Lecture Notes in Computer Science 1339

Edited by G. Goos, J. Hartmanis and J. van Leeuwen

Advisory Board: W. Brauer D. Gries J. Stoer

Springer
Berlin
Heidelberg
New York
Barcelona
Budapest
Hong Kong
London
Milan
Paris
Santa Clara
Singapore
Tokyo

Nabeel A. Murshed   Flávio Bortolozzi (Eds.)

Advances in Document Image Analysis

First Brazilian Symposium, BSDIA'97
Curitiba, Brazil, November 2-5, 1997
Proceedings

Springer

Series Editors

Gerhard Goos, Karlsruhe University, Germany

Juris Hartmanis, Cornell University, NY, USA

Jan van Leeuwen, Utrecht University, The Netherlands

Volume Editors

Nabeel A. Murshed
Flávio Bortolozzi
Pontifical Catholic University of Paraná
Lab. of Document Image Analysis and Neural Networks
R. Imaculada Conceicao, 1155, Curitiba, PR 80215-901, Brazil
E-mail: murshed@rla01.pucpr.br

Cataloging-in-Publication data applied for

Die Deutsche Bibliothek - CIP-Einheitsaufnahme

Advances in document image analysis : first Brazilian symposium ; proceedings /
BSDIA '97, Curitiba, Brazil, November 2 - 5, 1997. Nabeel A. Murshed ; Flávio
Bortolozzi (ed.). - Berlin ; Heidelberg ; New York ; Barcelona ; Budapest ;
Hong Kong ; London ; Milan ; Paris ; Santa Clara ; Singapore ; Tokyo :
Springer, 1997
  (Lecture notes in computer science ; Vol. 1339)
  ISBN 3-540-63791-5

CR Subject Classification (1991): I.4, I.5, I.7.2

ISSN 0302-9743
ISBN 3-540-63791-5 Springer-Verlag Berlin Heidelberg New York

Typesetting: Camera-ready by author
SPIN 10647919   06/3142 – 5 4 3 2 1 0   Printed on acid-free paper

Preface

Document Image Analysis has attracted, over the last few decades, many researchers from several scientific and technological fields. Many books and papers have been published on the subject, and several scientific meetings (conferences, workshops, etc.) have been organized annually to bring together researchers from around the world. These activities have made Document Image Analysis a well-established field of research that enjoys an enormous amount of attention. In Brazil and Latin America, activities in Document Image Analysis, are in their infancy. An invitation for participation is needed.

We were pleased to have the opportunity to organize the First Brazilian Symposium on Document Image Analysis (BSDIA'97). The symposium aims at promoting and motivating research and development in the fields of handwriting and graphics recognition, storage and information retrieval, workflow and document system architectures; and providing a forum for researchers, students, developers, and end-users, from academia and industry, to discuss and exchange ideas related to those areas. Within this context, we received 30 contributions from Brazil and abroad. A total of 19 contributions were selected for oral presentation at the symposium. Seven invited talks were included in these proceedings.

The symposium was organized by the Laboratory of Document Image Analysis and Neural Networks (LADIANN), of the Pontifícia Universidade Católica do Paraná; and was sponsored by CNPq, Equitel, IBM Itec, and Banco HSBC Bamerindus. We are grateful to our sponsors for their cooperation and support.

We are also grateful to the members of the program committee, who were also the invited speakers, and the participants for their support and active participation in the symposium. In addition, we are indebted to our faculty members and graduate and undergraduate students for their assistance in the organization of the symposium.

We hope that you will benefit from these proceedings.

November 1997
                                                        Nabeel A. Murshed
                                                        Flávio Bortolozzi

BSDIA'97 Committee

Symposium Chair
Flávio Bortolozzi, Brazil

Program Chair
Nabeel A. Murshed, Brazil

Program Committee

Abdel Belaid, France
Adnan Amin, Australia
Anselmo Chaves, Brazil
Carlos Heuser, Brazil
Celso A. Kaestner, Brazil
Ching Y. Suen, Canada
David Doerman, USA
Horst Bunke, Switzerland
Edgard Jamhour, Brazil

John C. M. Lee, Hong Kong
José Valdeni de Lima, Brazil
Keiiche Abe, Japan
Lee Luan Ling, Brazil
Paulo Cesar Masiero, Brazil
Rangachar Kasturi, USA
Rejean Plamondon, Canada
Robert Sabourin, Canada
Sebastiano Impedovo, Italy

Organized by:

Laboratory of Document Image Analysis and Neural Networks (LADIANN)
Department of Graduate Studies in Applied Informatics (PPGIA)
Pontifical Catholic University of Paraná (PUC-PR)
Brazil

Sponsored by:

CNPq, Equitel
IBM, Itec
Banco HSBC Bamerindus

Contents

Invited Talks

Oral Presentations

A. Low Level Processing

E. Application Systems

Handwriting Recognition:
State of the Art and Future Trends

G. Dimauro, S. Impedovo, G. Pirlo, A. Salzo

Dipartimento di Informatica, Università degli Studi di Bari
via Amendola 173, 70126 Bari, Italy

Abstract. This paper presents the State of the Art on handwriting recognition and points out some major future trends of scientific research in this field. Some of the most relevant results presented in the series of Workshops on Frontiers in Handwriting Recognition and Conferences on Document Analysis and Recognition are also discussed and an extensive bibliography of selected papers is reported.

1. Introduction

Handwriting recognition attracted the attention of researchers since the origin of computers. Nowadays, the technological progress made in the field of computer architectures and peripheral devices as well as the advances of scientific research make possible the development of new systems for handwriting recognition. In addition, the series of Workshops on Frontiers in Handwriting Recognition and the International Conferences on Document Analysis and Recognition provide continuously new stimuli to researchers, and point out that industrial applications of handwriting recognition can become a large business [1].

This paper presents the State of the Art in handwriting recognition and the new trends of scientific research in this field. Section 1 deals with the acquisition devices for handwriting recognition and reports a list of commercially diffuses scanners and graphic tablets. In Section 2 the most important preprocessing phases are discussed and some recent solutions are explained. Sections 4 and 5 report the results in the field of Character and numeral recognition and cursive word recognition respectively. The problem of automatic signature verification is discussed in Section 6. Section 7 highlights some of the most relevant future trends of scientific research in handwriting recognition: the use of contextual processing, the multi-expert approaches, and the CASE tools for systems design and development.

2. Data Acquisition Devices

Data acquisition devices for handwriting recognition systems can be divided in two main classes: off-line and on-line devices [2].

The most diffuse devices for off-line acquisition are the optical scanners which can be of four types: flatbed scanners, paper-fed scanners, handy scanners and drum scanners. Tables 1,2 3 and 4 report respectively some diffuse models of flatbed scanners, paper-fed scanners, handy scanners and drum scanners. Indeed, other scanners are also available for specific applications. Table 5 reports two scanners for check processing applications.

FLAT BED SCANNERS

MODEL	COMPANY	SCAN SPEED y	MAXIMUM RESOLUTION	MAX PAPER FORMAT
M3191F2	FUJITSU	2 sec/form	300 dpi	A4
M3096EX/GX	FUJITSU	5,6 sec/form	400 dpi	A3
M3098 EX/GX	FUJITSU	1,3 sec/form a 400 dpi	400 dpi	A4
M3099 A/G	FUJITSU	1 sec/form a 200 dpi	400 dpi	A3
SCANPARTNER ELECTRONIC OFFICE	FUJITSU	6 sec/form	400 dpi	A4
SCANMAKER IISP	MICROTEK	18 msec/line color	1200 dpi	A4
SCANMAKER III	MICROTEK	20 msec/line color	4800 dpi	A4
SCANMAKER E6	MICROTEK	42 sec. for A4 to color at 300 dpi	2400 dpi	A4
IMAGERY 1200 SP	TRUST	4 msec. for A4	9600 dpi	A4
IMAGERY 2400 SP	TRUST	3,6 msec. for A4	2400 dpi	A4
SPEEDY 8	UMAX	10 sec. for A4; 32 sec. for A4 at 300 dpi	6400 dpi	A4
PERSONAL SCANNER	UMAX	4,7 sec. for A4 3 sec. grey levels	4800 dpi	A4
VISTA COLOR PRO	UMAX	32 sec/form	800 dpi	A4
SUPER SPEEDY	UMAX	40 sec/form at 300 dpi	9600 dpi	A4
POWERLOOK II PROFESSIONAL	UMAX	7 sec.	9600 dpi	A4
SCANJET 4C	HEWLETT PACKARD	30 sec/form	600 dpi	A4
GT-8500	EPSON	17 sec/form	1600 dpi	A4
GT-9000	EPSON	2 msec/line for B/W 16 msec/line color	2400 dpi	A4
GT-5000	EPSON	11,5 msec/line	1200 dpi	A4
DUO SCAN	AGFA	10,5 msec/form	4000 ppi	A4
ARCUS II	AGFA	4 msec/line line art 15 msec/line color	3600 dpi	A4
SELECT SCAN	AGFA	2000 line/minute	4000 ppi	A4
SELECT SCAN PLUS	AGFA	2000 line/minute	8000 ppi	A4
HORIZON ULTRA	AGFA	0,8 msec/line	2540 ppi	A3
JX330	SHARP	13 sec/form	2400 ppi	A4

Table 1.

PAPERFED SCANNERS

MODEL	COMPANY	SCANSPEEDY	MAXIMUM RESOLUTION	PAPER FORMAT
PAGESCAN COLOR	LOGITECH	6 sec/form	400 dpi line/art 200 dpi color	21,6 x 35,6 cm
SCANJET 4S	HEWLETT PACKARD	7,5 sec/form	400 dpi line/art 200 dpi color	A4
PAGEREADER 800	HUSTEK	10 sec/form	400 dpi line/art 800 color	A4
GT 300	EPSON	15 sec/form	600 dpi line/art	A4
SCANPARTNER JUNIOR	FUJITSU	10 sec/form	300 dpi line/art	10,1 x 15,2 cm
PAGEWIZ	MICROTEK	12 sec/form	300 dpi line/art	A4

Table 2.

HANDY SCANNERS

MODEL	COMPANY	SCANSPEEDY	MAXIMUM RESOLUTION	SOFTWARE INTERFACE
SCANMAN COLOR	LOGITECH	Max 5 min. for text	400 dpi	TWAIN
SCANMAN COLOR PRO	LOGITECH	Max 5 min. for text	1600 dpi line/art 800 dpi color	TWAIN

Table 3.

DRUM SCANNERS

MODEL	COMPANY	MAXIMUM RESOLUTION	ROTATION SPEEDY	DEPTH
SCANMASTER 4500	HOWTEKJEI	4000 dpi	1200 turns/minute	12
SCREE DT-S1030AI	DIANIPPON	5200 dpi	1200 turns/minute	32
SCREE DT-S1045AI	DIANIPPON	8000 dpi	1200 turns/minute	32

Table 4.

CHECK SCANNERS

MODEL	COMPANY	SCANSPEEDY	MAXIMUM RESOLUTION	DEPTH
B1000	VISIONSHAPE	42 checks for min to 100 dpi	200 dpi	4 bit
BUIC 1000	DIGITALCHECK	42 checks for min to 100 dpi	200 dpi	4 bit

Table 5.

DIGITIZING TABLETS				
MODEL	COMPANY	RESOLUTION	ACTIVE AREA	FRESH RETRIEVEL
ACECAT II	ACECAD	up to 2000 lpi	5"x5"	NO
ACECAT III	ACECAD	up to 2540 lpi	5"x5"	NO
ACECAT 1212	ACECAD	up to 2540 lpi	12"x12"	NO
ACECAT 1812	ACECAD	up to 2540 lpi	18"x12"	NO
ARTPAD II	WACOM	2540 lpi	4"x5"	256
ARTZ II	WACOM	2540 lpi	6"x8"	256
ARTZ II	WACOM	2540 lpi	12"x18"	256
SUMMAPAD	SUMMAGRAPHICS	1016 ppi	4"x5"	256
SUMMA FLEX	SUMMAGRAPHICS	2540 lpi	6"x8"	256
SUMMA EXPRESSION	SUMMAGRAPHICS	2540 lpi	12"x12"	256
SUMMA EXPERT	SUMMAGRAPHICS	2540 lpi	12"x18"	256
DRAWINGSLATE II 32090	CALCOMP	100 lines/mm	6"x9"	256

Table 6.

The class of on-line acquisition devices includes both digitizing tablets and tablet-display devices. There are many types of digitizing tablets which use various technologies including electromagnetic/electrostatic, pressure sensitive and acoustic sensing in air medium types. Table 6 reports some models of commercially diffused digitizing tablets. However, it should be considered that, along with the recent advances in packaging technology, new low-priced unified tablet-display devices are available for on-line handwriting acquisition. They allow a very intelligent "paper and pencil-like" computer interface. A unified tablet-display device, which is roughly the size and the weight of a book, has an electronic stylus which provides electronic ink: an electronic tablet accurately captures the coordinates of pen-tip movements, electronic ink gives the trace of the pen-tip on the screen surface, giving an immediate feedback to the writer.

3. Preprocessing

The preprocessing algorithms provide the acquired data in a suitable form for further processing. In this phase the input image is generally cleaned from noise and errors caused by the acquisition process. A great number of well-defined algorithms for signal processing are currently used during the preprocessing phase. However, in handwriting recognition, the preprocessing deals with more specific problems than in other fields of pattern recognition.

The removal of the box in which off-line characters can be included has been faced by histogram techniques by Casey and Takahashi [3]. A more complex problem of text overlapping with non-textual strokes has been encountered by Govindaraju and Srihari [4] in reading handwritten addresses on mail pieces. Therefore, they address the problem of the identification of text from noise. This is not a trivial problem: when humans read a text with noise, they use a knowledge of text in perceiving part of the image. However, it has been argued that it is not necessary to use knowledge to perform segmentation. In fact, the "good" continuation principle can be applied to intersect boundaries to arrive at perceptually plausible segmentation. In practice the rule of "good" continuity is applied by following the path of minimal curvature change. After the strokes have been detected, they are identified according to simple rules based on the observation that variations in curvature are more present in textual strokes than in non-textual components. Therefore, non-textual components can be easily recognized and removed.

Another problem that arises in several applications of handwriting recognition is thinning. Thinning methods can be divided into two categories: pixelwise and non-pixelwise. Pixelwise methods consider the digitized image, representing the plain as a mesh, where each square element is a pixel. Pattern image is locally and iteratively processed, so the process is very slow and can lead to problems such as excessive erosion. An intelligent non-pixelwise methodology has been recently proposed by Nishida Suzuki and Mori [5]. They decompose a line picture into regular regions, singular regions and end-point regions, according to the classification of ordinary points, branch points and end-points. Each region can be represented by a graph node labelled with a number. Thus the nodes are connected to make a region adiacency graph. On the basis of this decomposition, the thinning is treated as an inference problem of singular regions from the adjacent regular regions. In fact, for a regular region, thin line representation can be allowed easily enough from the analysis of its contour.

Segmentation, which involves the extraction of the writing units from handwriting is another fundamental preprocessing phase. The units can be elementary strokes in character and numeral recognition or they can be entire characters in the recognition of entire words in cursive script recognition. Moreover, the segmentation processes use spatial information for off-line systems, but temporal information can also be used to detect the writing units in on-line handwriting recognition systems.

In the field of off-line character recognition, Casey-Takahashi suggest a supervised algorithm for segmenting handprinted characters of the NIST database [3]. Fenrich [6] proposes an unsupervised system for numeric string segmentation in which four multiple splitting methods are applied in sequence in order to achieve an acceptable segmentation: the histogram method, the upper-lower contour method, the upper-contour method, and the lower-contour method. It should be noted that these approaches are based on the fundamental consideration that the density of pattern pixels in those areas of the image containing significant patterns (characters, numerals etc.) is generally much greater than the density of pixels in areas which contain meaningless patterns (for instance ligatures). Moreover, as shown in the work of Liu and Tai [7] the importance of segmentation is crucial also in Chinese handwriting recognition since Chinese characters can be hierarchical decomposed in substructures called "radicals", strokes, and line-segments

Holt, Beglou and Datta [8] focus the problem of off-line cursive word recognition and point out that traditional approaches based on vertical segmentation in the direction of local minima of the upper profile of the word fail when slanted words are considered. The new approach they propose is based on segmentation boundaries rather than segmentation points. Each segmentation boundary is defined following clockwise the contour of the word from each local maximum to the corresponding local minimum breaking the word at an appropriate splitting point. Some heuristics for correct identification of local maximum and minimum points are also given to avoid unsatisfactory segmentation boundaries. The segmentation method of Hull, Ho, Favata, Govindaraju and Srihari for off-line cursive word recognition is integrated in the recognition process by a "divide and conquer" approach [9]. In the work of Fujisaki, Beigi, Tappert, Ukelson and Wolf [10] for on-line recognition, different segmentation techniques have been presented depending on the different types of writing.

In this field, in order to minimize the possible segmentation hypothesis in on-line applications, a new method has been proposed by Higgins and Ford to reduce deformations in handwriting [11]. Specifically, it works on the deformations of cusps, whereby a cusp becomes a loop or hump and vice-versa, because this deformation represents one of the main features of individuality between writers. So they propose a new method of consistent segmentation, flexible enough to be incorporated in many on line cursive script recognition systems. The method attempts to create a script sample considering the copy-book style that the writer learnt at school. Since there are a limited number of copy-book styles, if we can somehow map the graphical representation of the writing which appears on the page to the theoretical representation of the copy-book style, the great variety of personal styles may be considerably reduced.

4. Character and Numeral Recognition

Two methods are generally adopted for character and numeral recognition [12]: pattern matching and structural analysis. Pattern matching techniques may be classified depending on the features used (gray level, stroke density, stroke direction, relation between segments) and the method of feature extraction (projections, peripheral, local, structure integration and expansion functions). Structural analysis methods can be classified in accordance with the relative weight of bottom-up or top-down processing and with the techniques for stroke segment extraction. Some of the most important applications of structural methods concern the recognition of Chinese Characters (Kanji). The recognition of Kanji characters involves several problems in contrast to alphanumeric character recognition: (P1) Complicated interrelation between line segments and redundancy, (P2) Occurrence of a great number of categories, (P3) Existence of mutually resembling categories. In this field, problem P3 is contradictory to problem P1: while in pattern matching methods it is difficult to overcome problem P3, the structural analysis methods are impeded by problem P1.

The structural decomposition approaches pose other difficult problems: while in off-line character recognition the set of strokes, which compose a character, can be obtained only by taking into account static information, in on-line character recognition systems dynamic information can also be considered. So, in order to identify the strokes, while off-line character recognition systems use control points like end-points, cross-points, bend-points, local extrema in x and y direction and so on, in on-line systems many authors consider as strokes each part of the character between a pen-down and the successive pen-up. The classification is then performed using reference look-up tables or decision tree classifiers. Moreover, in on-line applications, the problem of the number and order of the strokes must be faced for character decomposition.

A structural decomposition procedure for a stroke-order free, on-line Chinese character recognition system which uses a dynamic matching method has been proposed by Shiau, Kung, Hsieh, Chen and Kao [13]. Specifically, they use a hierarchical representation of feature strings. A character description procedures include a radical item for indexing the radicals' structural description procedures. This allows one to avoid a complex stroke-by-stroke description of whole characters. In fact, the structural description procedures try to decompose the character into radicals, and then to decompose each radical into strokes. However, if strokes between any two radicals cannot be separated, different procedures for direct character decomposition into stroke domain are used.

The "enhanced loci" algorithm proposed by Downton, Tregidgo, Leedham and Hendrawan [14], for every white point of the background of the character, vertical and horizontal vectors are generated and the presence of intersection of vectors with the character is used to obtain a feature code.

Often, also hybrid approaches are used for the recognition of characters and numerals. They are usually based on statistical and structural information of the patterns. The hybrid approach for handwritten numeral recognition proposed by Wang, Nagendreprasad, Gupta [15] is based on histograms and a relaxation matching algorithm for classification. A histogram consists of a vector of numbers where each number is obtained as the sum of the number of pixels in a particular direction. The set of histograms for a digit consits of four vectors: vertical histogram, horizontal histogram, right diagonal histogram and left diagonal histogram. Thus the i-th element in the vertical histogram is the sum of the pixels of the pattern in the i-th column of the bitmap. The other histograms are similarly formed. The next stage of this system is based on a neural network that uses the histograms for performing the recognition. The rejects from this stage are fed to a recognition algorithm based on Freeman's octal chain code primitives. From each character, an encoding of a string of primitives is extracted. The relaxed matching is used to compare the elements of this string with the elements of the dictionary strings.

The need to obtain a very high recognition rate has led to the development of sophisticated post-processing systems for error correction. In this field, an expert system for error rectification of handprinted character recognition systems is proposed by Nakano [16].

A new system for automatic error correction has been proposed by Seino, Tanabe and Sakay [17]. It uses linguistic post-processing based on word occurrence probability, by which mistakes are detected and corrected automatically. In fact, current OCRs are not sufficiently accurate for business use. So, in order to increase the accuracy to a business use level, the Knowledge Processing method has been adopted which detects reading mistakes and corrects them with linguistic information.

The growth in knowledge of character shapes and their recognition during the last thirty years poses new questions today. One of the most important is whether in character recognition, man still remains better than the machine or not. The need to obtain machines capable of matching human performance in character recognition has led to the investigation of human capabilities in recognizing patterns.

Dimauro, Impedovo and Pirlo [18] investigate the behaviour of man in recognizing handwritten characters. For this purpose, from an input mono-stroke numeral, a set of plane curves is obtained which includes a numeral similar to the original one and meaningless curves. In fact, under certain conditions the input numeral can be described through three Fourier descriptors in a three-dimensional discrete space of complex values. In this way, through slight variations of the phase of one of more of the Fourier descriptors describing the original numeral, it is possible to obtain a set of curves very similar to each other, but not always recognizable by man as belonging to the class of the original handwritten numeral. Humans are then used to recognize the set of generated curves. The results shown that each subject needs a short personal learning phase in order to eliminate his uncertainty in recognizing patterns. The duration and development of this learning change depends on the subject and all subjects, after a short learning phase, show a tendency to reject more curves.

Differences between computer and human recognition of handprinted characters have also been investigated by Suen, Guo and Li [19]. They propose a domain decomposition method and a computer algorithm to investigate the recognition rates of patterns based on their distinctive parts, and to find all the potential candidates when there is uncertainty or confusion. Taking into consideration a sample of patterns with 89 most common styles of alphanumeric handprints, the recognition rate of the different parts of a characters is evaluated. This research is of great interest where only parts of the pattern are available for recognition due to missing in writing, gambling, poor scanning, noise or various kinds of distorsion. On the other hand, it may also lead to concentrate the efforts of the recognition algorithms on the most distinctive parts of the characters for more effective and efficient recognition.

5. Cursive Word Recognition

Undoubtedly, cursive word recognition is the most interesting and attractive part of handwriting recognition. In this field, on line acquisition is the most profitable to get a flexible model, because the temporal or dynamic components of the drawing can be registered. Thus, one of the greatest advantage of on-line devices is that the process itself yields a natural segmentation of the shape into strokes. Off-line recognition appears, and is, more difficult, not only because of the loss of dynamic information, but also because the scripted sheet reading camera process brings additional noise on the remaining information.

One of the major difficulties of the word recognition descends from the great variability observed in different samples of script issued from the same writer over time or from different writers. So it is difficult to find a reliable description of a word able to represent all the admitted occurrences of the input shape.

Consequently, in the field of off-line recognition of handwritten words, several works have been devoted up to now to word description techniques using the structural approach. At the purpose, many researchers divide the regular part from the singular part of the trace before performing coding [9, 20]. The concept of regular and singular class of features is widely used by many groups in the world: the regular class is made of features, of which the iconic representation is periodic or quasi periodic; the singular class is made of features which are the accidents of the regular features. Examples of regular features are textures and lines. Examples of singular features are edges for textures, extremities, crossing, cusps and forks for line pictures.

In the work of Simon and Baret [20] a dictionary is used which contains a set of codes for the words used in the field of bank cheques.In this approach, after preprocessing, the regular part of the word is extracted. This part, which is also called the axis of the word, is defined as the shortest path from one extremity (left) to the other (right) and that remains in the body of the word. The singularities are then derived by complementation, i.e. by taking off the axis from the word image. The symbolic description chain reports detailed information about word shape. Among the others it reports the interruptions, points or accent above the word, ascender or descender (with a height value from 0 to 9), extremities, direction of straight segments in the word, direction sector orthogonal to the chord of a curve in the word, loops and so on. The matching procedure is carried out starting from the analysis of anchors in the chain that define robust features and then using a dynamic matching for the other parts of the code.

The technique proposed by Hull, Ho, Favata, Govindaraju and Srihari [9] performs word recognition without explicit segmentation of the word. Specifically this approach, developed for cursive word recognition in postal addresses, extracts a chain code from the contour of the whole word and then uses this code to derive singular features. Furthermore, the algorithm of Hull, Ho, Favata, Govindaraju and Srihari is based on an analysis of writing styles. Each image is assigned to the following categories of writing: discretely printed, fully cursive and broken cursive. Discretely printed is formed by printing nearly every character in the word even if some of those characters touched one another, fully cursive is formed by a single continuous motion of the writing instrument, broken cursive is formed by more than one writing motion and it contains at least one cursively written component spanning more than one character.

Camillerapp, Lorette, Menier, Oulhadj and Pettier [21] propose a system for off-line handwriting recognition based on a structural approach. Each word is represented by its graph model deduced directly from the grey-level image by detecting specific primitives along the baseline of the word. Recognition is performed at the word level: each model in the dictionary is compared to the entity to be recognized. The likelihood is evaluated as the difference between similarity and dissimilarity rates. Tree comparison is based on dynamic programming and may be viewed as a two-dimensional adaptation of the string to string problem.

A new method based on a syntactic description of the words for automatic recognition of off-line Arabic cursive handwritten words is also proposed by Zahour,

Taconet and Faure [22] and Taconet, Zahour and Faure [23]. They evaluate word geometric characteristics as end-points, nodes, and sometimes sharp point etc. to define a graphical complexity index for off-line drawing. The word complexity index is defined as the double of the difference between the number of branches in the word and the number of nodes in the word. The complexity index is rather stable for one writer only, for whom, however, a preliminary vocabulary learning is required. Furthermore it has the advantage of being completely independent from geometrical parameters such as lengths of branches, curvature intensity, position with respect to the basic line etc., and of being a global word index. Unfortunately when a cursive word is written also using discretely printed characters, its complexity index can significantly change. Another measure of complexity is given by the estimated number of main pen-up and pen-down movements. This estimation is usually done by counting the number of transitions between the lower basic line and the upper basic line of the word. This counting can be done practically by counting the number of crossing points with a medium line between the two basic lines. A variation of this technique can be done by counting the number of connected components situated between the medium line and the upper basic line, and having at least one crossing point with it. The existence of word indexes can be useful for fast detection of a reduced list of candidate words on which further, more accurate procedures must be performed.

In the field of on-line recognition of run-on handwriting the results have mainly been obtained using unified tablet-displays such as a paper-like computer interface. Fujisaki, Beigi, Tappert, Ukelson and Wolf [10] developed a system that classifies strokes, generates character hypotheses, by means of a hypothesis generator, and verifies them by means of a hypothesis tester to estimate the most suitable character sequence for each word.

Schomaker and Teulings [24] consider the stroke-based approach for cursive script recognition versus the character-based approach. In the first approach, after stroke feature vector quantization by a Kohonen self-organizing network, a symbolic classification procedure is performed on strings of symbolic stroke codes to identify characters. This approach yields a significant, correct classification of neatly written on-line cursive script, but it lacks the generalisation capability that is necessary to display invariance in the geometric distortions and variabilities inherent to cursive script. In the second approach the handwritten input word is first segmented into characters on the basis of the minima in the tangential velocity. Feature vectors describing complete characters are used to train a Kohonen network. The character-based recognizer yields a better performance at the cost of a larger amount of computation, due to the high dimensionality of the feature vector.

Moreover, several classification techniques have been tested: models based on nearest neighbour matching, rule-based recognition and the Markov model. For instance Camillerapp, Lorette, Menier, Ouladhj and Pettier [21] use an ergodic Markov model including 8 states to represent a word. The model can be considered a fully connected graph in which each node represents a state corresponding to one direction of chain code, and each arc is associated with a change in the direction of the pen tip trajectory. Markov models of different n orders have also been investigated. In these models the probability of there being a particular change in the direction depends only on the n states which have been encountered previously. In this case a full-word approach is considered to overcome the problem of segmentation.

6. Signature Verification Systems

Signature verification involves monetary transactions, legal certifications, automatic control of physical entry to a protected area, and so on. In this field the results have mainly been obtained with on-line signature verification systems, where dynamical information from the input process is available [25].

Recently, research in on-line signature verification has been devoted to the selection of better features conveying more information about the writing process. Yoshimura and Yoshimura [26] propose the use of function features based on the direction of pen movements during signing. They point out the effectiveness of the new features for signature verification. Moreover, they show that by using functions as discriminating features, the elastic matching procedure assures significant advantages in obtaining a similarity measure, but that this is generally not very efficient in terms of computational speed. In order to avoid time-consuming matching procedures, Dimauro, Impedovo and Pirlo [27] use specific Fourier descriptors that contain information on both the shape and dynamics of the writing process of signature strokes. In order to reduce the effect of variability in signing, they propose a knowledge-base oriented to the specific set of components of the signature, as detected in the training phase of the system. The procedure for automatic clustering of the components of each signer, is based on an improved k-means clustering algorithm which uses topological features. For each class of components, some parameters are computed and used in the stroke-verification process. The overall verification of the signature is then performed using the verification responses obtained from each component of the signature.

Moreover, three types of matching procedures can be adopted for automatic signature verification [28]: *wholistic matching*, *regional matching* and *multiple-regional matching* procedures. When *wholistic matching* procedures are adopted the input signature is matched sequentially against each reference signature and the signatures are considered as a whole. When *regional matching* is used the input signature is matched against each reference signature by comparing couples of corresponding strokes of the two specimens. Of course, also in this case the reference signatures are sequentially used in the verification process. When the *multiple-regional matching* is used, each stroke of the input signature is verified individually by using all the corresponding strokes of the reference signatures. Successively, the responses obtained at stroke level are combined in order to achieve a verification response for the entire signature.

In recent research, new aspects of signature verification have also been considered. In the work of Congedo, Dimauro, Impedovo and Pirlo [29], a measure of the stability of a dynamical signature is proposed based on dynamic non linear time-warping procedure. This information is used to select the near-optimal set of signature to be used as reference for a SVS according to a specific representation space (position, displacement, velocity, acceleration and so on). Brittan and Fairhurst [30] presented a conceptual architecture of the image classifier and its matching for signature verification. They develop a hierarchically structured, or multi-layer, classifier architecture, where a series of classifiers are cascaded one after the other.

By placing a number of contraints on each of the classifiers, a significantly higher level of performance is achieved compared with implementation on a single classifier.

7. Future Trends in Handwriting Recognition

The enormous effort of Scientific Community in the field of handwriting recognition is producing a new scenario extremely interesting and profitable. From one side, current handwriting recognition systems can achieve performance very high and in some cases even better than human. From the other side, many applications still exist in which higher performances are required. This is the case of tax form and bank check recognition systems, for which a very reliable response is required even if handwritten characters of very degraded quality and with strong shape distortions are considered [31].

Therefore, new issues have been recently arised as new challenges for scientists in the field of handwriting recognition. Among others, they are the use of contextual processing, the multi-expert approaches, and the use of CASE tools for system development.

7.1 Contextual Processing for handwriting recognition

In a man-machine interaction, based on handwriting recognition, the use of different sources of knowledge (morphological, linguistic and pragmatic) and of a hierarchy of contexts at different levels (for strokes, letters, bigrams, trigrams, words, sentences) make the recognition much more efficient [32]. Hull [33] points out that several different types of contextual contraints can be considered in handwritten text recognition: graphical orthographical, statistical, structural-syntactic and structural-semantic.

In the system of Fujisaki, Beigi, Tappert, Ukelson and Wolf [10] for run-on cursive script recognition, two types of linguistic constraints are used. A filter based on acceptable character-type transition checks transition of character types over a given hypothesis path, and, if a transition is infrequent or prohibited, a penality score is considered. For instance, a hypothesis path containing a transition from a lower to an uppercase character is penalized. Another filter validates character sequences by character tri-grams. If a sequence of three letter transition is infrequent or prohibited in the character tri-gram data base, a penality score is added to the path. In the work of Schomaker and Teulings [24] knowledge of the handwritten script is used to perform a top-down process for recognizing ambiguous shapes,while they observe that bottom-up information is usually sufficient to recognize unambiguous patterns.

Finally, it must be pointed out that some questions still remain open. They are related to the combination of knowledge from different sources, the discrimination between correct and incorrect recognition, the detection of errors and the automatic acquisition of knowledge.

7.2 Expert Combination

Expert combination is a diffuse strategy to improve the performance of individual classifiers. More specifically, it has been recently observed that the combination of the decisions of several classifiers provide more reliable results [34].

Therefore, many combination methods have been recently proposed which can be classified into three catagories depending on the quantity of information they combine [35]: *abstract-level*, *ranked-level* and *measurement-level* combination methods.

Abstract-level combination methods are used when each classifier outputs a simple class label. The majority vote is probably the simplest combination method of this category. It assigns to each class a score equal to the number of classifiers for which the class is the top candidate. The final classification response is the class with the highest score [36,37]. Another important example of abstract-level method is the Behaviour-Knowledge Space (BKS) [38]. The BKS method is based on a k-dimensional space, where k is the number of of classifier. In this space each dimension corresponds to the decision of one classifier. The unit which is the intersection of the classifier decisions of the current input is called the focal unit (F.U.). In each unit is stored three kinds of data: (1) Total number of meaning samples, (2) The best representative class, (3) The number of incoming samples belonging to the class. The BKS method operates in two stages: in the "learning" stage the BKS is constructed using the learning samples; in the "operation" stage, according to the decision taken from each individual classifier a focal unit is selected and the final decision is taken.

Ranked-level combination methods are used when each classifier outputs a ranked list of class labels ordered according to the degree of membership of the input pattern. Two effective methods are those based on the Borda Count function [39] and on logistic regression [40].

When each classifier also provides a measure of the degree of membership of the input pattern to each class *Measurement-level* combination methods are used. A measurement-level combination method based on the Bayes theory is presented in [41]. In [42] the Dempster-Shafer theory of evidence is used to integrate information obtained by multiple classifiers. Also neural networks [43] and fuzzy systems [44] can be used to combine experts at the measurement level.

Recently, the problem of classifier combination has been considered also from other points of view. Specifically, from the consideration that in classifier combination, the degree of recognition accuracy improvement depends on how much classifiers complement each other, in [45] a new technique is presented to develop complementary classifiers. In [46] and [47], the problem of classifier combination is addressed in terms of the performance of the combination method depending on the correlation among individual classifier. At the purpose a suitable index of similarity measuring the stochastic correlation among classifiers is introduced and a systematic analysis of each combination method is performed [46]. This analysis provides new tools for the dynamic selection of individual classifiers and combination methods, and more generally of the multi-expert topology (fully serial, fully parallel, hybrid, etc.) [48].

7.3. CASE Tools for System Development

The large variety of algorithms today available for the several phases of the pattern recognition process imposes the use of Computer Aided Software Engineering (CASE) tools to quickly prototype and maintain the systems.

Currently, many CASE tools are today available for this purpose, like the SPW (Signal Processing Worksystem) and the Khoros system, and efforts in using them for system development are in progress in various Universities.

For instance, Khoros [49] is a software integration development environment that emphasizes information processing and data exploration. It provides a series of programming toolkits to quickly prototype new software and offers the possibility to interact on one or more of the following levels: Application End-User level, Visual Programmer level, Toolbox Programmer level, Infrastructure level.

The Application End-User utilizes one of the Khoros interactive, stand-alone information processing and data visualization applications.

The Visual Programmer encapsulates a data flow visual program into a stand-alone application by using the visual programming language *"Cantata"*. Here, the data operators and visualization operators are available as "icons" or "glyphs" and are grouped together by function into *toolboxes*. The Visual Programmer writes a program by selecting operators from the toolboxes via the menu and icons and by connecting them in order to form a data flow graph.

The Toolbox Programmer level comprises some software development tools for creating new toolboxes and new operators. Specifically, the tool *"Craftsman"* is used to create or manage a toolbox, the tool *"Composer"* is used to create, edit and manage a software object or operator and the tool *"Guise"* is used to create and interactively design a GUI (Graphical User Interface) for the operators.

The fourth level provides high-level programming libraries that form the infrastructure of the Khoros system.

The Khoros object oriented view of software development provides a methodology for creating reusable libraries so that the need to write new custom code is limited to the functional details of an algorithm. The basic ideas of the Khoros system are the *software object* and the *toolbox object*.

A *software object* is composed of source code, documentation, configuration files and user interface specifications.

A *toolbox object* is a collection of software objects which are characteristic of a given application domain. The fundamental motivation for using toolboxes is the need to reduce and manage complexity. Many factors contribute to complexity of a software system: amount of source code, dependencies between components, number of people involved with the software development.

Recently, the Khoros system has been successfully used by Dimauro, Impedovo, Pirlo and Salzo for designing and developing a first prototype of the Italian bankcheck processing system [50]. Since several algorithms were available for each phase of the bankcheck processing, they were grouped into different toolboxes (i.e. preprocessing, layout analysis, courtesy amount recognition, legal amount recognition, signature verification) and divided into subcategories in order to make more understandable their function and easier their access. Then, they were interactively combined within the workspace of the *"Cantata"* visual programming environment in order to form a data flow diagram. The prototype for bankcheck processing system, obtained using the Khoros CASE tool, results flexible, extendible and easy to manage. In fact it is very simple to configure a new system simply changing the disposition of the operators into the workspace or selecting new ones from the specific application toolboxes. No compiling is required since the glyphs are dynamically connected [51,52].

Conclusion

Handwriting recognition has had an enormous development in the recent years as shown by the interest towards the series of Workshops and Conferences in this field.

From the analysis of some relevant results presented on the recent literature on handwriting recognition, this paper reports the State of the Art in this field and points out the main trends of research in this field.

References

1. S. Impedovo, "Frontiers in Handwriting Recognition", in "Fundamentals in Handwriting Recognition", S. Impedovo (ed.), NATO-ASI Series, Springer-Verlag Publ., Berlin, 1994,

2. S. Impedovo, L. Ottaviano, S. Occhinegro, "Optical Character Recognition -A Survey", *International Journal of Pattern Recognition and Artificial Intelligence* 5 (1,2), 1-24 (1991).

3. R.G. Casey, H. Takahashi, "Experience in segmenting and classifying the NIST data base", in "From Pixels to Features III - Frontiers in Handwriting Recognition", S. Impedovo, J.C. Simon (eds.), Elsevier 1992, pp. 5-16.

4. V. Govindaraju, S.N. Srihari, "Separating handwritten text from interfering stroke", in "From Pixels to Features III - Frontiers in Handwriting Recognition", S. Impedovo, J.C. Simon (eds.), Elsevier 1992, pp.17-28.

5. H. Nishida, T. Suzuki, S. Mori, "Thin line representation from contour representation of handprinted characters", in "From Pixels to Features III - Frontiers in Handwriting Recognition", S. Impedovo, J.C. Simon (eds.), Elsevier 1992, pp. 29-40.

6. R. Fenrich, "Segmentation of automatically located handwritten numeric strings", in "From Pixels to Features III - Frontiers in Handwriting Recognition", S. Impedovo, J.C. Simon (eds.), Elsevier 1992, pp. 47-60.

7. Y.-J. Liu, J.-W. Tai, "An on-line Chinese character recognition system for handwritten in Chinese calligraphy", in "From Pixels to Features III - Frontiers in Handwriting Recognition", S. Impedovo, J.C. Simon (eds.), Elsevier 1992, pp. 87-99.

8. M.J.J. Holt, M.M. Beglou, S. Datta, "Slant-independent letter segmentation for off-line cursive script recognition", in "From Pixels to Features III - Frontiers in Handwriting Recognition", S. Impedovo, J.C. Simon (eds.), Elsevier 1992, pp. 41-46.

9. J.J. Hull, T.K. Ho, J. Favata, V. Govindaraju, S.N. Srihari, "Combination of segmentation-based and wholistic handwritten word recognition algorithms", in "From Pixels to Features III - Frontiers in Handwriting Recognition", S. Impedovo, J.C. Simon (eds.), Elsevier 1992, pp. 261-272.

10. T. Fujisaki, H.S.M. Beigi, C.C. Tappert, M. Ukelson, C.G. Wolf, "On-line recognition of unconstrained handprinting: a stroke based system and its evaluation", in "From Pixels to Features III - Frontiers in Handwriting Recognition", S. Impedovo, J.C. Simon (eds.), Elsevier 1992, pp. 297-312.

11. C.A. Higgins, D.M. Ford, "A new segmentation method for cursive script recognition", in "From Pixels to Features III - Frontiers in Handwriting Recognition", S. Impedovo, J.C. Simon (eds.), Elsevier 1992, pp. 75-86.

12. K. Yamamoto, H. Yamada, T. Saito, "Current state of recognition method for Japanese characters and database for research of handprinted character recognition", in "From Pixels to Features III - Frontiers in Handwriting Recognition", S. Impedovo, J.C. Simon (eds.), Elsevier 1992, pp. 105-116.

13. S.L. Shiau, S.J. Kung, A.J. Hsieh, J.W. Chen, M.C.Kao, "Stroke-order free on-line Chinese character recognition by structural decomposition method", in "From Pixels to Features III - Frontiers in Handwriting Recognition", S. Impedovo, J.C. Simon (eds.), Elsevier 1992, pp. 117-128.

14. A.C. Downton, R.W.S. Tregidgo, C.G. Leedham, Hendrawan, "Recognition of handwritten British postal addresses", in "From Pixels to Features III - Frontiers in Handwriting Recognition", S. Impedovo, J.C. Simon (eds.), Elsevier 1992, pp. 129-144.

15. P.S.P. Wang, M.V. Nagendraprasad, A.Gupta, "A neural net based 'hybrid' approach to handwritten numeral recognition", in "From Pixels to Features III - Frontiers in Handwriting Recognition", S. Impedovo, J.C. Simon (eds.), Elsevier 1992, pp. 145-154.

16. Y. Nakano, "Advanced application systems for handwritten character recognition", in "From Pixels to Features III - Frontiers in Handwriting Recognition", S. Impedovo, J.C. Simon (eds.), Elsevier 1992, pp. 185-190.

17. K. Seino, Y. Tanabe, K. Sakai, "A linguistic post processing based on word occurrence probability", in "From Pixels to Features III - Frontiers in Handwriting Recognition", S. Impedovo, J.C. Simon (eds.), Elsevier 1992, pp. 191-200.

18. G. Dimauro, S. Impedovo, G. Pirlo, "Uncertainty in the recognition process: some considerations on human variable behaviour", in "From Pixels to Features III - Frontiers in Handwriting Recognition", S. Impedovo, J.C. Simon (eds.), Elsevier 1992, pp. 215-222.

19. C.Y. Suen, J. Guo, Z.C. Li, "Computer and human recognition of handprinted characters by parts", in "From Pixels to Features III - Frontiers in Handwriting Recognition", S. Impedovo, J.C. Simon (eds.), Elsevier 1992, pp. 223-236.

20. J.C. Simon, O.Baret, "Cursive words recognition", in "From Pixels to Features III - Frontiers in Handwriting Recognition", S. Impedovo, J.C. Simon (eds.), Elsevier 1992, pp. 241-260.

21. J. Camillerapp, G. Lorette, G. Menier, H. Oulhadj, J.C. Pettier, "Off-line and on-line methods for cursive handwriting recognition", in "From Pixels to Features III - Frontiers in Handwriting Recognition", S. Impedovo, J.C. Simon (eds.), Elsevier 1992, pp. 273-288.

22. A. Zahour, B. Taconet, A. Faure, "Machine recognition of arabic cursive writing", in "From Pixels to Features III - Frontiers in Handwriting Recognition", S. Impedovo, J.C. Simon (eds.), Elsevier 1992, pp. 289-296.

23. B. Taconet, A. Zahour, A. Faure, "A new global off-line recognition method for handwritten words", in "From Pixels to Features III - Frontiers in

Handwriting Recognition", S. Impedovo, J.C. Simon (eds.), Elsevier 1992, pp. 327-338.

24. L.R.B. Schomaker, H.-L. Teulings, "Stroke-versus character-based recognition of on-line, connected cursive script", in "From Pixels to Features III - Frontiers in Handwriting Recognition", S. Impedovo, J.C. Simon (eds.), Elsevier 1992, pp. 313-326.

25. G. Dimauro, S. Impedovo, G. Pirlo, "Algorithms for Automatic Signature Verification", in "Handbook of Character Recognition and Document Image Analysis",H.Bunke,P.S.P.Wang(eds.),World Scientific Publ.1997,pp.605-621.

26. I. Yoshimura, M. Yoshimura, "On line signature verification incorporating the direction of pen movement - An experimental examination of the effectiveness", in "From Pixels to Features III - Frontiers in Handwriting Recognition", S. Impedovo, J.C. Simon (eds.), Elsevier 1992, pp. 353-362.

27. G. Dimauro, S. Impedovo, G. Pirlo, "A stroke-oriented approach to signature verification", in "From Pixels to Features III - Frontiers in Handwriting Recognition", S. Impedovo, J.C. Simon (eds.), Elsevier 1992, pp. 371-384.

28. G.Pirlo, "Algorithms for Signature Verification", in *Fundamentals in Handwriting Recognition*, S. Impedovo ed., Springer Verlag, Berlin, 1994, pp. 433-454.

29. G. Congedo, G. Dimauro, S. Impedovo, G. Pirlo, " A new methodology for the measurement of local stability in dynamical signatures", Proc. of the *Fourth International Workshop on Frontiers in Handwriting Recognition*, Taipei, Taiwan, 1994, pp- 135-144.

30. P. Brittan, M.C. Fairhurst, An approach to handwritten signature verification using a high performance parallel architecture, in "From Pixels to Features III - Frontiers in Handwriting Recognition", S. Impedovo, J.C. Simon (eds.), Elsevier 1992, pp. 385-390.

31. S. Impedovo, H. Bunke, P.S.P. Wang (eds.), *Automatic Bankcheck Processing*, IJPRAI, World Scientific Publ.,Singapore, 1997, Vol.11,No.4 &5.

32. I.J. Evett, C.J. Wells, F.G. Keenan, T. Rose, R.J. Whitrow, "Using linguistic information to aid handwriting recognition", in "From Pixels to Features III", S.Impedovo and J.C.Simon eds., Elsevier 1992, pp. 339-348.

33. J.J.Hull, "Language-level syntactic and semantic constraints applied to visual word recognition", in *Fundamentals in Handwriting Recognition*, S. Impedovo ed., Springer Verlag, Berlin, 1994, pp. 289-312.

34. C.Y.Suen, C. Nadal, T.A. Mai, R. Legault, L. Lam, "Recognition of totally unconstrained handwritten numerals based on the concept of multiple experts", Proc. IWFHR-1, 1990, Montreal, Canada, pp. 131-143.

35. Ley Xu, Adam Krzyzak, Ching Y-Suen, "Methods of Combining Multiple Classifiers and Their Applications to Handwriting Recognition", *IEEE Transaction on Systems, Man and Cybern.*, Vol. 22, N. 3, 1992, pp. 418-435.

36. F. Kimura, Z. Chen and M. Shridhar, "An integrated character recognition algorithm for locating and recognizing zip codes", Proc. U.S. Postal Service Advanced Technology Conf., Nov. 1990, pp. 605-619.

37. L. Lam and C.Y. Suen, "Increasing Experts for Majority Vote in OCR: Theoretical Considerations and Strategies", Prof. IWFHR-4, Taipey, Taiwan, 1994, pp. 245-254.

38. Huang, C.Y. Suen, "An Optimal Method of Combining Multiple Classifiers for Unconstrained Handwritten Numeral Recognition", Proc. of IWFHR-3, Buffalo, NY, 1993, pp. 11-20.

39. T.K.Ho, J.J. Hull, S.N. Srihari, "Combination of Structural Classifiers", IAPR Workshop on Syntactic and Structural Pattern Recognition, Murray Hill, New Jersey, June 13-15, 1990, pp. 123-136.

40. T.K. Ho, J.J. Hull, S.N. Srihari, "Decision Combination in Multiple Classifier Stsrems", *IEEE Trans. on Pattern Analysis Machine Intelligence*, Vol. 16, No. 1, Jan. 1994, pp. 66-75.

41. N. Gorsky, "Practical Combination of Multiple Classifiers", in *Progress in Handwriting Recognition*, A.C. Downton and S. Impedovo (eds.), World Scientific Publ., Singapore, 1997, pp.277-284.

42. E. Mandler, J. Schurmann, "Combining the classification results of independent classifiers based on the Dempster-Shafer theory of evidence", Pattern Recognition Artificial Intelligence,1988,pp.381-393.

43. Y.S.Huang, K.Liu, C.Y. Suen, "A Neural Network Approach for Multi-classifier recognition systems", Proc. of IWFHR-4, 1994, pp.235-244.

44. F. Yamaoka, Y. Lu, A. Shaout and M. Shridhar, "Fuzzy integration of classification results in a Handwritten Digit Recognition System", Prof. of IWFHR-4, Taipei, Taiwan, 1994, pp. 255-264.

45. T. Kawatani, H. Shimizu, "Complementary Classifier Design Using difference principal components", Proc. of ICDAR '97, IEEE Press., 1997, pp. 875-880.

46. G. Dimauro, S. Impedovo, G. Pirlo and A. Salzo, "A New Methodology for the Evaluation of Combination Processes", in *Progress in Handwriting Recognition*,A.C.Downton-S.Impedovo (eds.),WSP, Singapore, 1997.

47. J.Kim, K. Seo, K. Chung, "A Systematic Approach to classifier Selection on Combining Multiple Classifiers for Handwritten Digit Recognition", Proc. of ICDAR '97, IEEE Press., 1997, pp. 459-462.

48. A.F.R. Rahman, M.C. Fairhurst, "Introducing New Multiple Expert Decision Combination Topologies: A Case Study using Recognition of Handwritten Characters", Proc. of ICDAR '97, IEEE Press., 1997, pp. 886-891.

49. K.Konstantinides, J.Rasure, "The Khoros Application Development Environment for Image and Signal Processing". IEEE Journal of Image Processing, 1993.

50. G. Dimauro, S. Impedovo, G. Pirlo, A. Salzo, "Bankcheck recognition systems: re-engineering the design process". In *Progress in Handwriting Recognition*, A.C. Downton and S. Impedovo (eds.), World Scientific Publ., Singapore, 1997, pp. 419-425.

51. G. Dimauro, S. Impedovo, G. Pirlo, A. Salzo, "Automatic Bankcheck Processing: A New Engineered System". In *Automatic Bankcheck Processing*, IJPRAI, S. Impedovo, P.S.P. Wang and H.Bunke (eds.), World Scientific Publ., Singapore, 1997, Vol. 11., No. 4, pp. 1-38.

52. G. Dimauro, S. Impedovo, G. Pirlo, A. Salzo, "A Multi-Expert Signature Verification System for Bankcheck Processing". In *Automatic Bankcheck Processing*, IJPRAI, S. Impedovo, P.S.P. Wang and H.Bunke (eds.), World Scientific Publ., Singapore, 1997, Vol.11., No. 5.

Crucial Combinations for the Recognition of Handwritten Letters

C. Y. Suen

Centre for Pattern Recognition and Machine Intelligence
Concordia University,
1455 de Maisonneuve Blvd. West, Montreal
Quebec H3G 1M8, Canada

Z. C. Li

Department of Applied Mathematics
National Sun Yat-sen University,
Kaohsiung, Taiwan 80424

Abstract

Crucial features are important for the identification of patterns. Great efforts have been devoted to discover the most distinctive features. This paper presents the ideas of splitting patterns into 4- or 6- parts, integration for recognition, and selection of crucial combinations. The crucial combinations are explored with particular reference to 26 handprinted letters of the English alphabet. This paper proposes a new and simple algorithm to perform the crucial combinations. Also, the largest confusion regions and algorithms for finding them are provided. The most useful crucial combinations and the largest confusion combinations are listed in this paper.

Key words: Character recognition, handwritten characters, alphabetic handprints, recognition rate, regional decomposition method, part combination, crucial part, crucial combination, confusion combination.

1   Introduction

In [1], the crucial combinations and largest confusion regions are proposed with strict analysis, and the computer experiments are carried out for alphanumerals in six-partitions. In this paper, we will focus on exploitation of the crucial combinations and the largest confusion regions for 26 English letters only. Note that the capital letters in handwriting are often used in formal documents, titles of papers, book chapters, store and building names, etc. Hence it is also worthwhile to discover their distinct features. Note also that many reports on handwriting numerals and alphanumerals have appeared , but not on letters only. Let us consider the samples in Table 1, which represents 60 most common styles of 26 letter handprints in the North America, where those with "*" are the Canadian standard characters [5]. If the individual writing style and letters are identified, the models, M_{60-60} and M_{60-26}, are referred to in this paper respectively. In [1], the recognition rates are first evaluated for parts and combinations, then the perfect combinations can be found by $\rho = 100\%$. Based on the perfect combinations, the crucial combinations are then sought by the computer algorithms. In this paper, for seeking the crucial combinations and the largest confusion combinations, we may identify directly all perfect combinations without counting their recognition rates. This simplification of algorithms benefits character recognitions.

We consider 4- and 6- partitions, the combinations indicated by part numbers are shown in Table 2. To complement with the computer results for 36 alphanumerals in 6-partitions in [1], we will mainly provide the different results for 26 letters in 4-partitions.

In the next section, we introduce the crucial combinations and the largest confusion regions, and describe their algorithms. In the last section, the important results are given, accompanied with some necessary pattern analysis.

2 Crucial Combinations and their Simplified Algorithms

Let a pattern x_i be divided into M basic parts:

$$x_i = \{x_{i1}, x_{i2}, \ldots, x_{iM}\}, \quad M \geq 1. \tag{2.1}$$

Suppose that for part j there exists a set of independent bases $\{\phi_k^{(j)}\}$ such that each part of patterns belongs to one and just one basis, e.g., the k_jth: $x_{ij} = \phi_{k_j(i)}^{(j)}$. Hence the integer $k_j(i)$ denotes the belonging relation, thus leading to an integer vector $\vec{z}_i = \{z_{i1}, z_{i2}, \ldots, z_{iM}\}^T$, where $z_{ij} = k_j(i)$. As a consequence, we obtain a classification matrix for N patterns, $y_1, y_2, \ldots, y_N$ (see [2])

$$Z = \begin{bmatrix} z_{11} & z_{12} & \cdots & z_{1M} \\ z_{21} & z_{22} & \cdots & z_{2M} \\ \cdots & & & \\ z_{N1} & z_{N2} & \cdots & z_{NM} \end{bmatrix}. \tag{2.2}$$

In this paper, M = 4 or 6, N = 60.

The basic classification matrix Z is beneficial to combination recognition, because it reduces considerably the CPU time. For details readers may refer to the regional decomposition method [2].

2.1 Perfect Combinations

Perfect combinations occur if the pattern or character can be identified uniquely. In other words, under the given model, no other patterns (or characters) have the same parts or combinations. Denote the 60 patterns in Table 1 by

$$x_1, x_2, \ldots, x_{60} \tag{2.3}$$

and 26 letters by

$$C_1, C_2, \ldots, C_{26}. \tag{2.4}$$

	Char		Char		Char		Char
A_1^*	A	F_3^*	F	M_1	M	T	T
A_2	A	G_1	G	M_2^*	Ṁ	U_1	U
B_1	B	G_2	G	M_3	M	U_2	U
B_2	B	G_3	G	M_4	M	U_3^*	U
B_3^*	B	G_4	G	N	N	V	V
C_1	C.	G_5^*	G	O_1	O	W_1	W
C_2^*	C	H	H	O_2^*	Ø	W_2^*	W
D_1	D	I_1^*	I	P_1	P	W_3	W
D_2	D	I_2	I	P_2^*	P	W_4	W
D_3^*	D	J_1	J	Q	Q	X_1	X
E_1	E	J_2^*	J	R_1^*	R	X_2^*	X
E_2	E	K_1^*	K	R_2	R	Y_1^*	Y
E_3^*	E	K_2	K	S_1	S	Y_2	Y
F_1	F	K_3	K	S_2	S	Z_1^*	Z
F_2	F	L	L	S_3^*	S	Z_2	Z

Table 1 Patterns of 60 most common styles of letter handprints where "*" denotes Canadian standard characters.

N	Combinations
1	1 0 0 0
2	2 0 0 0
3	3 0 0 0
4	4 0 0 0
5	1 2 0 0
6	1 3 0 0
7	1 4 0 0
8	2 3 0 0
9	2 4 0 0
10	3 4 0 0
11	1 2 3 0
12	1 2 4 0
13	1 3 4 0
14	2 3 4 0
15	1 2 3 4

1	2
3	4

4-partitions

Table 2a   15 combinations for 4-partitions.

6-partitions

N	Combinations	N	Combinations
1	1 0 0 0 0 0	33	2 3 5 0 0 0
2	2 0 0 0 0 0	34	2 3 6 0 0 0
3	3 0 0 0 0 0	35	2 4 5 0 0 0
4	4 0 0 0 0 0	36	2 4 6 0 0 0
5	5 0 0 0 0 0	37	2 5 6 0 0 0
6	6 0 0 0 0 0	38	3 4 5 0 0 0
7	1 2 0 0 0 0	39	3 4 6 0 0 0
8	1 3 0 0 0 0	40	3 5 6 0 0 0
9	1 4 0 0 0 0	41	4 5 6 0 0 0
10	1 5 0 0 0 0	42	1 2 3 4 0 0
11	1 6 0 0 0 0	43	1 2 3 5 0 0
12	2 3 0 0 0 0	44	1 2 3 6 0 0
13	2 4 0 0 0 0	45	1 2 4 5 0 0
14	2 5 0 0 0 0	46	1 2 4 6 0 0
15	2 6 0 0 0 0	47	1 2 5 6 0 0
16	3 4 0 0 0 0	48	1 3 4 5 0 0
17	3 5 0 0 0 0	49	1 3 4 6 0 0
18	3 6 0 0 0 0	50	1 3 5 6 0 0
19	4 5 0 0 0 0	51	1 4 5 6 0 0
20	4 6 0 0 0 0	52	2 3 4 5 0 0
21	5 6 0 0 0 0	53	2 3 4 6 0 0
22	1 2 3 0 0 0	54	2 3 5 6 0 0
23	1 2 4 0 0 0	55	2 4 5 6 0 0
24	1 2 5 0 0 0	56	3 4 5 6 0 0
25	1 2 6 0 0 0	57	1 2 3 4 5 0
26	1 3 4 0 0 0	58	1 2 3 4 6 0
27	1 3 5 0 0 0	59	1 2 3 5 6 0
28	1 3 6 0 0 0	60	1 2 4 5 6 0
29	1 4 5 0 0 0	61	1 3 4 5 6 0
30	1 4 6 0 0 0	62	2 3 4 5 6 0
31	1 5 6 0 0 0	63	1 2 3 4 5 6
32	2 3 4 0 0 0		

Table 2b   63 combinations for 6-partitions.

We may express their relations by

$$C_j = X_i, \quad 1 \leq j \leq 26, \quad 1 \leq i \leq 60, \tag{2.5}$$

where the ith pattern also means the jth letter. In M_{60-60}, the part j of x_i is a perfect part if only

$$z_{ij} \neq z_{kj}, \quad \forall k, \quad k \neq i, \tag{2.6}$$

where z_{ij} are the matrix entries given in (2.2). In fact, when the values of z_{ij} are different from other z_{kj}, the jth part does not resemble any part of other patterns. This also implies that part j is a perfect part of pattern i.

When 26 letters are our targets to be identified, we do not distinguish the writing styles. Hence the perfect parts in M_{60-26} can be determined if

$$z_{ij} \neq z_{kj}, \quad \forall k, \quad C_k \neq C_i, \tag{2.7}$$

where C_k and C_i denote the letters represented by x_k and x_i respectively (see [4]).

We can easily prove by [3] that the recognition rates

$$\rho_i j = 100\% \quad \text{in } M_{60-60} \text{ and } M_{60-26}. \tag{2.8}$$

Moreover, a letter has a perfect part if this part is perfect for all its writing styles. Hence we may easily conclude which part is perfect in 26 letters, based on the perfect parts of 60 patterns.

As to the combinations in M_{60-60}, a combination is perfect if this combination does not have any resembling combinations of other patterns.

Let us now describe the above more clearly. Denote an integer set

$$K = \{1, 2, \cdots, M\}, \tag{2.9}$$

and a subset of r-combinations

$$K_r = \{k_1, k_2, \cdots, k_r\} \quad 1 \leq k_j \leq M, \tag{2.10}$$

where $\{k_j\}$ are distinct positive integers. Then

$$K_r \subseteq K, \quad r \leq M. \tag{2.11}$$

Now consider the combination K_r instead of basic parts in [2–4]. Two patterns x_i and x_k are confused with each other in the group K_r if all parts of K_r, i.e. $\forall j \in K_r$, are confused with each other. We may represent this by

$$\prod_{\forall j \in K_r} N_d(z_{ij}, z_{kj}) = 1, \tag{2.12}$$

where $N_d(z_{ij}, z_{kj})$ is the identifying function given by

$$N_d(z_{ij}, z_{kj}) = \begin{cases} 1 & \text{if} \quad z_{ij} = z_{kj}, \\ 0 & \text{if} \quad z_{ij} \neq z_{kj}. \end{cases} \tag{2.13}$$

Let ρ_{i, K_r} denote the recognition rate of combination K_r. Then the perfect combination K_r in M_{60-60} and M_{60-60} occurs if

$$\prod_{\forall j \in K_r} N_d(z_{ij}, z_{kj}) = 0, \quad \forall k, \quad k \neq i \tag{2.14}$$

and

$$\prod_{\forall j \in K_r} N_d(z_{ij}, z_{kj}) = 0, \quad \forall k, \quad C_k \neq C_i. \tag{2.15}$$

in fact, Eq. (2.14) leads to at least $\exists j \in K_r$ such that

$$N_d(z_{ij}, z_{kj}) = 0, \quad \forall k, \quad k \neq i. \tag{2.16}$$

This implies that combination K_r of x_i is distinct from that of x_k due to the non-resembling part j. Note that Eq. (2.15) is for the comparisons of different letters only. Eqs. (2.6) and (2.7) can also be represented by (2.14) and (2.15). Also we can show from [3]

$$\rho_{i, K_r} = 100\%. \tag{2.17}$$

Since the perfect combinations and the confusion combinations are complement to each other, a combination is a confusion if it is not perfect.

2.2 Crucial Combinations and the Largest Confusion Regions

Since crucial parts are special characteristics of crucial combinations, we use combinations to include parts as well. Simply speaking, crucial combinations are the roots or bases of perfect combinations, the minimal distinct features of a pattern. Therefore, a crucial combination must be a perfect combination. A combination that does not include any crucial combination must be a confusion. Strict definitions of crucial combinations can also be found in [1]. The crucial combinations are most significant in letter recognition, and in other pattern recognition. Suppose that the crucial combinations are selected from the perfect combinations, the ratios of their numbers can then be defined by

$$R_{cru.} = \frac{I_{cru.}}{I_{perf.}} \tag{2.18}$$

providing an efficiency of crucial combinations. The complete perfect combinations can be easily restored from the crucial combinations [1].

On the other hand, the largest confusion regions are also the roots or bases of confusion combinations. A combination that is not included into any largest confusion combination must also be a perfect combination. The largest confusion regions may be selected from the algorithms given in Section 2.3 below. The ratios to the confusion combinations are defined by

$$R_{conf.} = \frac{I_{Larg.}}{I_{conf.}} \tag{2.19}$$

where $I_{Larg.}$ and $I_{conf.}$ are the numbers of the largest confusion regions and the confusion combinations. Evidently,

$$I_{Larg.} + I_{conf.} = 2^M - 1, \tag{2.20}$$

where M = 4 or 6,

$$I_{Larg.} + I_{conf.} = 15 \text{ or } 63. \tag{2.21}$$

2.3 Algorithm for Crucial Combinations

We may design the following simple algorithm consisting of four steps.

- **Step I.** For the given set F of all perfect combinations obtained in Section 2.1, we partition F into M subsets G_r with length $1, 2, \ldots, M$

$$\{G_1\}, \{G_2\}, \ldots, \{G_M\}. \tag{2.22}$$

- **Step II.** Select as the crucial combinations all perfect combinations with the *shortest* combination length, e.g., one or two or more if no perfect part exists. Denote by r_m the length of combinations, then

$$r_M = 1 \ or \ 2 \ (\ even \ \geq 2). \tag{2.23}$$

- **Step III.** Determine if each combination with longer length in $\{G_r\}$, $r = r_m + 1$ is not a crucial combination. This can be done by checking whether or not this combination includes one of the crucial combinations already selected. Otherwise, this combination stands as a new, crucial combination.

- **Step IV.** Stop when $r_m = M$; otherwise let

$$r_m \Leftarrow r_m + 1 \tag{2.24}$$

and return to Step III.

2.4 Algorithm for the Largest Confusion Requires

An algorithm is given below to seek set R of the largest confusion regions from set F of perfect combinations.

- **Step I.** Partition the complement set $\bar{F}$ of F into M groups with the same length from 1 to M:

$$\{B_1\}, \{B_2\}, \ldots, \{B_M\}. \tag{2.25}$$

- **Step II.** Select first all confusion combinations with the *longest* length as the largest confusion regions, and denote the maximal length by r_M. For M-partition, we have

$$r_M = M - 1, \ or \ r_M = M \tag{2.26}$$

While $r_M = M$ indicates the existing assemble pairs (see [4]).

- **Step III.** Determine if each combination with shorter length in $\{\bar{F}\}$, $r = r_M - 1$ is included in one of the largest confusion regions already selected. If not, this combination stands as a new, largest confusion region.

- **Step IV.** Stop when $r_M = 1$; otherwise let

$$r_M \rightarrow r_M - 1, \tag{2.27}$$

and return to Step III.

Note that the difference in Algorithms in 2.3 and 2.4 lies in the selecting order: one from short to long pattern lengths; the other from long to short.

To close this section, we consider the cases when the perfect combinations are not complete. This may fit some practical application. First, we use all existing perfect combinations to obtain the "temporarily" crucial combinations, called the set $\{G_r\}$. Once a new perfect combination K_r occurs, we ignore K_r if

$$\exists G_r, \ such \ that \ G_r \subseteq K_r. \tag{2.28}$$

Otherwise set up this perfect combination K_r to be a new crucial combination added into set $\{G_r\}$; also we must discard any temporarily crucial combinations G_r if

$$G_r \subset K_r, \tag{2.29}$$

because K_r is the new perfect (i.e. crucial) combination.

Note that the above implementation is valid only under the given models, e.g. either M_{60-60} or M_{60-26}. If some new patterns are added, i.e., the models M_{n-k} are changed, and then the entire perfect combinations as well as their crucial combinations may vary. In this case, we should seek all the perfect combinations and their crucial combinations from the very beginning of algorithms in Sections 2.3 and 2.4.

3 Pattern Analysis

In this section, computer experiments are conducted to demonstrate the significance of crucial combinations and largest confusion regions, as well as their new simplified algorithms. Analysis on the resulting data from the methods of this paper will produce many interesting and fascinating conclusions in handprinted letters and their recognition. Comparisons are also made to lead to important conclusions in letters recognition.

Choose a sample set of 60 patterns illustrated in Table 1, which contains 26 English letters, where those with '*' are the Canadian Standard characters. Note that the 60 patterns of Table 1 are a subset of 89 alphanumeric patterns in [1–4]. In this paper, the 36 alphanumerals are also provided purely for comparison. We use four-partitions and six-partitions as given in Table 2. There are 15 different combinations for four-partitions and 63 different combinations for six-partitions, where the nonzero integers denote the combined parts.

By means of the computer algorithms in Sections 2.3 and 2.4, the crucial combinations and the largest confusion regions have been identified. Table 3 lists the results of 26 letters for 4-partitions. In order to compare them clearly, we have also provided the results of M_{89-36} in Table 4. Comparing Table 4 with Table 3 shows that the number of perfect combinations increases from 232 to 264; but the number of confusion combinations decreases from 158 to 126. It is interesting to note that the numbers of crucial combinations in M_{60-26} and M_{89-36} are the same 77; but the number of largest confusion regions decreases from 58 to 51. Moreover, we can see that all the letters have larger numbers of perfect combinations in M_{60-26} than those in M_{89-36}. The crucial combinations of two models, letters "I" and "O" do not have perfect combinations in M_{89-36}, but they do have in M_{60-26}. We note that under the circumstance of only letters or numerals, the resemble pairs of entire patterns will no longer exist. Therefore, all the patterns must have perfect combinations in M_{60-26}.

For 60 writing styles of 26 letters, we list the results for 4-partitions in Table 5. Comparing Table 5 with Table 3, the number of perfect combinations of any letter is larger than those of all its writing styles, obviously. This conforms again the analysis made in the previous paper [1]. For six-partitions, we also provide the results in Table 6 for comparison.

In order to display a clear view on the graphical structure of combinations, for 4-partitions we sketch in Fig. 1 all the crucial combinations and the largest confusion regions by the rectangles, where the black parts represent the existing combined parts and the white parts represent the non-existing parts. Also, for 6-partitions, we sketch in Fig. 2 all the crucial combinations and the largest confusion regions. These results are important to character analysis and identification. First, we note that all crucial combinations of 26 letters in 4-partitions have length ≤ 2. For 6-partitions, all crucial combinations have length ≤ 3. This can be interpreted by the fact that at most half of pattern areas cover the distinct characteristics of the characters.

Take character "Q" in Fig. 1 as an example. We find just one crucial part, #4, and one largest confusion region (#1&#2&#3). This can be interpreted by the fact that "Q" has the special stroke in the bottom right, which is very unique. Any combinations including the crucial part #4 can be identified as letter "Q" itself without any mistakes. Furthermore, since any combination including the crucial part #4 must also be a perfect combination, thus we can restore 8 perfect combinations. Therefore, we may store just one crucial part #4, and recover all perfect and confusion combinations. All those distinct features of handprinted character can be completely understood, simply stored and easily identified, even saving a lot of time. Again from Fig. 1, we can see that

Characters		Cru.	Perf.	Ratio	Larg.	Conf.	Ratio
1	A	3	14	21.43	1	1	100.00
2	B	5	10	50.00	3	5	60.00
3	C	3	11	27.27	2	4	50.00
4	D	3	11	27.27	2	4	50.00
5	E	3	11	27.27	2	4	50.00
6	F	2	6	33.33	2	9	22.22
7	G	3	11	27.27	2	4	50.00
8	H	3	13	23.08	2	2	100.00
9	I	4	9	44.44	2	6	33.33
10	J	3	13	23.08	2	2	100.00
11	K	3	13	23.08	2	2	100.00
12	L	3	11	27.27	2	4	50.00
13	M	2	12	16.67	1	3	33.33
14	N	4	9	44.44	2	6	33.33
15	O	2	6	33.33	2	9	22.22
16	P	2	6	33.33	2	9	22.22
17	Q	1	8	12.50	1	7	14.29
18	R	4	8	50.00	4	7	57.14
19	S	3	13	23.08	2	2	100.00
20	T	4	9	44.44	2	6	33.33
21	U	4	9	44.44	3	6	50.00
22	V	3	14	21.43	1	1	100.00
23	W	2	10	20.00	2	5	40.00
24	X	3	7	42.86	2	8	25.00
25	Y	2	6	33.33	2	9	22.22
26	Z	3	14	21.43	1	1	100.00
Sum		77	264	29.17	51	126	40.48

Table 3 Crucial combinations and largest confusion regions of 26 letters in $M_{60\text{-}26}$ for 4-partitions.

Characters		Cru.	Perf.	Ratio	Larg.	Conf.	Ratio
1	A	3	14	21.43	1	1	100.00
2	B	4	9	44.44	3	6	50.00
3	C	3	11	27.27	2	4	50.00
4	D	3	11	27.27	2	4	50.00
5	E	3	11	27.27	2	4	50.00
6	F	2	6	33.33	2	9	22.22
7	G	2	10	20.00	2	5	40.00
8	H	4	12	33.33	3	3	100.00
9	I	0	0	0.00	1	15	6.67
10	J	4	12	33.33	3	3	100.00
11	K	4	12	33.33	3	3	100.00
12	L	5	10	50.00	3	5	60.00
13	M	3	11	27.27	2	4	50.00
14	N	4	9	44.44	2	6	33.33
15	O	0	0	0.00	1	15	6.67
16	P	2	6	33.33	2	9	22.22
17	Q	1	8	12.50	1	7	14.29
18	R	4	8	50.00	4	7	57.14
19	S	3	8	37.50	3	7	42.86
20	T	4	9	44.44	2	6	33.33
21	U	4	9	44.44	3	6	50.00
22	V	3	13	23.08	2	2	100.00
23	W	2	10	20.00	2	5	40.00
24	X	3	7	42.86	2	8	25.00
25	Y	2	6	33.33	2	9	22.22
26	Z	5	10	50.00	3	5	60.00
Sum		77	232	33.19	58	158	36.71

Table 4 Crucial combinations and largest confusion regions of 26 letters in M_{89}. $_{36}$ for 4-partitions.

Patterns		Cru.	Perf.	Ratio	Larg.	Conf.	Ratio
1	A_1	5	62	8.06	1	1	100.00
2	A_2	4	53	7.55	3	10	30.00
3	B_1	6	40	15.00	4	23	17.39
4	B_2	6	40	15.00	4	23	17.39
5	B_3	8	48	16.67	3	15	20.00
6	C_1	5	31	16.13	2	32	6.25
7	C_2	9	49	18.37	2	14	14.29
8	D_1	6	52	11.54	3	11	27.27
9	D_2	6	52	11.54	3	11	27.27
10	D_3	7	53	13.21	3	10	30.00
.	.	.	.	.	.	.	.
.	.	.	.	.	.	.	.
.	.	.	.	.	.	.	.
46	T	4	36	11.11	2	27	7.41
47	U_1	7	44	15.91	3	19	15.79
48	U_2	9	50	18.00	4	13	30.77
49	U_3	10	51	19.61	4	12	33.33
50	V	5	59	8.47	2	4	50.00
51	W_1	5	62	8.06	1	1	100.00
52	W_2	5	61	8.20	2	2	100.00
53	W_3	5	62	8.06	1	1	100.00
54	W_4	4	58	6.90	2	5	40.00
55	X_1	9	49	18.37	2	14	14.29
56	X_2	8	45	17.78	2	18	11.11
57	Y_1	6	57	10.53	2	6	33.33
58	Y_2	5	47	10.64	2	16	12.50
59	Z_1	5	62	8.06	1	1	100.00
60	Z_2	5	61	8.20	2	2	100.00
Sum		308	2917	10.56	140	863	16.22

Table 5 Crucial combinations and largest confusion regions of 60 patterns in $M_{60\text{-}26}$ for 4-partitions.

Characters		Cru.	Perf.	Ratio	Larg.	Conf.	Ratio
1	A	4	53	7.55	3	10	30.00
2	B	6	40	15.00	4	23	17.39
3	C	3	28	10.71	2	35	5.71
4	D	6	52	11.54	3	11	27.27
5	E	5	37	13.51	4	26	15.38
6	F	6	42	14.29	3	21	14.29
7	G	1	32	3.12	1	31	3.23
8	H	9	49	18.37	5	14	35.71
9	I	4	36	11.11	2	27	7.41
10	J	2	48	4.17	1	15	6.67
11	K	8	48	16.67	3	15	20.00
12	L	5	50	10.00	2	13	15.38
13	M	4	54	7.41	2	9	22.22
14	N	10	51	19.61	4	12	33.33
15	O	4	23	17.39	3	40	7.50
16	P	4	36	11.11	3	27	11.11
17	Q	1	32	3.12	1	31	3.23
18	R	4	32	12.50	4	31	12.90
19	S	3	56	5.36	1	7	14.29
20	T	4	36	11.11	2	27	7.41
21	U	6	41	14.63	3	22	13.64
22	V	5	59	8.47	2	4	50.00
23	W	6	57	10.53	3	6	50.00
24	X	6	42	14.29	2	21	9.52
25	Y	4	46	8.70	2	17	11.76
26	Z	5	61	8.20	2	2	100.00
Sum		125	1141	10.96	67	497	13.48

Table 6 Crucial combinations and largest confusion regions of 26 letters in M_{60-26} for 6-partitions.

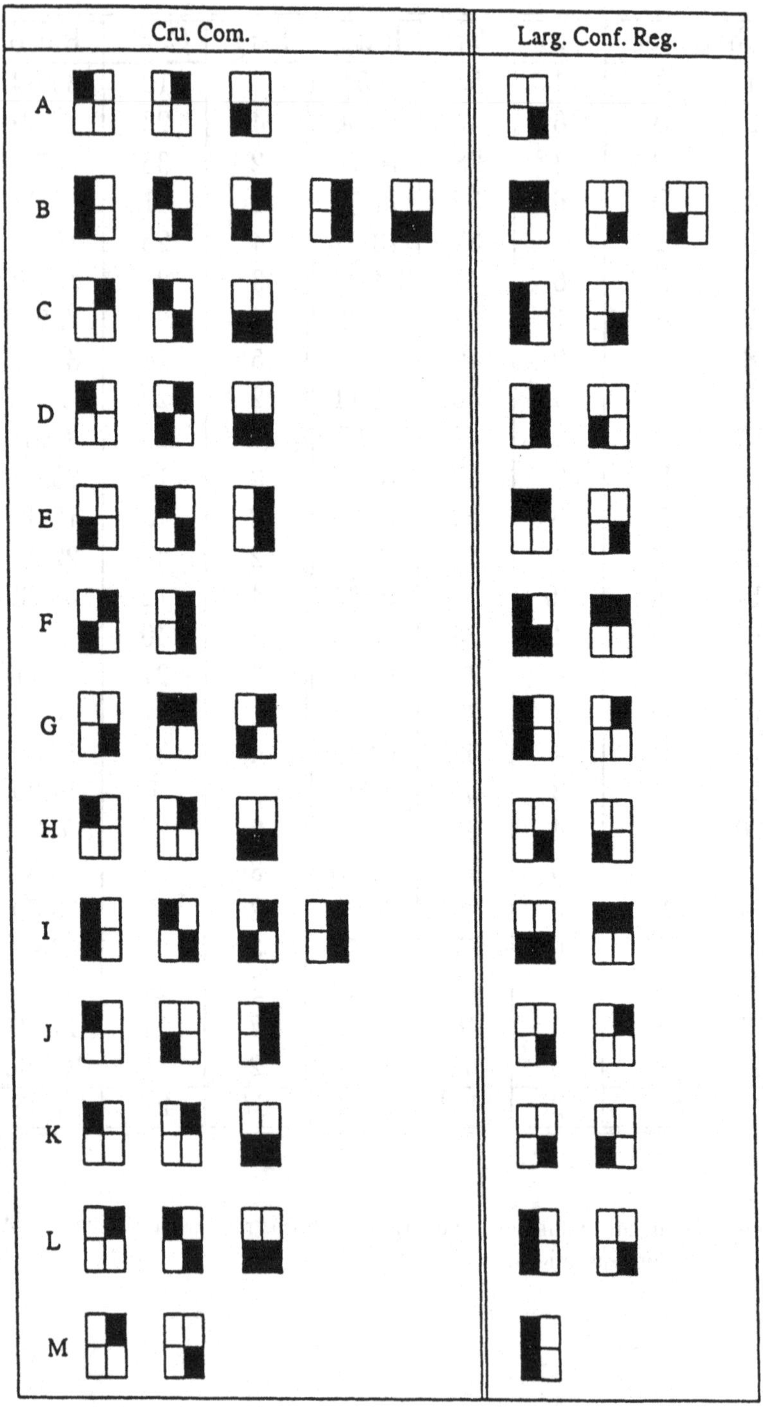

Fig. 1   Basic crucial combinations and largest confusion regions for 4-partitions in $M_{60\text{-}26}$.

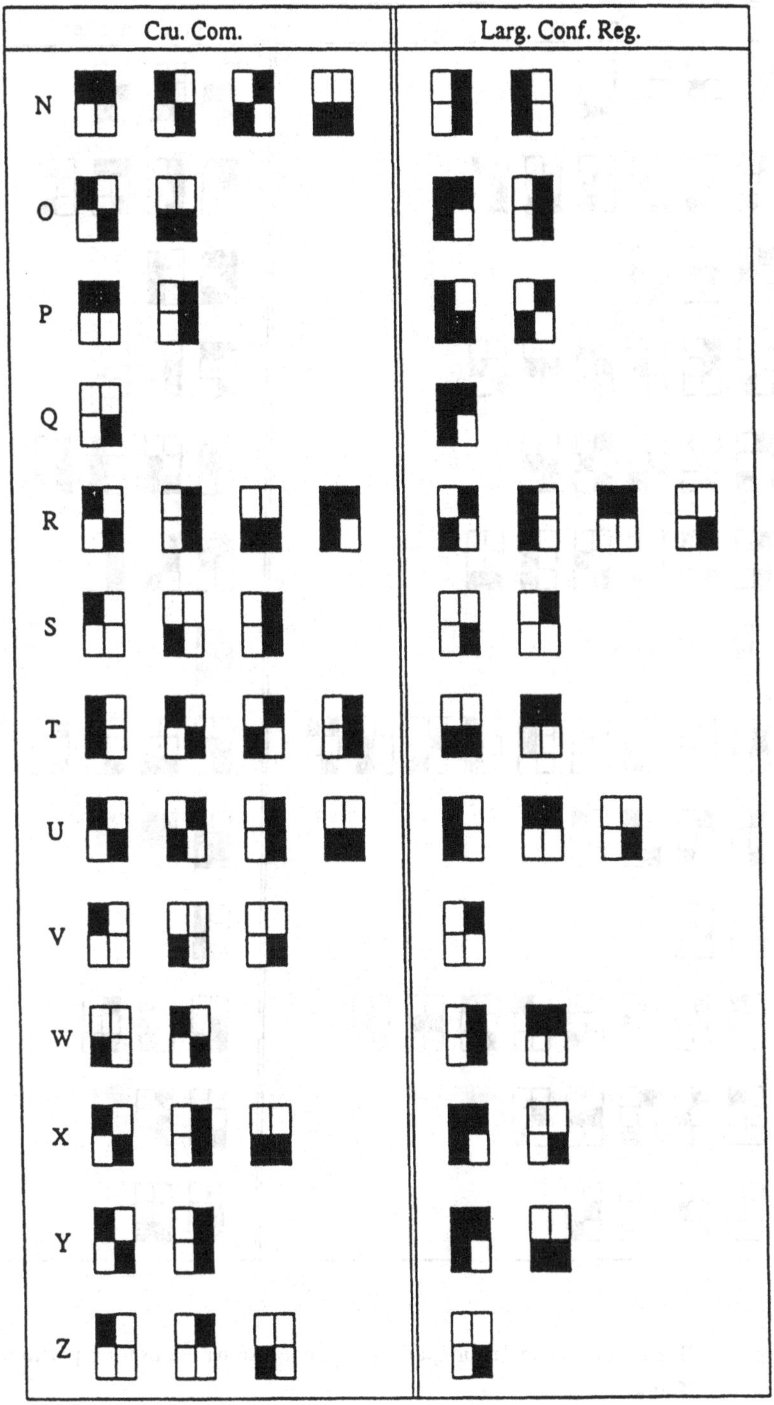

Fig. 1 (cont'd) Basic crucial combinations and largest confusion regions for 4-partitions in M_{60-26}.

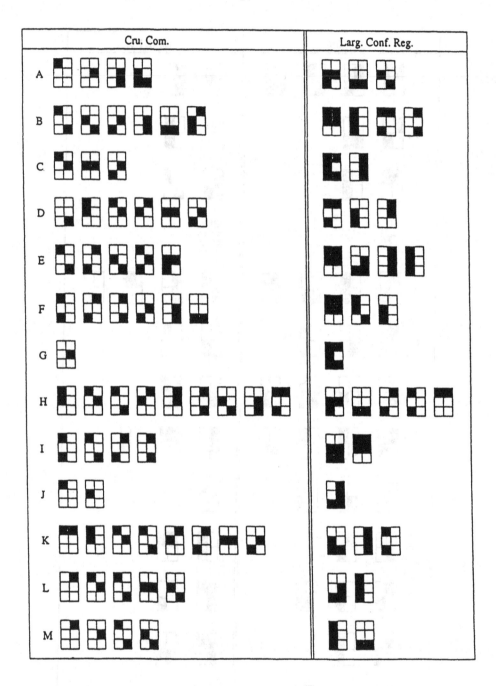

Fig. 2 Basic crucial combinations and largest confusion regions for 6-partitions in $M_{60\text{-}26}$.

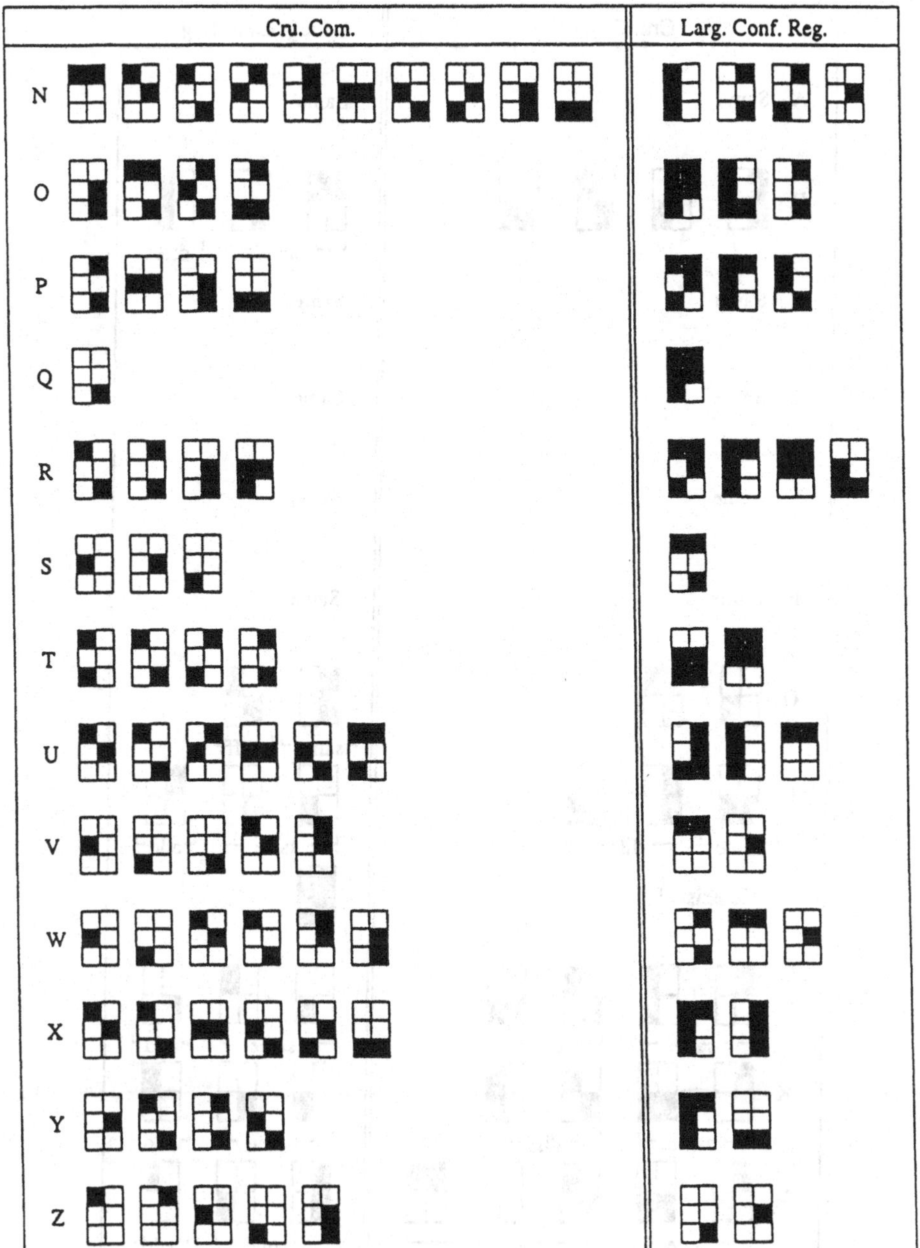

Fig. 2 (cont'd) Basic crucial combinations and largest confusion regions for 6-partitions in M_{60-26}.

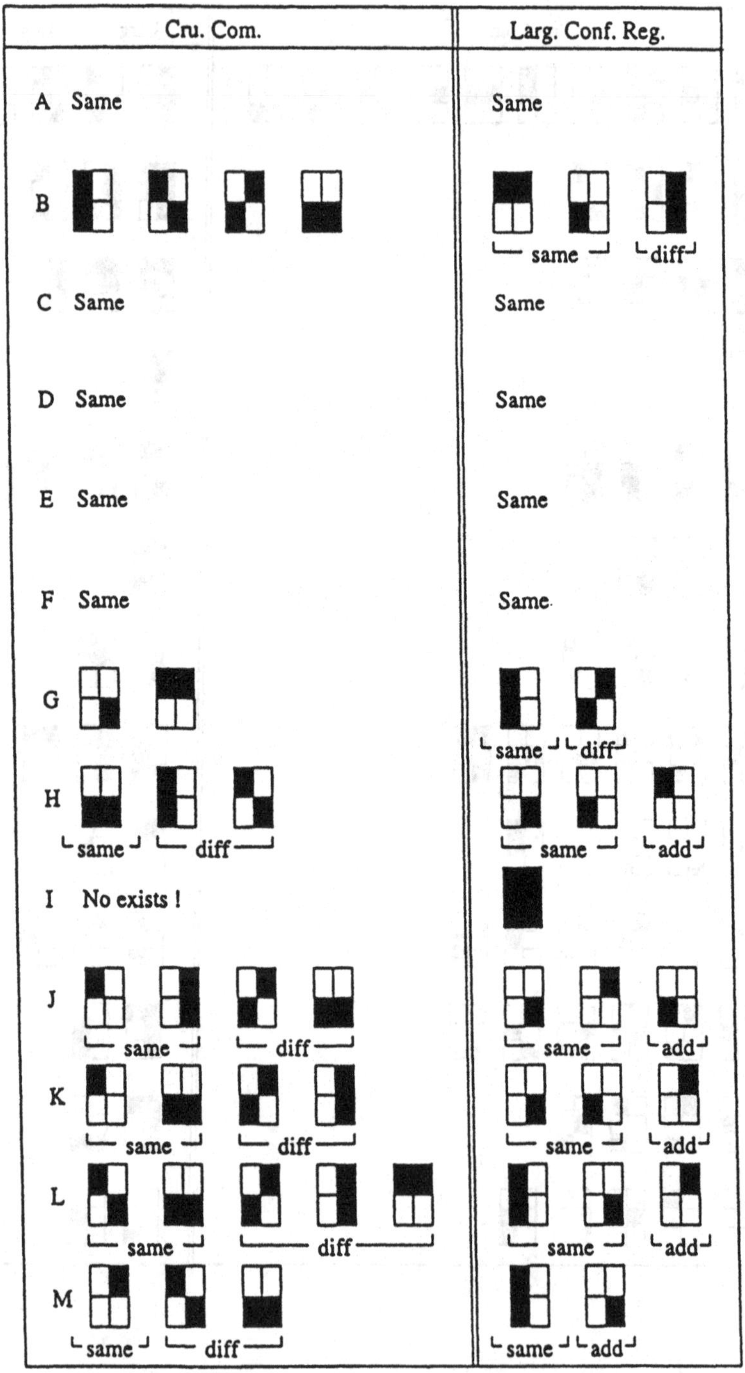

Fig. 3 Basic crucial combinations and largest confusion regions for 4-partitions in $M_{89\text{-}36}$.

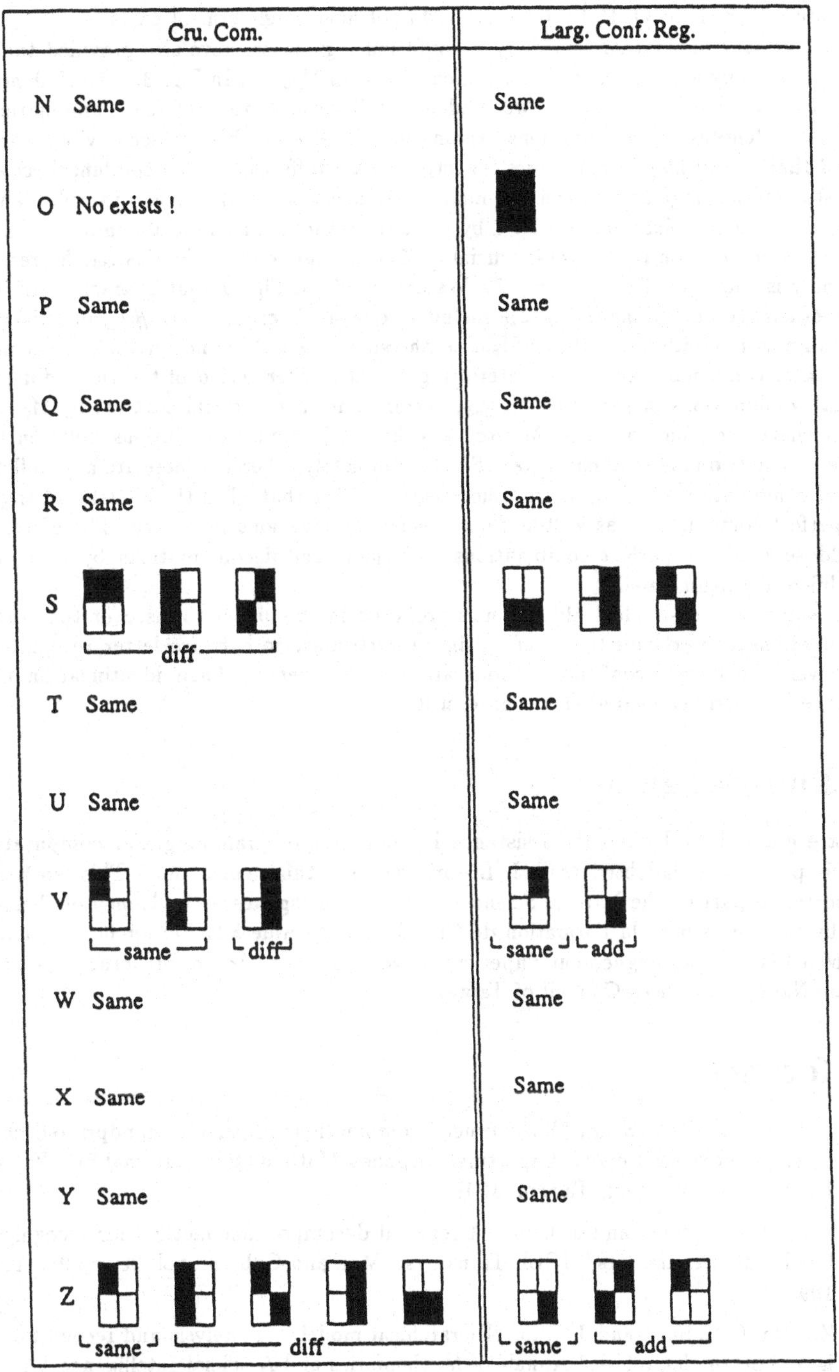

Cru. Com.
Larg. Conf. Reg.
N Same
Same
O No exists !
P Same
Same
Q Same
Same
R Same
Same
S
diff
T Same
Same
U Same
Same
V
same
diff
same
add
W Same
Same
X Same
Same
Y Same
Same
Z
same
diff
same
add

Fig. 3 (cont'd) Basic crucial combinations and largest confusion regions for 4-partitions in $M_{89\text{-}36}$.

characters B, F, I, N, O, P, R, T, U, X, Y do not have single crucial parts.

For crucial combinations and largest confusion regions, we have also provided the results in Fig. 3 by M_{89-36}, as well as a comparison with M_{60-26}. In Fig. 3, "same" denotes the same combinations between two models; "diff" denotes the different combinations; and "add" denotes the combinations existing only in M_{89-36}. We can see obviously from Fig. 3 that nearly half of characters (or letters) have different crucial combinations and largest confusion regions between two models. Take character "B" as an example. There are five crucial combinations in Fig. 1 by M_{60-26}; but four same crucial combinations by M_{89-36}. Combination (#2&#4) is crucial in M_{60-26}; but confused in M_{89-36}. Moreover, we analysis the case of character "L". We can see from Fig. 1 that character "L" has three crucial combinations in M_{60-26} including one single crucial part #2 and two combinations with length two. But in M_{89-36}, the single part #2 is not crucial; it becomes the largest confusion region. It is interesting that the intersection of the three different crucial combinations in Fig. 3 by M_{89-36} become one single crucial part #2 by M_{60-26}. Other characters, such as J, K, M, etc., also have the similar conclusions between two models. The two models of character "S" are completely different, there are no duplicate crucial combination and largest confusion regions. Note that when the model is changed, the perfect combinations as well as their crucial combinations may vary. Therefore, we should seek all the perfect combinations and their crucial combinations by using the simplified algorithms again.

In summary, a s-implied algorithm is proposed in this paper to discover the crucial combinations of parts for 26 letters, using 4-partitions, to enhance letter recognition. Moreover, the largest confusion regions are also discovered. Their identification may enhance handwritten character recognition [6–15].

Acknowledgements

We are grateful to J. Guo for assistance in computer programming and computation in this paper; and indebted to H.J. Li for preparing this manuscript. This work was supported in part by the Natural Sciences and Engineering Research Council of Canada, and by the Fonds pour la Formation de Chercheurs et l'Aide à la Recherche of Québec, the Ministère de l'Enseignement Superieur et de la Science (Action Structurantes), and by the National Sciences Council of Taiwan.

References

[1] Z. C. Li and C. Y. Suen, "Basic crucial combinations of parts to handprinted characters," Technical Report, Department Applied Mathematics, National Sun Yat-sen University, Kaohsiung, Taiwan, 1997.

[2] Z.C. Li, C.Y. Suen and J. Guo, "A regional decomposition method for recognising handprinted characters," IEEE Trans Syst. Man and Cybern., vol. 25, pp.988-1010, 1995.

[3] Z.C. Li, C.Y. Suen and J. Guo, "Hierarchical models for analysis and recognition of handwritten characters," Annals of Mathematics and Artificial Intelligence Vol. 10, pp.149-174, 1994.

[4] C.Y. Suen, J. Guo and Z.C. Li, "Analysis and recognition of alphanumeric handprints by parts," IEEE Trans. on Sys. Man and Cybern., Vol.24, pp.614-631, 1994.

[5] C.Y. Suen et al, Canadian standard alphanumeric character set for handprinting, Z243.34–M1983, Canadian Standards Association, Toronto, 1983.

[6] C.Y. Suen (Ed.), *Frontiers in Handwriting Recognition*, CENPARMI, Concordia University, 1990.

[7] P.S. P. Wang (Ed.), *Character & Handwriting Recognition: Expanding Frontiers*, World Scientific, Singapore, 1991.

[8] P.A. Koles, "Clues to letters recognition implications for the design of characters," J. Typographical Research, Vol. 3, no. 2, pp.145-168, 1968.

[9] S. Impedovo and G. Dimauro, "An interactive system for the selection of handwritten numeral classes," Proc. 10th International Conference on Pattern Recognition, Atlantic City, New Jersey, June 1990, pp. 563–566.

[10] C. Nadal, R. Legault and C.Y. Suen, "Complementary algorithms for the recognition of totally unconstrained handwritten numerals," Ibid, pp. 443–449.

[11] V. K. Govindan and A. P. Shivaprasad, "Character recognition - a review," Pattern Recognition, Vol. 23, pp.671-683, 1990.

[12] J. Mantas, "An overview of character recognition methodologies," Pattern Recognition, Vol. 19, pp.425-430, 1986.

[13] R. Kasturi and L. O'Gorman, "Document image analysis: a bibliography," Machine Vision and Applications, An International Journal, Vol. 5, pp.231-243, 1992.

[14] H. Bunke and A. Sanfeliu, *Syntactic and Structural Pattern Recognition Theory and Applications*, World Scientific, Singapore, 1990.

[15] R. Lindwurm, T. Breuer and K. Kreuzer. "Multi-expert system for handprint recognition," Proc. 5th Int. Workshop on Frontiers in Handwriting Recognition, pp. 125-129, Sept. 1996.

Recognition of Printed and Handwritten Arabic Characters

Adnan Amin

School of Computer Science and Engineering, University of New South Wales,
Sydney New South Wales 2052, Australia

Abstract. Within the image processing arena, the field of pattern recognition or matching has been the focus of extensive research. Many researchers have attempted to solve the problem of pattern recognition in general, while many others have been interested in the specific problem of character recognition because of its potential application. In addition, the automated recognition and processing of (Hand-) printed characters is also the basic function that should be given attention in order to improve man-machine communication.

A large number of research papers and reports have already been published on Latin, Chinese and Japanese characters. However, although almost a third of a billion people worldwide, in several different languages, use Arabic characters for writing, little research progress, in both on-line and off-line, has been achieved towards the recognition of Arabic characters. This is a result of the lack of adequate support in terms of funding, and other utilities such as Arabic database, dictionaries, etc.. The main objectives of this paper are: to identify the major problems related to handwritten and printed Arabic characters; to present a general panorama on the research technique in the domain of Arabic character recognition and, in particular, to present some different systems that we have developed during the past two decades in both on-line and off-line recognition

1 Introduction

For the past three decades there has been increasing interest among researchers in problems related to machine simulation of the human reading process. Intensive research has been carried out in this area with a large number of technical papers and reports in the literature devoted to character recognition. This subject has attracted immense research interest not only because of the very challenging nature of the problem, but also because it provides the means for automatic processing of large volumes of data in postal code reading [1, 2], office automation [3, 4], and other business and scientific applications [5–7].

Much more difficult, and hence more interesting to researchers, is the ability to automatically recognize handwritten characters [8–10]. The complexity of the problem is greatly increased by the noise problem and by the almost infinite variability of handwriting as a result of the mood of the writer and the nature of the writing. Analysing cursive script requires the segmentation of characters within the word and the detection of individual features. This is not a problem unique to computers; even human beings, who possess the most efficient optical reading device (eyes), have difficulty in recognizing some cursive scripts and have an error rate of about 4% in reading tasks in the absence of context [11].

The different approaches covered under the general term character recognition fall into either the on-line or off-line category, each having its own hardware and recognition algorithms.

In on-line character recognition systems, the computer recognizes the symbols as they are drawn [12–16]. The most common writing surface is the digitizing tablet, which typically has a resolution of 200 points per inch and a sampling rate of 100 points per second, and deals with one-dimensional data.

Off-line recognition is performed after the writing or printing is completed. Optical Character Recognition, OCR [17–21], deals with the recognition of optically processed characters rather than magnetically processed ones. In a typical OCR system, input characters are read and digitized by an optical scanner. Each character is then located and segmented and the resulting matrix is fed into a preprocessor for smoothing, noise reduction, and size normalization. Off-line recognition can be considered the most general case: no special device is required for writing and signal interpretation is independent of signal generation, as in human recognition.

Many papers have been concerned with the recognition of Latin, Chinese and Japanese characters. However, although Arabic characters are used in several widespread languages, little research has been conducted toward the automatic recognition of Arabic characters because of the strong cursive nature of its writing rules.

The objectives of this paper are to identify the problems related to handwritten Arabic characters, and describe different methods for the recognition of handwritten Arabic characters for both on-line and off-line recognition. The remainder of this paper is organized as follows: Section 2 reviews some of the basic characteristics of Arabic writing. Section 3 covers different approaches for segmentation and feature extraction, and presents various methods adopted for recognition in both on-line and off-line systems. Finally, concluding remarks are given in Sec. 4.

2. General Characteristics of the Arabic Writing System

A comparison of the various characteristics of Arabic, Latin, Hebrew and Hindi scripts are outlined in Table 1. Arabic is written from right to left. Arabic text (machine printed or handwritten) is cursive in general and Arabic letters are normally connected on the base line. This feature of connectivity will be shown to be important in the segmentation process. Some machine printed and handwritten texts are not cursive, but most Arabic texts are, and thus it is not surprising that the recognition rate of Arabic characters is lower than that of disconnected characters such as printed English.

Arabic writing is similar to English in that it uses letters (which consist of 29 basic letters), numerals, punctuation marks, as well as spaces and special symbols. It differs from English, however, in its representation of vowels since Arabic utilizes various diacritical markings. The presence and absence of vowel diacritics indicates different meanings in what would otherwise be the same word. For example, مـ درسة

is the Arabic word for both "school" and "teacher". If the word is isolated, diacritics are essential to distinguish between the two possible meanings. If it occurs in a sentence, contextual information inherent in the sentence can be used to infer the appropriate meaning. In this paper, the issue of vowel diacritics is not treated, since it is more common for Arabic writing not to employ these diacritics. Diacritics are only found in old manuscripts or in very confined areas.

Table 1. Comparison of various scripts.

Characteristics	Arabic	Latin	Hebrew	Hindi
Justification	R-to-L	L-to-R	R-to-L	L-to-R
Cursive	Yes	No	No	Yes
Diacritics	Yes	No	No	Yes
Number of vowels	2	5	11	–
Letters shapes	1 – 4	2	1	1
Number of letters	28	26	22	40
Complementary characters	3	–	–	–

The Arabic alphabet is represented numerically by a standard communication interchange code approved by the Arab Standard and Metrology Organization (ASMO). Similar to the American Standard Code for Information Interchange (ASCII), each character in the ASMO code is represented by one byte. An English letter has two possible shapes, capital and small. The ASCII code provides separate representations for both of these shapes, whereas an Arabic letter has only one representation in the ASMO table. This is not to say, however, that the Arabic letter has only one shape. On the contrary, an Arabic letter might have up to four different shapes, depending on its relative position in the text. For instance, the letter (ع A'in) has four different shapes: at the beginning of the word (preceded by a space), in the middle of the word (no space around it), at the end of the word (followed by a space), and in isolation (preceded by an unconnected letter and followed by a space). These four possibilities are exemplified in Fig.1.

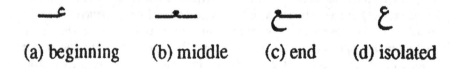

(a) beginning   (b) middle   (c) end   (d) isolated

Fig. 1. Different shapes of the Arabic letter 'A'in (a) beginning (b) middle (c) end (d) isolated.

In addition, different Arabic characters may have exactly the same shape, and are distinguished from each other only by the addition of a complementary character. Complementary characters (A portion of a character that is needed to complement an Arabic character) are positioned differently, for instance, above, below or within the confines of the character. Fig. 2 depicts two sets of characters, the first set having five characters and the other set three characters. Clearly, each set contains characters which differ only by the position and/or the number of dots associated with it. It is worth noting that any erosion or deletion of these complementary characters results in a misrepresentation of the character. Hence, any thinning algorithm needs to efficiently deal with these dots so as not to change the identity of the character.

(a)   (b)

Fig. 2. Arabic characters differing only with regard to the position and number of associated dots.

Arabic writing is cursive and is such that words are separated by spaces. However, a word can be divided into smaller units called subwords (see Appendix). Some Arabic characters are not connectable with the succeeding character. Therefore, if one of these characters exists in a word, it divides that word into two subwords. These characters appear only at the tail of a subword, and the succeeding character forms the head of the next subword. Fig. 3 shows three Arabic words with one, two, and three subwords. The first word consists of one subword which has nine letters; the second has two subwords with three and one letter, respectively. The last word contains five subwords, each consisting of only one letter.

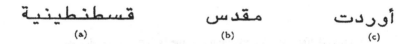

(a)   (b)   (c)

Fig. 3. Arabic words with constituent subwords.

Arabic writing can be, in general, classified into typewritten (Naskh), handwritten (Ruq'a) and artistic (or decorative Calligraphy, Kufi, Diwani, Royal, and Thuluth) styles as shown in Fig. 4. Handwritten and decorative styles usually include vertical combinations of charactes called ligatures. This feature makes it difficult to determine the boundaries of the characters. Furthermore, characters of the same font have different sizes (i.e. characters may have different widths even though the two characters have the same font and point size). Hence, word segmentation based on a fixed size width cannot be applied to Arabic.

Fig. 4. Different styles and Fonts for the Writing of Arabic Text

3. Recognition of Arabic Characters

There are two strategies which have been applied to printed and handwritten Arabic character recognition. These can be categorized as follows:

(i) Holistic strategies in which the recognition is globally performed on the whole representation of words and where there is no attempt to identify characters individually. These strategies were originally introduced for speech recognition and can fall into two categories:

    (a) Methods based on distance measurements using Dynamic Programming [22, 23].

    (b) Methods based on a probabilistic framework (Hidden Markov Models) [24–27].

(ii) Analytical strategies in which words are not considered as a whole, but as sequences of small size units and the recognition is not directly performed at word level but at an intermediate level dealing with these units, which can be graphemes, segments, pseudo-letters, etc. [28, 29].

This section covers different techniques used for the recognition of Arabic characters in both on-line and off-line systems.

3.1. On-Line Recognition Systems

An on-line character recognition system typically uses a magnetic graphic tablet as the main input device. Such a tablet operates through a special pen in contact with the surface of the tablet which emits the coordinates of the plotted points at a constant frequency. Breaking contact prompts the transmission of a special character. Thus, recording on the tablet produces strings of coordinates separated by signs indicating when the pen has ceased to touch the tablet surface.

On-line recognition has several interesting characteristics. First, recognition is performed on one-dimensional data rather than two-dimensional images as in the case of off-line recognition. The writing line is represented by a sequence of dots whose location is a function of time. This has several important consequences:

- The writing order is available and can be used by the recognition process.
- The writing line has no width.
- Temporal information, like velocity can also be taken into consideration.
- Additionally, penlifts can be useful in the recognition process.

The IRAC (Interactive Recognition of Arabic Character) system [30] adopted a structural classification method for recognizing on-line, handwritten isolated Arabic characters. The system consists of three major steps. First, preprocessing using an algorithm inspired by Berthod and Jancenne [31] which accomplished two objectives: the reduction of the number of points and the correction of any deformation. Features such as the shape of the main stroke, the number of strokes, the characteristics relative to the group of dots and the presence of an eventual "zigzag" (which can be either isolated, e.g. ك or attached, e.g. س) are then extracted from the character. If these features do not permit the recognition of a character by a simple dichotomic consultation of the dictionary, backtracking is performed in order to correct the shape of the main stroke. Finally, secondary features of the main stroke such as frame size, the start point and the curvature are extracted using a distance function in order to remove the ambiguity and provide an exact match.

Amin [32] proposed a system for on-line Arabic word recognition. The hand drawing is directly segmented into characters on the basis of certain heuristic criteria.

For every word, the characters are connected to each other by horizontal segments from right to left. Statistically, these connections appear:

(i)   almost always after an intersection point,

(ii)   often after a cusp point,

(iii)   sometimes simply after a change of curvature.

These points serve remarkably as separators in the segmentation process and directly permit one to obtain a list of the characters of the word component. However, since the separators are determined by statistical considerations, the segmentation has to proceed through successive essays (i.e. tentative tries), and the character recognition module is responsible for validating each essay. Hence, the two modules (segmentation and recognition) are highly interactive.

Each class of separator is affected by a priority. The intersection point has the highest priority and the cusp point has the lowest priority. An Arabic handwritten word is segmented into characters by using the separators according to their priority and their appearance in the word. With the help of each separator, portions of the handwriting are extracted and transmitted to the character recognition module for identification. If no character has been recognized, the corresponding segmentation is cancelled and a new tentative try is carried out with the separator of the next lower priority. For example, assume E is the end of the last character identified. If the portion of handwriting between E and the first intersection point is not identified as a character, then the second essay will be carried out on the portion between E and the next cusp point.

The character recognition module is similar to the system which recognizes isolated characters; this system is in fact a further development of the method for the recognition of isolated Arabic characters, described in [30].

Finally, three hypotheses at most, i.e. the three best score candidates provided by the character recognition module, are associated with each of the characters which are extracted from the handwritten Arabic word by the segmentation module. The set of hypotheses form a lattice (Fig. 5). Identification of the word consists of traversing the lattice to find the path of the best score corresponding to a word in the dictionary. Binary diagrams [33] are used for resolving ambiguities and for eliminating all candidates which are not present in the dictionary of known words.

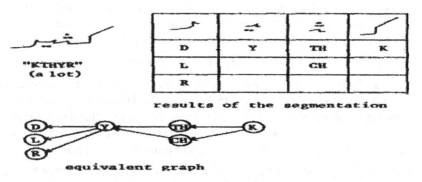

Fig. 5. An example of segmenting the Arabic word and the equivalent graph

In [34] two methods were presented for recognizing cursive Arabic words. The first is a syntactical method based on the segmentation of words into primitives. These primitives are: open and closed curves, vertical/horizontal stroke, cusp point, inflection point, group of dots, shape of the secondary stroke and pencil lif.

The syntactic representations of the characters are merged into a tree structure whose arcs are labelled with primitive names. Recognition consists of finding a path through the tree, using the primitive classes of the description of the character to be recognized as instructions for choosing the arcs to take. The terminal node of the path gives the name of the recognized character. The recognition of words works from the list of primitives provided by the segmentation process. It consists of finding all combinations of successive characters utilizing binary diagrams to eliminate all ineligible combinations of letters.

The second method of [35] is global without segmentation into characters. It uses some properties of Arabic writing, in which seven characters cannot have a left connection. These characters are:

This property causes words to be decomposed into pieces (Fig. 3 (c)) whenever one of these characters occurs.

This method uses the notion of stroke instead of character. A vector defining the main parameters of a word (number of secondary strokes, size and position of groups of dots, number of intersection points, number of cusp points) makes it possible to recognize a word and each of its constituent strokes. Whenever the same vector yields several possibilities, they are classified according to a score computed from secondary parameters of each stroke (the start point, form and angular variation of the main stroke).

Furthermore, to enhance the recognition rate, a syntactical and semantical analyzer that verifies the grammatical structure and the meaning of the Arabic sentence is used [35].

El-Sheikh and El-Taweel [36] proposed a system for recognizing properly segmented handwritten Arabic characters. Four groups of characters are identified depending on the position of the character within a word (isolated, beginning, middle or at the end). Moreover, each group is further classified into four subgroups depending on the number of strokes (one, two, three, or four) in the character. The classification of characters within each subgroup is done by means of some features, such as the existence of a maximum or a minimum on either the horizontal or vertical direction of the main stroke, the ratio between length and width, the type of secondary stoke and other features.

Al-Emami and Usher [37] presented a system for on-line recognition of handwritten Arabic words. Words were entered via a graphic tablet and segmented into strokes based on the method proposed by Belaid et al. [38]. In the preliminary learning process, specifications of the strokes of each character are fed to the system, while in the recognition process, the parameters of each stroke are found and special rules are applied to select the collection of strokes that best match the features of one of the stored characters. However, few words were used in the learning and testing processes, which makes the performance of the system questionable.

3.2. Off-Line Recognition Systems

Off-line character recognition systems typically use a scanner as the main input device. Off-line recognition can be considered as the most general case: no special device is required for writing and signal interpretation is independent of signal generation, as in human recognition.

3.2.1. Word segmentation

The segmentation phase is a necessary step in recognizing printed Arabic text. Any error in segmenting the basic shape of Arabic characters will produce a different representation of the character component.

Two techniques have been applied for segmenting machine printed and handwritten Arabic words into individual characters: implicit and explicit segmentations.

(i)   Implicit segmentation (straight segmentation): In this technique, words are segmented directly into letters. This type of segmentation is usually designed with rules that attempt to identify all the character's segmentation points.

(ii) Explicit segmentation: In this case, words are externally segmented into pseudo-letters which are then recognized individually. This approach is usually more expensive due to the increased complexity of finding optimum word hypotheses.

In all Arabic characters, the width at a connection point is much less than the width of the beginning character. This property is essential in applying the baseline segmentation technique [39, 40]. The baseline is a medium line in the Arabic word in which all the connections between the successive characters take place. If a vertical projection of bi-level pixels is performed on the word (Eq. (3.1)),

$$v(j) = \sum_i w(i, j) \qquad (3.1)$$

where $w(i, j)$ is either zero or one and i, j index the rows and columns, respectively, the connectivity point will have a sum less than the average value (AV) (Eq. 3.2)

$$AV = (1 / Nc) \sum_{j=1}^{Nc} Xj \qquad (3.2)$$

and where Nc is the number of columns and Xj is the number of black pixels of the jth column.

Hence, each part with a sum value much less than AV should be a boundary between different characters. However if the histogram produced from the vertical projection does not follow the condition of Eq. (3.3), the character remains unsegmented, as illustrated in Fig. 6.

By examining Arabic characters , it is found that the distance between successive peaks does not exceed one third the width of the Arabic character. That is

$$|d_k| < d_l / 3 \qquad\qquad (3.3)$$

where d_k is the distance between kth peak and peak $k+1$, and d_l is the total width of the character.

Fig. 6. An example of segmentation of the Arabic word كثير into characters (a) Arabic word (b) histogram (c) word segmented into characters.

Moreover, at the end of a word or a subword, Eq. (3.4) is also to hold.

$$L_{k+1} > 1.5 * L_k \qquad\qquad (3.4)$$

where L_k is the kth peak in the histogram. This rule is brought to bear because of the inter-connectivity of Arabic characters and their shapes at the end of a word.

An example of a histogram of a word that should be segmented into five parts based on the average value appears in Fig. 7. However, when the last rule is applied only the first four segmentation locations should be chosen, hence reducing the number of characters to four since both conditions are to hold simultaneously.

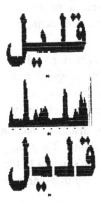

Fig. 7. An example of the Arabic word قليل and its segmentation into characters (a) Arabic word (b) histogram. (c) word segmented into characters.

This approach depends heavily on a predefined threshold value related to the character width. Moreover, this approach will not work effectively for skewed images.

Almuallim and Yamaguchi [41] proposed a structural recognition technique for Arabic handwritten words. Their system consists of four phases. The first is preprocessing, in which the word is thinned and the midline of the word is detected. Since it is difficult to segment a cursive word into letters, words are segmented into separate strokes and classified as complementary characters, strokes with a loop and strokes without a loop. These strokes are then further classified using their geometrical and topological properties. Finally, the relative positions of the classified strokes are examined, and the strokes are combined in several steps into the string of characters that represents the recognized word. System failures in most cases were due to incorrect segmentation of words.

Segmentation is also achieved by tracing the outer contour [42] of a given word and calculating the distance between the extreme points of intersection of the contour with a vertical line. The segmentation is based on a horizontal scan from right to left of the closed contour using a window of adjustable width w. For each position of the window, the average vertical distance h_{av} is calculated across the window. At the boundary between two characters, the following conditions should be met:

(i) $h_{av} < T$. In this case, a silence region is detected, which means that the average vertical distance over the window should be less than a certain preset threshold T.

(ii)   Detected boundaries should lie on the same horizontal line (the base line).

(iii)   No complementary characters should be located (above or below the base line) at a silence region.

Readjustment of parameters w and T as well as backtracking may occur if segmentation leads to a rejected character shape.

El-Khaly and Sid-Ahmed [43] segment a thinned word into characters by following the average baseline of the word and detecting when the pixels start to go higher or lower than it.

Shoukry [44] used a sequential algorithm based on the input-time tracing principle which depends on the connectivity properties of the acquired text in the binary image domain. This algorithm bears some resemblance to an algorithm devised by Wakayama [45] for the skeletonization of binary pictures.

Kurdy and Joukhadar [46] use the upper distance function of the subword, which is the set of the highest points in each column. They assign to each point of the function a token name by comparing the point's height to the height and token name of the point on its right. Using a grammar, they then parse the sequence of tokens of a subword to find the connection points.

Finally, Amin and Al-Sadoun [47, 48] adopted a new technique for segmenting Arabic text. The algorithm can be applied to any font and it permits the overlay of characters. There are two major problems with the traditional segmentation method which depends on the baseline:

(i)   Overlapping of adjacent Arabic characters occurs naturally, see Fig. 8 (a). Hence, no baseline exists. This phenomenon is common in both printed and handwritten Arabic text.

(ii)   The connection between two characters is often short. Therefore, placing the segmentation points is a difficult task. In many cases, the potential segmentation points will be placed within a character rather than between characters.

The word in Fig. 8 (a) was segmented utilizing a baseline technique. Fig. 8 (b) shows the proper segmentation and the result of the new segmentation method is shown in Fig. 8 (c).

The new technique can be divided into four major steps. First is the digitization step in which the original image is transformed into a binary image utilising a scanner (300 dpi). Second, there is a preprocessing step in which the Arabic word is thinned

using a parallel thinning algorithm. Third, the skeleton of the image is traced from right to left using a 3 * 3 window and a binary tree is constructed. The Freeman code [49] is used to describe the skeleton shape. Finally, the binary tree is segmented into subtrees such that each subtree describes a character in the image.

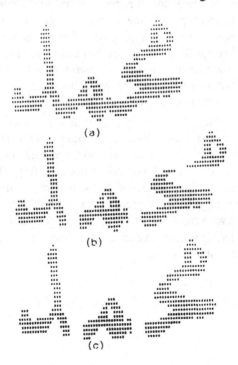

Fig. 8. Example of an Arabic word ـحـد and different techniques of the segmentation.

3.2.2. Feature extraction and recognition

It is known that features represent the smallest set that can be used for discrimination purposes and for a unique identification for each character. Features can be classified into two categories:

(i)   Local features which are usually *geometric* (e.g. concave/convex parts, type of junctions: intersections/T-junctions/endpoints etc.).

(ii)   Global features which are usually *topological* (connectivity, number of connected components, number of holes, etc.) or *statistical* (Fourier transform, invariant moments, etc.).

Parhami and Taraghi [50] presented a technique for the automatic recognition of machine printed Farsi text (which is similar to Arabic text). The authors first segment the subword into characters by identifying a series of potential connection points on the baseline at which line thickness changes from or to the thickness of the baseline. Although they also have some rules to keep characters at the end of a subword intact, they segment some of the wider characters (e.g. س) into up to three segments. Then they select twenty features based on certain geometric properties of the Farsi symbols to construct a 24 bit vector that is compared with entries of a table where an exact match is checked first. The system is heavily font dependent, and the segmentation process is expected to give incorrect results in some cases.

Table lookup is used for the recognition of isolated handwritten Arabic characters [51]. In this approach, the character is placed in a frame which is divided into six rectangles and a contour tracing algorithm is used for coding the contour as a set of directional vectors by using a Freeman code. However, this information is not sufficient to determine Arabic characters, therefore extra information related to the number of dots and their position is added. If there is no match, the system will add the feature vector to the table and consider that character as a new entry.

Amin and Masini [39] adopted a structural approach for recognizing printed Arabic text. Words and subwords are segmented into characters using the base-line technique. Features such as vertical and horizontal bars are then extracted from the character using horizontal and vertical projections (Fig. 9). Four decision trees, chosen according to the position of the character within the word which was computed by the segmentation process, have been used. The structure of four decision trees allows a rapid search for the appropriate character. Furthermore, trees are utilized in distinguishing characters that have the same shape but appear in different positions within a word.

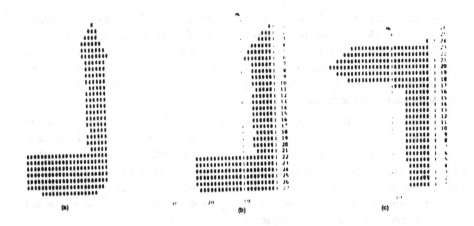

Fg. 9. Vertical and horizontal scanning of the character ل a) character ل (b) horizontal scanning (c) vertical scanning.

Amin and Mari [40] proposed a new technique for multifont Arabic text which includes character and word recognition. A character is divided into many segments by a horizontal scan process (Fig. 10). In this way, segments are connected to form the basic shape of the character. Segments not connected with any other segment are considered to be complementary characters. By using the Freeman code [49], the contour detection process is applied to these segments to trace the basic shape of the character and generate a directional vector through a 2 * 2 window. A decision tree is then used for the recognition of the characters. Finally, a Viterbi algorithm [52] is used for Arabic word recognition to enhance the recognition rate. The main advantage of this technique is to allow an automatic learning process to be used.

Segment 1	Segment 2	Segment 3

Fig. 10. Major segments of character

The study reported in [53, 54] utilizes moment invariant descriptors to recognize the characters. Other techniques include a set of Fourier descriptors from the coordinate sequences of the outer contour which is used for the recognition [42]. Also, in [55] each character is assigned a logical function where characters are preclassified into four groups depending on the existence of certain pixels in a specified location of the image.

In [56] table lookup is adopted for the recognition of isolated Arabic characters. In this approach, the character is placed in the frame window and divided into small windows to extract some features. These features include end points, intersection points, corners, and the relationship between the length and width of the window frame. Characters are identified by an association between feature points and their locations within the window frame. The recognition is achieved by finding a match between unknown characters and entries in a lookup table.

To enhance the recognition rate of an OCR system, some characteristic morphological properties of the Arabic language can be used. Amin and Al-Fedaghi [6] describe a method for spell correction of Arabic words. They correct spelling errors and complete words that have some unrecognized characters using an algorithm that depends on the frequencies of roots and patterns in Arabic.

Amin and Al-Sadoun [57, 58] proposed a structural approach for recognizing handwritten Arabic characters. The binary image of the character is first thinned using a parallel thinning algorithm and then the skeleton of the image is traced from right to left using 3 * 3 window in order to build a graph to represent the character. Features like straight lines, curves and loops are then extracted from the graph. Finally, a five layer artificial neural network is used for the character classification. Each character is classified in term of the segments used in the system such as dot, hamza, line, curve and loop. The relationships between the segments are encoded in the object inter-relationship matrix. The overall design of the input layer uses 150 neurons. Fig. 11 illustrates an example of the character representation using this input layer design.

Altuwaijri and Bayoumi [59] introduced a system for recognizing printed Arabic words using artificial Neural Networks (NN). The system can be described into three different steps: first the Arabic input word is segmented into characters using a similar approach as [39]. Next, six moments are used for extracting features from the segmented characters feeding it the neural network. Finally, a multi-layer perceptron network with back–propagation learning with one hidden layer is used to classify the character.

Finally, Al-Badr and Haralick [60] proposed a system to recognize machine printed Arabic words without prior segmentation by applying mathematical morphology operations on the whole page to find the locations where shape primitives are present. They then combine those primitives into characters and print out the character identities and their location on the page.

Segment 1 (1 Dot Above)
1 0 0 0 0 1 0 0 1 0 X X
Segment 2 (East Curve Small Upper)
0 0 0 1 0 1 0 0 0 0 0 1
Segment 3 (East Curve Large)
0 0 0 1 0 1 0 0 0 1 0 0

Segment inter-relationship matrix
(2,1) Relationship 1 0 0 0 0 0 0 0
(3,1) Relationship 1 0 0 0 0 0 0 0
(3,2) Relationship 1 0 0 0 0 0 0 1

Fig. 11. Complete representation of an Arabic character for the neural network input layer

4. Conclusion

This paper presented the problems related to printed and handwritten Arabic characters, and much of the important research work was briefly described in an attempt to present the current status of Arabic character recognition research. This is still an open research area and there is no commercial Arabic OCR system available yet. This is because of the segmentation problem, which is in fact similar to the segmentation of cursive script in many languages, and because of the complexity of Arabic characters. Moreover, all the algorithms presented in this paper deal with unvocalized text and the recognition of vowel diacritics is an extremely important research area in the Arabic language.

As stated previously, no vital computational techniques in this area have yet been fully explored. As such, this field is of importance for future research.

References

1. L. D. Harmon, Automatic recognition of printed and script, *Proc. IEEE,* **60,** 10, 1165–1177, 1972.

2. A. A. Spanjersberg, Experiments with automatic input of handwritten numerical data into a large administrative system, *IEEE Trans. Man Cybern.* **8** (4), 286–288, 1978.

3. L. R. Focht and A. Burger, A numeric script recognition processor for postal zip code application, *Int. Conf. Cybernetics and Society,* 486–492, 1976.

4. J. Schuermann, Reading Machines, *6th Int. Conf on Pattern Recognition,* Munich, Germany, 1031–1044, 1982.

5. R. Plamondon and R. Baron, On-line recognition of handprint schematic pseudocode for automatic Fortran code generator, *8th Int. Conf on Pattern Recognition,* Paris, 741–745, 1986

6. A. Amin and S. Al-Fedaghi, Machine recognition of printed Arabic text utilising a natural language morphology, *Int. J. of Man-Machine Studies* **35**(6), 769–788, 1991.

7. D. Guillevic and C. Y. Suen, Cursive script recognition: A fast reader scheme, *2nd Int. Conf. on Document Analysis and Recognition*, Japan, 311–314, 1993.

8. M. K. Brown and S. Ganapathy, Preprocessing technique for cursive script word recognition, *Pattern Recogn.* **19**(1), 1–12, 1983.

9. R. H. Davis and J. Lyall, Recognition of handwritten characters a review, *Image and Vision Computing* **4**(4), 208–218, 1986.

10. E. Lecolinet and O. Baret, *Cursive Word Recognition: Methods and Strategies, Fundamentals in Handwriting Recognition*, ed. S. Impedovo, 235–263, 1994.

11. C. Y. Suen, R. Shingal and C. C. Kwan, Dispersion factor: A quantitative measurement of the quality of handprinted characters, *Int. Conf. of Cybernetics and Society*, pp. 681–685, 1977.

12. A. Amin and A. Shoukry , Topological and statistical analysis of line drawing, *Pattern Recogn. Lett.* **1**, 365–374, 1983.

13. J. Kim and C. C. Tappert, Handwriting recognition accuracy versus tablet resolution and sampling rate, 7th *Int. Conf. on Pattern Recognition*, Montreal, 917–918, 1984.

14. J. R. Ward and T. Kuklinski, A model for variability effects in handprinted with implication for the design of handwritten character recognition system, *IEEE Trans. Man Cybern.* **18**, 438–451, 1988.

15. F. Nouboud and R. Plamondon, On-line recognition of handprinted characters: Survey and beta tests, *Pattern Recogn.* **25**(9), 1031–1044, 1990.

16. C. Tappert, C. Y. Suen and T. Wakahara, The state of the art in One–line Handwriting Recognition, *IEEE Trans. Pattern Anal Machine Intell.* **PAMI–12**, 787–808, 1990.

17. C. Y. Suen, M. Berthod and S. Mori, Automatic recognition of handprinted characters, the state of the art, *Proc. IEEE* **68** (4), 469–483, 1980.

18. J. R. Ullmann, Advance in character recognition, *Application of Pattern Recognition*, ed. K. S. Fu, 197–236, 1982.

19. V. K. Govindan and A. P. Shivaprasad, Character recognition A– review, *Pattern Recogn.* **23** (7), 671–683, 1990.

20 S. Impedovo, L. Ottaviano and S. Occhinegro, Optical character recognition– A survey, *Int. J. of Pattern Recogn. and Artif. Intell.* **5** (1&2) 1–24, 1991.

21. S. Srihari, From pixel to paragraphs: the use of models in text recognition, *Second Annual Symp on Document Analysis and Information Retrieval*, Las Vegas, USA, 47–64, 1993.

22. M. Khemakhem, Reconnaissance de caracters imprimes par comparaison dynamique, These de Doctorate de 3 e'me cycle, University of Paris XI, 1987.

23. M. Khemakhem and M. C. Fehri, Recognition of Printed Arabic charcters by comparaison dynamique, Proc. First Kuwait Computer Conference, pp. 448–462, 1989.

24. H. Y. Abdelazim and M. A. Hashish, Interactive font learning for Arabic OCR, Proc. First Kuwait Computer Conference, 464-482, 1989.

25. H. Y. Abdelazim and M. A. Hashish, Automatic recognition of handwritten Hindi numerals, Proc. of the 11th National Computer Conference, Dhahran, 287-299, 1989.

26. Z. Emam and M. A. Hashish, Application of Hidden Markov Model to the recognition of isolated Arabic word, Proc. of the 11th National Computer Conference, Dhahran, 761-774, 1989.

27 R. Schwartz, C. LaPre, J. Makhoul, C. Raphael, and Y. Zhao. Language independent ocr using a continuous speech recognition system. 13th International Conference on Pattern Recognition, vol. C, pages 99-103, Vienna, Austria,1996.

28. B. Al-Badr and S. Mahmoud, Survey and bibliography of Arabic optical text recognition, Signal Processing 41 , 49-77, 1995.

29. A. Amin, Arabic Character Recognition. Handbook of Character Recognition and Document Image Analysis edited by H Bunke and P S P Wang, May 1997.

30. A. Amin, A. Kaced, J. P. Haton and R. Mohr, Handwritten Arabic characters recognition by the IRAC system, *5th Int. Conf. on Pattern Recognition,* Miami, USA, 729–731, 1980.

31. M. Berthod and P. Jancenn, Le pretraitement des traces manuscrits sur une tablette graphique, *2eme, congres AFCET-INRIA,* Toulouse, France, 195–209, 1979.

32. A. Amin, Machine recognition of handwritten Arabic word by the IRAC II system, *6th Int. Conf. on Pattern Recognition,* Munich, 34–36, 1982.

33. E. M. Riseman and R. W. Ehrich, Contextual word recognition usig binary diagrams, *IEEE Trans. Computer* c–20 (4), 397–403, 1971.

34. A. Amin, G. Masini and J. P. Haton, Recognition of handwritten Arabic words and sentences, *7th Int. Conf. on Pattern Recognition,* Montreal, 1055–1057, 1984.

35. A. Amin, IRAC: Recognition and understanding systems, *Applied Arabic Linguistic and Signal and Information Processing,* ed. R. Descout, Hemisphere, New York, 159–170, 1987.

36. T. S. El-Sheikh and S. G. El-Taweel, Real-time Arabic handwritten character recognition, *Pattern Recogn.* 23 (12), 1323–1332, 1990.

37. S. Al-Emami and M. Usher, On-line recognition of handwritten Arabic characters, *IEEE Trans. Pattern Anal Machine Intell.* PAMI-12, 704–710, 1990.

38. A. Belaid and J. P. Haton, A syntactic approach for handwritten mathematical formula recognition, *IEEE Trans. Pattern Anal. Machine Intell.* **PAMI-6**. 105–111, 1984.

39. A. Amin and G. Masini, Machine recognition of muti-fonts printed Arabic texts, *8th Int. Conf. on Pattern Recognition,* Paris, 392–395, 1986.

40. A. Amin and J. F. Mari, Machine recognition and correction of printed Arabic text, *IEEE Trans. Man Cybern.* **9**(1), 1300–1306, 1989.

41. H. Almuallim and S. Yamaguchi, A method of recognition of Arabic cursive handwriting, *IEEE, Trans. Pattern Anal. and Machine Intell.* **PAMI-9**, 715–722, 1987.

42. T. El-Sheikh and R. Guindi, Computer recognition of Arabic cursive script, *Pattern Recogn.* **21** (4), 293–302, 1988.

43. F. El-Khaly and M. Sid-Ahmed, Machine recognition of optically captured machine printed Arabic text, *Pattern Recog.* **23** (11), 1207–1214, 1990.

44. A. Shoukry, A sequential algorithm for the segmentation of typewritten Arabic digitized text, *Arabian J. Sc. and Eng.* **16**(4), 543–556, 1991.

45. T. Wakayama, A core-line tracing algorithm based on maximal square moving, *IEEE Trans. Pattern Analysis and Machine Intell.* **PAMI-4**, 68–74, 1982.

46. B. M. Kurdy and A. Joukhadar, Multifont recognition system for Arabic characters, *3rd Int. Conf. and Exhibition on Multi-lingual Computing (Arabic and Roman Script),* U.K, 731–739, 1992.

47. A. Amin and H. Al-Sadoun, A segmentation technique of Arabic text, *11th Int. Conf. on Pattern Recognition,* 441–445, 1992.

48. H. B. Al-Sadoun and A. Amin, A new structural technique for recognizing printed Arabic text, *Int. J. of Pattern Recognition and Artif. Intell.* **9**(1), 101–125, 1995.

49. H. Freeman, On the encoding of arbitrary geometric configuration, *IEEE. Trans. Electronic Comp.* **EC-10**, 260–268, 1968.

50. B. Parhami and M. Taraghi, Automatic recognition of printed Farsi texts, *Pattern Recog.* **14**,(6), 395–403, 1981.

51. S. Saadallah and S. Yacu, Design of an Arabic character reading machine, *Proc. of computer Processing of the Arabic language,* Kuwait, 1985.

52. D. Forney, The Viterbi algorithm, *Proc. IEEE* **61** (3), 268–278, 1973.

53. S. El-Dabi, R. Ramsis and A. Kamel, Arabic charcter recognition system: Statistical approach for recognizing cursive typewritten text, *Pattern Recogn.* **23** (5), 485–495, 1990.

54. H. Al-Yousefi and S. S. Udpa, Recognition of Arabic characters, *IEEE Trans. Pattern Anal. and Machine Intell.* **PAMI-14**, 853–857, 1992.

55. A. Nouh, A. Ula and A. Sharaf-Edin, Boolean recognition technique for typewritten Arabic character set, *Proc. 1st King Saud Univ. Symp. on Computer Arabization*, Riyadh, 90–97, 1987.

56. K. Jambi, Arabic charcter recognition: Many approaches and one decade, *Arabian J. Sc. Eng.* **16** (4), 499–509, 1991.

57. A. Amin and H. Al-Sadoun, Handprinted Arabic character recognition system, *12th Int. Conf. on Pattern Recognition*, 536–539, 1994.

58. A. Amin and H. Al-Sadoun, Handprinted Arabic character recognition system using an arficial neural network, *Pattern Recognition* **29** (4), 663–675, 1996.

59. M. Altuwaijri, and M. Bayoumi, Arabic Text Recognition Using Neural Networks, International Symposium on Circuits and Systems - ISCAS'94. pp. 415-418, 1994.

60. B. Al-Badr and R. Haralick, Segmentation-free word recognition with application to Arabic, *3rd Int. Conf. on Document Analysis and Recognition*, Montreal, 355–359, 1995.

Objective Evaluation of the Discriminant Power of Features in an HMM-Based Word Recognition System

A. El-Yacoubi[1,2], M. Gilloux[2], R. Sabourin[1,3] and C.Y. Suen[1]

[1] *Centre for Pattern Recognition and Machine Intelligence*
Department of Computer Science, Concordia University
1455 de Maisonneuve Boulevard West, Montréal, Canada H3G 1M8

[2] *Service de Recherche Technique de La Poste*
Department Reconnaissance, Modélisation Optimisation (RMO)
10, rue de l'ile Mabon, 44063 Nantes Cedex 02, France

[3] *Ecole de Technologie Supérieure*
Laboratoire d'imagerie, de vision et d'intelligence artificielle (LIVIA)
1100 Notre-Dame Ouest, Montréal, Canada H3C 1K3

Abstract

This paper describes an elegant method for evaluating the discriminant power of features in the framework of an HMM-based word recognition system. This method employs statistical indicators, entropy and perplexity, to quantify the capability of each feature to discriminate between classes without resorting to the result of the recognition phase. The HMMs and the Viterbi algorithm are used as powerful tools to automatically deduce the probabilities required to compute the above-mentioned quantities.

Keywords: Handwritten Word Recognition, Feature Extraction, Discriminant Power of Features, Hidden Markov Models.

1. Introduction

Usually, a word recognition system is described according to the following scheme (Fig. 1):

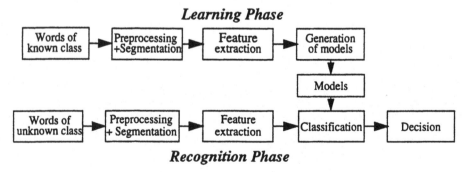

Fig. 1. Architecture of a word recognition system

The first two steps are optional, and may be required or not depending on the complexity of the problem to be solved. For instance, in the case of automatic reading of legal amounts in a cheque processing system, the vocabulary of possible words is around 30. This led many researchers to model each word as a whole [Gilloux92] [Paquet93] [Gorski94] [Knerr96] [Suen96], dropping in this way the segmentation phase. When dealing with the problem of automatic sorting of handwritten addresses, the recognition of city names [Gilloux95] [Cohen94] [Caesar94] [Shrihari95] or street names [ElYacoubi95] involves the use of large vocabularies. Therefore, a segmentation of words into smaller units such as letters is necessary, to reduce the set of models to a reasonable size. On the other hand, due to the presence of guide lines on cheques, the slant of the writing baseline isn't as frequent or important as it is in the latter application.

Unlike preprocessing and segmentation, the feature extraction phase is always present in a word recognition system. The aim of this phase is to reduce redundancy in the input image while preserving the meaningful information for recognition. Among the desired characteristics, the features should be easy to extract, robust to noise, insensitive to intra-class variability and sensitive to inter-class variability.

To judge the quality of the features, most approaches try to measure their effect on recognition rates. However, these methods are not reliable since they depend strongly on the recognition model that uses them. In this paper, we present an approach that carries out an objective evaluation of the discriminant power of features, without resorting to the results of the recognition phase. This approach is based on some statistical indicators that are well known in information theory, such as entropy and perplexity. To this purpose, we use hidden Markov models [Bahl83] [Cox88] [Poritz88] [Rabiner89] [Lee91] along with the Viterbi algorithm [Forney73] to label each feature extracted from the input image by one of the classes considered in our word recognition system: letters or half letters. This allows us to automatically determine the various probabilities needed to compute the quantities mentioned above.

The organization of this paper is as follows. Section 2 gives an overview of our word recognition system, including steps of preprocessing, segmentation, feature extraction and handwritten word modeling. In section 3, we present our method of evaluating the discriminative power of features. Section 4 describes the experiments we performed to validate our approach. Finally, section 5 gives some concluding remarks and perspectives.

2. System Overview

Our system is designed to recognize unconstrained handwritten words (or sequence of words) such as those encountered in handwritten postal addresses. Therefore, we must cope with all writing styles: handprinted, cursive or mixed (Fig. 2). In this section, we present the outline of our approach. A more detailed description can be found in [ElYacoubi96].

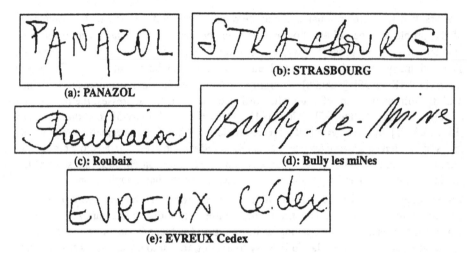

Fig. 2. Variability of handwriting styles: Handprinted (**a**) and (**b**), Cursive (**c**) and Mixed (**d**) and (**e**).

Markovian modeling assumes that each input word image is represented as a sequence of observations. These features should be statistically independent, once the underlying hidden state chain is known. To satisfy the latter requirement, we first apply to the word image a preprocessing phase, aimed at removing of information that is not meaningful to recognition. Then, we perform segmentation and feature extraction processes to transform the input image into an ordered sequence of symbols.

2.1. Preprocessing

As we said before, the goal of preprocessing is to reduce irrelevant information, such as noise and writing slant, that increases the complexity of the recognition task. In our system, we use four preprocessing steps to normalize word images: baseline slant normalization, character skew correction, lower case letter area normalization when dealing with cursive words, and finally smoothing. We give below a brief description of this system. For more details see [ElYacoubi94].

Our approach for baseline slant normalization consists of two steps. The first estimates roughly the baseline as the line that best fits with the lower contour minima (using the least square method) after having filtered those minima corresponding to downstrokes. The writing baseline is then normalized by simply replacing the (x,y) coordinates of each pixel of the bitmap by the new coordinates (x',y') given by: $x' = x$ and $y' = y - x.tan\alpha$ where α is the angle between the horizontal axis and the detected line (Fig. 3b). In the second step, a second model taking as input the normalized image approximates the writing slant as the succession of line segments linking together the lower contour minima (after filtering). The definitive normalization is then done on each zone vertically delimited by two successive minima, using the same transform as before, the angle of which varies dynamically depending on the slant of the line linking the current pair of minima (Fig. 3c).

To normalize the slant of characters, we sample the word image contour at a rate of 8 pixels per step and we compute the slope of each segment linking two successive samples. The horizontal or nearly horizontal segments are not considered. The average character slant is then estimated from the mean of the slope of the remaining segments (Fig. 3d).

The aim of the third preprocessing is to normalize the height of the lower-case letter area (or the core region) of cursive words. The technique we developed is similar to the second step of the first preprocessing described above. After having detected the upper contour maxima, we filter those belonging to ascenders or capitalized letters. The lower-case letter area normalization is then carried out on each zone vertically delimited by two successive maxima, using a nonlinear transform of this zone's pixels, which sets the two maxima to the same height, while keeping the baseline horizontal (Fig. 3e).

Finally, we smooth the output image to remove some noise that appears at the borders of the word image due to the various geometric transformations made in preprocessing (Fig. 3f). The overall processing time is around one second in average on a Sun Sparcstation IPX.

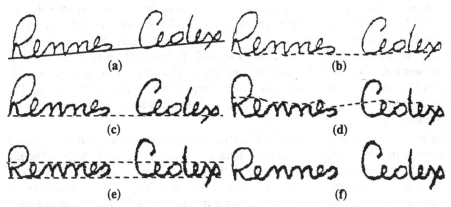

Fig. 3. Preprocessing steps applied to word images: (a) Original image, (b) and (c) Baseline slant normalization, (d) character slant normalization, (e) Lower-case letter area normalization, (f) Definitive image after smoothing.

2.2. Segmentation

The segmentation process attempts to split each word image into its component letters. Since this operation is difficult due to ambiguity between letters, we choose a segmentation algorithm that is not perfect but will produce a deliberately high number of segmentation points, offering in this way several segmentation options, the best one being validated during recognition. This algorithm, highly inspired by the approach developed by M. Leroux [Leroux 91], merely detects lower contour minima of words. Then each minimum is validated as a segmentation point or not, according to some rules that roughly impose a vertical cut of a shape without crossing any loop, while minimizing the vertical transition histogram of the word image. As a result, our segmentation algorithm splits a word (or a sequence of words) into *graphemes*, each of which may be a correctly segmented, an under-segmented or an oversegmented letter (Fig. 4).

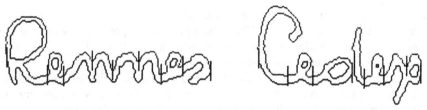

Fig. 4. Segmentation of words into letters or pseudo-letters.

2.3. Feature Extraction

The sequence of segments obtained by the segmentation process is transformed into a sequence of symbols by considering two sets of features. The first set, of size 27, is based on global features such as loops, ascenders and descenders. These characteristics are encoded in various ways according to their size and position. Then, each different combination of these features within a segment is encoded by a different symbol. For example, in (Fig. 5a), the first segment is encoded by symbol "L" reflecting the existence of a large ascendent and a loop located above the core region. The second segment is encoded by symbol "o" indicating the presence of a small loop within the body of the writing. The third segment is represented by symbol "–", which encodes shapes without any interesting feature, *etc*.

The second feature set is based on the analysis of the bidimensional contour transition histogram of each segment in the horizontal and vertical directions. After a filtering phase, the histogram values may be 2, 4 or 6. We focus only on the median part of the histogram, which represents the stable area of the letter. In each direction, we determine the dominant transition number (2, 4 or 6). Each different pair of dominant transition numbers is then encoded by a different symbol or class. After having created some further subclasses, this coding leads to a set of 14 symbols. (Fig. 5b) shows three examples of such a coding. Letters B, C and O, whose pairs of dominant transition numbers are (6,2), (4,2) and (4,4), are encoded by symbols called "B", "C" and "O" respectively.

We also have five segmentation features that try to reflect the way segments are linked together. These features consist of the space between two segments, in the case of disconnected letters, and of the segmentation point vertical position, in the case of connected letters. The space is encoded in three ways, according to its relative width with respect to the average segment width, while the second feature is encoded into two different symbols according to its relative vertical position compared to the word baseline. Finally, the feature extraction process represents each word image by two symbolic descriptions of equal length, each consisting of an alternating sequence of symbols encoding a segment shape and of symbols encoding the segmentation point associated to this shape (Fig. 5c).

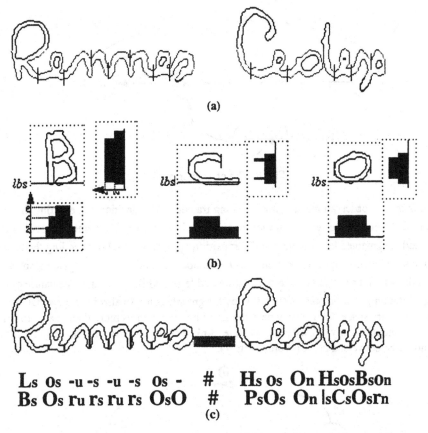

Fig. 5. Feature extraction phase: (a) first type of feature, (b) second type of feature, (c) representation of the image as a pair of descriptions, each consisting of an alternating sequence of shape symbols (large size) and segmentation symbols (small size).

2.4. Handwritten Word Modeling

When considering a large vocabulary such as those involved in the recognition of city names, it is not possible to build a different model for each word. Therefore, our handwriting modeling is carried out at the letter level. As our segmentation process may produce either a correct segmentation of a letter, an oversegmentation of a letter into two segments, or a letter omission, we built a five state HMM having three paths to take account of these configurations (Fig. 6).

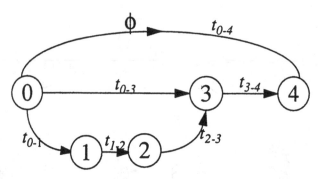

Fig. 6. The character model

In this model, observations are emitted along transitions. Transition t_{04}, emitting the null symbol Φ, models the letter omission case. Transition t_{03} emits a symbol encoding the correctly segmented letter shape, while transition t_{34} emits a symbol encoding the nature of the segmentation point associated with this shape. Transitions t_{01} and t_{23} are associated with the left and right parts of an oversegmented letter, while t_{12} models the nature of the segmentation point that gave rise to this oversegmentation. We also have a special model for inter-word space in the case where the input image contains more than one word (Fig. 5). This model simply consists of two states linked by two transitions, modeling a space or no space between a pair of words (Fig. 7).

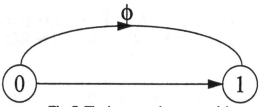

Fig. 7. The inter-word space model

In the learning phase, since we have the exact labelling of each word image, the word model is built by concatenating the appropriate elementary letter models (Fig. 8). This learning is performed using Viterbi training [Rabiner93] [Cox88]. After initializing the model parameters with random values, each word is matched against its associated feature sequence *via* the Viterbi algorithm. According to the current model, observation sequences of the training set are segmented into states by recovering the optimal alignment path. The re-estimations of the new parameters of the HMMs are then directly obtained by examining the number of transitions between each pair of states and the number of observation symbols for each state (or transition). This procedure is repeated until the increase in the global probability of observing feature sequences falls below a small fixed threshold. This strategy allows the word models, for a sufficient learning database, to detect the regularity of the observed phenomena, so that they are able to learn precise letter models without any need of hand-segmenting feature sequences into subsequences that correspond to letters (manual labelling) as in [Chen94] [Knerr96] [Kim97].

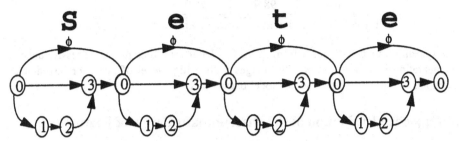

Fig. 8. The word model for word "Sete" in the learning phase. Note that the final state (4) of a letter model serves as the initial state (0) of the following model.

In the recognition phase, the word model is still built by concatenating letter models. However, since no information is available on the style in which an unknown word has been written, a letter model here actually consists of two models in parallel, associated with the upper and lower case modes of writing a letter (Fig. 9). As a result, two consecutive letter models are now linked by four transitions associated with the various ways two consecutive letters may be written: uppercase-uppercase (UU), uppercase-lowercase (UL), lowercase-uppercase (LU) and lowercase-lowercase (LL). The probabilities of these transitions are estimated from the same learning database which served for HMM parameter estimation. To complete our model description, we should add that each emitted observation along a transition corresponding to a shape is in fact a combination of two observations, related to the two sets of features we are using.

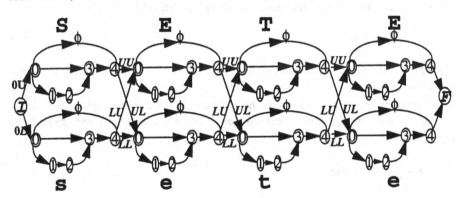

Fig. 9. The word model in recognition

The word recognition system described above was trained on a learning set of 11,410 unconstrained handwritten French city names extracted manually from real mail envelopes, and was tested on a different set of 4,313 images. The recognition rates (RR) obtained with this model for lexicons of size 10, 100 and 1,000 and without rejection are given in Table 1.

Lexicon	10	100	1000
RR	98.9	95.0	86.3

Table 1. Recognition rates obtained with our recognition model

The system errors come mainly from splitting a letter into three pieces, excessive under-segmentation in a word, ambiguity, images with a high level of noise, or insufficient examples to reliably estimate some model parameters.

3. Objective Evaluation of the Discriminant Power of Features

In order to compare the performance of several feature sets, it is useful to have a measure of the intrinsic information carried by features. Most approaches, however, use the recognition module to indirectly evaluate a feature set by comparing the recognition rates obtained by using that set. These techniques are not reliable since they strongly depend on the robustness of the recognition phase. Moreover, they are highly dependent of the lexicon of the possible words. For instance, if the length (number of characters) of words in this lexicon shows a high variation, the recognition is much easier regardless of the discriminant power of the features employed. Finally, we cannot rely on such techniques to judge individually each feature; all that we can have is unquantified information about the global discriminant power of the feature set. In this section, we describe a new way to evaluate the quality of features without resorting to the results of the recognition phase. The technique we developed is based on statistical indicators, entropy and perplexity, well known in the field of information theory. These measures were introduced to speech recognition [Bahl83] to evaluate the difficulty of the task of recognizing words in a specific application. In this case, the associated entropy is defined as:

$$H = -\sum_{w} p(w) \cdot \log_2 p(w) \tag{3.1}$$

where $p(w)$ is the *a priori* probability of word w related to the language of the considered application.

In our problem, the goal is to evaluate the difficulty of the task at a higher level, that is after the feature extraction phase. This leads us to define the *conditional entropy H(f)* of a given feature f:

$$H(f) = -\sum_{i=1}^{N_c} p(c_i|f) \cdot \log p(c_i|f) \tag{3.2}$$

where c_i are the classes considered in our modeling, and N_c is the number of classes (letters or half letters, as we will see in next section).

$H(f)$, measured in bits, quantifies the capability of feature f to discriminate between the classes c_i. The maximum of $H(f)$ is $\log N_c$. It occurs when

$$p(c_i|f) = \frac{1}{N_c} \qquad \forall c_i \tag{3.3}$$

In this case, feature f does not contribute any information to discriminate between classes, and can be considered as useless. On the other hand, $H(f)$ reaches its minimum, which is 0, when there exist one class c_i such that:

$$p(c_i|f) = 1 \quad \text{and} \quad p(c_j|f) = 0 \qquad \forall j \neq i \tag{3.4}$$

The *conditional perplexity* $P(f)$ of a feature f is related to its entropy $H(f)$ by the following definition:

$$P(f) = 2^{H(f)} \tag{3.5}$$

The advantage of using perplexity *versus* entropy is that it varies between 1 and N_c. Thus, it can be directly compared to the size of the vocabulary.

The *global entropy* is obtained by carrying out a sum over all the features weighted by their *a priori* probability:

$$H = \sum_{j=1}^{N_f} p(f_j) \cdot H(f_j) \tag{3.6}$$

where N_f is the size of the feature set. H allows us to quantify globally the discriminant power of the feature set. The corresponding *global perplexity* P is given by:

$$P = 2^H \tag{3.7}$$

This quantity also varies between 1 and N_c, and reaches its maximum when for each feature f_j, we have

$$p(c_i|f_j) = \frac{1}{N_c} \qquad \forall i \in [1, N_c] \tag{3.8}$$

4. Experiments

Considering the modeling in our system, each feature related to a segment (or grapheme) may be generated by a letter, left half part of a letter or right half part of a letter. Therefore, the number of classes to be discriminated by features is three times the number of possible letters. The latter, whose size is 86, include upper-case letters, lower-case letters, numerals and some other special characters such as "/", "-", etc. Moreover, some letters such as n, u, v, w, N, U, V, W that are likely to be regularly split into two or more segments are represented each by two "letters", e.g. n_1 and n_2 for letter n.

To evaluate the entropies and perplexities presented in section 3, we must know the probability $p(c_i|f_j)$ for each pair (c_i, f_j). The natural way to do this is to label manually each segment or grapheme by the associated correct class (letter or half letter). However, this operation is not desirable, because it costs a lot economically and requires important human means. To overcome this problem, we use the HMMs along with the Viterbi algorithm to automatically label each segment with the appropriate class. The idea is that at the end of the training process described in section 2, each word HMM (sequence of character HMMs) is matched with its associated feature sequence. Then using the *backtracking* procedure of the Viterbi algorithm, we recover the best path in the model to label each feature (segment) in an automatic way. Figure 10 shows an example of such a technique.

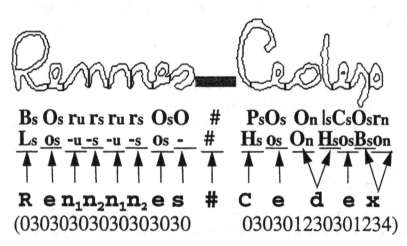

Fig. 10. Recovering the best path (and state sequence) through the feature sequence and using it to label each feature with its corresponding letter or pseudo-letter.

When applied to our learning database containing 11,410 word images, this technique allowed us to build a labelled grapheme database of 161,000 samples (letters or half letters). Note that the best path recovered for labelling purposes turned out to be almost always correct. This is due to the fact that our character and word models are well adapted to the data that they are supposed to represent. The quantities $p(c_i|f_j)$ are then computed as the occurrence frequency of the pair (c_i, f_j) over the number of samples associated with feature f_j. Quantities $p(f_j)$ are estimated in the same way by their occurrence frequency in the grapheme database.

Having determined the required probabilities, we computed the perplexities associated with the three types of features that could be used in our system: the first and second feature sets, and the third set that results from the combination of the first two. Table 2 and Table 3 show the perplexity corresponding to the 11 best features of the first set and the second set respectively, in other words features that have the 11 lowest perplexities.

f	z	G	0	o	O	L	Q	K	b	d	s
P	10.3	10.4	12.2	12.9	15.2	15.7	16.0	18.9	23.2	23.7	23.7

Table 2. The 11 best features of the first set with their associated perplexities

f	E	A	r	O	B	C	P	Q	W	m	|
P	10.5	19.7	22.6	23.3	23.3	28.1	28.6	30.9	31.6	32.0	42.2

Table 3. The 11 best features of the second set with their associated perplexities

Note that as expected for the first feature set, the feature with the highest perplexity was symbol "−" (49.7) representing shapes that do not contain any loops, ascenders or descenders.

Now, according to equations (3.6) and (3.7), we computed the global perplexities associated with the three feature sets mentioned above. We obtained

$$P_1 = 34.06 \qquad P_2 = 30.27 \qquad P_3 = 18.90 \tag{3.9}$$

These results can be interpreted by saying that the three feature sets we tested bring down the perplexity (complexity) of the recognition task from $N_c=258$ to 34.06, 30.27 and 18.90 respectively. The theoretical goal is to obtain a perplexity equal to 1, in which case the recognition problem is solved even before the recognition phase. However, since the features are used for word recognition and not for character recognition, the perplexity is not required to be too close to 1, because the word recognition module can still integrate other powerful knowledge sources such as context and redundancy, supplied by the order in which features appear in the sequence extracted from the input image, the length of words, the language model, *etc.*

On the other hand, we see that the perplexity decreases dramatically when we use the combination of our two feature sets as a new feature set. This proves that to some extent, the two types of features we are considering complement each other. Note that segmentation features were not considered in the results given above. Their integration would have certainly led to better results. Nevertheless, these features were very useful in helping the word model detect successfully the optimal character boundaries in the training phase, used to label the extracted graphemes from segmented input images.

To study the relationship between perplexity and recognition rate, we used three models for testing [ElYacoubi96]. The first two use respectively the first and second set of features (we will refer to these sets as E_1 and E_2), while the third model is the one described in section 2 using the combination of E_1 and E_2 as a feature set. The tests were carried out on the same learning and testing databases described before. The results obtained are given in Table 4.

lexicon	10	100	1000
with E_1	96.4%	85.4%	66.7%
with E_2	96.5%	87.2%	67.8%
with $E_1 \times E_2$	98.9%	95.0%	86.3%

Table 4. Recognition rates obtained respectively with first and second feature sets and a combination of the two.

From these results and the perplexity values given in equation (3.9), we can observe that there is a clear correlation between recognition rate and perplexity. The smaller the perplexity is, the better recognition rate is obtained.

5. Conclusion

We present in this paper an objective method to evaluate the discriminant power of features in an HMM-based word recognition system, making use of two well-understood indicators in information theory, entropy and perplexity. We showed that the HMMs allow us to compute automatically all the required probabilities to determine the above quantities. Our method can be considered as a powerful tool to judge or compare the quality of several feature sets without resorting to the results of the recognition phase. An interesting issue is to compute the global perplexity of the whole feature sequence or to derive it from the individual grapheme perplexities. This will allow us to evaluate indirectly the quality and the complexity of an input word image to be recognized. Unfortunately, this is a hard task in practice.

References

[Bahl83] Bahl L., Jelinek F., Mercer R. "A maximum likelihood approach to speech recognition", IEEE Transactions on Pattern Analysis and Machine Intelligence (PAMI) 5:179-190, 1983.

[Caesar94] Caesar T., Gloger J.M., Kaltenmeier A., Mandler E., "Handwritten Word Recognition Using Statistics," The Institution of Electrical Engineers. Printed and published by the IEE, Savoy Place, London WC2R 0BL, UK, 1994.

[Chen94] Chen M.Y., Kundu A., Zhou J., "Off-Line Handwritten Word Recognition Using a Hidden Markov Model Type Stochastic Network," IEEE, TPAMI, Vol. 16, No. 5, May 1994, pp. 481-496.

[Cohen94] Cohen E., Hull J.J., Srihari S.N., "Control Structure for Interpreting Handwritten Addresses," PAMI, Vol. 16, NO. 10, October 1994.

[Cox88] Cox S.J., "Hidden Markov Models for Automatic Speech Recognition: Theory and Application," Br Telecom Technol J Vol. 6 No. 2, April 1988.

[ElYacoubi94] El-Yacoubi A., Bertille J.M., Gilloux M., "Towards a more effective handwritten word recognition system," Proceedings of the Fourth International Workshop on Frontiers in Handwriting Recognition (IWFHR-IV), pp. 378-385, Taiwan, December 1994.

[ElYacoubi95] El-Yacoubi A., Bertille J.M., Gilloux M., "Conjoined Location and Recognition of Street Names Within a Postal Address Delivery Line," ICDAR'95, pp. 1024-1027, Vol. 2, Montréal, August 1995.

[ElYacoubi96] A. El-Yacoubi. "Modélisation Markovienne de l'écriture manuscrite. Application à la reconnaissance des adresses postales," PhD thesis, Université de Rennes I, France, 1996.

[Forney73] Forney G.D., "The Viterbi Algorithm," Proceedings of the IEEE, Vol. 61, NO. 3, March 1973.

[Gillies92] Gillies A.M., "Cursive Word Recognition Using Hidden Markov Models," Proc. of the 5th US-Postal Service Advanced Technology Conf., pp. 557-562, Nov 30-Dec 2, 1992.

[Gilloux92] Gilloux M., Leroux M. "Recognition of cursive script amounts on postal cheques," Proc. of the 5th USPS Advanced Technology Conf., pp. 545-556, 1992.

[Gilloux95a] Gilloux M., Leroux M., Bertille J-M., "Strategies for Cursive Script Recognition Using Hidden Markov Models," Machine Vision and Applications, Vol. 8, pp. 197-205, 1995.

[Gorski94] Gorski N.D., "Experiments with Handwriting Recognition using Holographic Representation of Line Images," Pattern Recognition Letters 15, pp. 853-859, 1995.

[Kim95] Kim G., Govindaraju V., "Handwritten Word Recognition for Real-Time Applications," International Conference on Document Analysis and Recognition (ICDAR), Vol. 1, pp. 24-27, August 14-16, 1995.

[Kim97] Kim G., Govindaraju V., "A Lexicon Driven Approach to Handwritten Word Recognition for Real-Time Applications," PAMI, Vol. 19, No. 4, April 1997, pp. 366-379.

[Knerr96] Knerr S., Baret O., Price D., Simon J.C., "The A2iA Recognition System for Handwritten Checks," Proceedings of Document Analysis Systems, Malvern, Pennsylvania, October 14-16, 1996, pp. 431-494.

[Lee91] Lee K.F., Hon H.W, Hwang M.Y., Huang X., "Speech Recognition Using Hidden Markov Models: A CMU Perspective," Elsevier Science Publishers B.V. (North-Holland), 1991, pp. 497-508.

[Leroux91] Leroux M., "Reconnaissance de Textes Manuscrits à Vocabulaire Limité avec Application à la Lecture Automatique des Chèques," PhD thesis, Université de Rouen, 1991.

[Paquet93] Paquet T., Lecourtier Y., "Recognition of Handwritten Sentences Using A Restricted Lexicon," Pattern Recognition, Vol. 26, NO. 3, pp. 391-407, 1993.

[Poritz88] Poritz A.B., "Hidden Markov Models: A Guided Tour," Proceedings of the IEEE International Conference on Acoustics, Speech, and Signal Processing (ICASSP'88), pp. 7-13, 1988.

[Rabiner93] Rabiner L., Juang B-H, "Fundamentals of Speech Recognition," Prentice Hall Signal Processing Series, pp. 382-384, 1993.

[Rabiner89] Rabiner L.R., "A tutorial on Hidden Markov Models and Selected Applications in Speech Recognition," Proc. IEEE, 77 (2): 257-286, February 1989.

[Shrihari95] Shrihari S.N, Shin Y-C, Ramanaprasad V, Lee D-S, "Name and Address Block Reader System for Tax Form Processing," International Conference on Document Analysis and Recognition (ICDAR), Vol. 1, pp. 5-10, August 14-16, 1995.

[Simon92] Simon J.C., Baret O., "Cursive Word Recognition," From Pixels to Features III. Frontiers in Handwriting Recognition, S. Impedovo and J.C. Simon Editors (North-Holland), pp. 241-260, 1992.

[Suen96] Suen C.Y., Lam L., Guillevic D., Strathy N.W., Cheriet M., Said J.N., Fan R., "Bank Check Processing System," International Journal of Imaging Systems and Technology, Vol. 7, 392-403 (1996).

Effect of Variability on Letters Generation with the Vectorial Delta-Lognormal Model

Wacef Guerfali & Réjean Plamondon

Laboratoire Scribens
École Polytechnique de Montréal
C.P. 6079, Succursale Centre-Ville, Montréal (Québec) Canada H3C 3A7
e-mail: guerfali@scribens.polymtl.ca or rejean@scribens.polymtl.ca

Abstract:

One of the primary goals of handwriting modeling is to understand how humans represent, control and generate compex movements. Moreover handwriting modeling has been also used for practical applications such as handwriting analysis and recognition. Numerous models used to date were not strong enough to explain and support some fundamental results about biomechanical or neurophysiological systems and neither practical enough to be used for accurate handwriting generation in the kinematic and the spatial domains. The vectorial delta-lognormal model has been shown, in the past few years, to answer to these two paradigms showing accuracy for simple movements simulation and flexibility for letters and cursive handwriting generation. In this paper we show how this model can help to study handwriting variability, particularltly, the effects of the fluctuations of the commands and of the neuromuscular effectors, on the movement generated. Some characters models are proposed with examples of the variability effects. A parametric representation of allographs can then be used to represent basic shapes and some models of distortion, to generate a variety of prototypes.

Keywords:

Handwriting Models, Vectorial Delta-Lognormal Model, Handwriting generation, Handwriting variability, Models of letters distortion.

1 Introduction

One of the difficulties found in both validating and training of handwriting recognizer systems is related to the use of large databases of different characters [GUY-96]. An alternative for large scale data collection is the use of handwriting generation models to automatically generate a wide variety of different letters from a small database of letter prototypes. Many authors have proposed in the past few years several interesting approaches for handwriting generation that can be considered for this kind of tasks (see Plamondon & Maarse for models review before 89 [PLA-89] and [BUL-93][MOR-94][STE-94][SIN-94] for more recent models). One of the most complete approach proposed to date, the vectorial delta-lognormal model [PLA-95c][GUE-96], is summarized in this paper. This model has found its origin in fundamental studies of simple rapid movements, where it helped to explain several psychophysical phenomena related to 2D movements [PLA-93a,b][PLA-95a,b].

In this paper, we show how the vectorial delta-lognormal model is used to represent and generate, with a small set of parameters, prototypes of allographs that can be considered as the basic or the "ideal" representations of the specific shape and the kinematics of a handwriting movement. Intra and inter writer variability can then be modeled by the density function associated to each random variable of the model parameters [PLA-96c].

The paper presents in its first part a short overview of the vectorial delta-lognormal model. The second part proposes a strategy for letter model generation, and an analysis-by-synthesis algorithm is presented to show how spatio-temporal characteristics of a reference subject can be used for letter modeling. The third part of the paper suggests some strategies for the study and the description of letter variability. Three groups of parameters are used for the analysis of handwriting fluctuations. First the kinematic command parameters, second the static command parameters and third the effectors parameters that describe the timing properties of the neuromuscular systems involved in the generation of each particular stroke.

2 The vectorial delta-lognormal model

The vectorial delta-lognormal model is used for describing complex 2D movements based on the kinematic theory recently proposed by R. Plamondon [PLA-93a,b][PLA-95a,b]. According to the kinematic theory, simple human movements can be described in the velocity domain by the response of a synergistic action of an agonist and an antagonist neuromuscular network [PLA-93a][PLA-95a]. The resulting curvilinear velocity V(t) for a single stroke along a circular path obeys to the delta-lognormal law and is described by equations 1 and 2 [PLA-93a][PLA-95a].

$$V(t) = D_1\, \Lambda(t;\, t_0, \mu_1, \sigma_1^2) - D_2\, \Lambda(t;\, t_0, \mu_2, \sigma_2^2) \tag{1}$$

$$\textit{where} \quad \Lambda(t;\, t_0, \mu_p, \sigma_i^2) = \frac{1}{\sqrt{2\pi}\, \sigma_i^2\, (t-t_0)}\, e^{-\frac{1}{2\sigma_i^2}(\ln(t-t_0)-\mu_i)^2} \tag{2}$$

The parameters D_1 and D_2 in equation 1 represent respectively the amplitude of the agonist and the antagonist commands, while the parameters μ_i and σ_i reflect the logtime delay and the logresponse time that mainly describe the timing properties of agonist and the antagonist subsystems used to generate a particular movement. Under these conditions, the velocity profile of a neuromuscular synergy, in response to a synchronous impulse commands occurring at t_0, is described by the weighted difference of two lognormals curves [PLA-93a][PLA-95a].

To describe two dimensional movements, the velocity of each single stroke is considered as a vector whose magnitude is described by equation 1 and direction with respect to an arbitrary reference as a function of time by equation 3. Each stroke starts with an initial orientation θ_0 and a constant curvature C_0 [PLA-95c][GUE-95ab].

$$\theta(t) = \theta_0 + C_0 \int_{t_0}^{t} V(\tau) \, d\tau \qquad (3)$$

Complex movements such as fluent cursive handwriting, can then be generated by time overlapping two or more velocity vectors representing individual strokes to generate any complex handwriting shape. The instantaneous module and direction of an n strokes velocity vector are described by equations 4 and 5 [PLA-96b][GUE-96].

$$V(t) = \sqrt{\left(\sum_{i=1}^{n} V_i(t) \, \cos(\theta_i(t))\right)^2 + \left(\sum_{i=1}^{n} V_i(t) \, \sin(\theta_i(t))\right)^2} \qquad (4)$$

$$\theta(t) = \arctan\left(\frac{\sum_{i=1}^{n} V_i(t) \, \sin(\theta_i(t))}{\sum_{i=1}^{n} V_i(t) \, \cos(\theta_i(t))}\right) \qquad (5)$$

Using equations 4 and 5, the complete kinematics of a 2D trajectory can be recovered as well as the pentip position x(t) and y(t), with respect to an arbitrary posture $P_0(x_0, y_0)$ and starting time t=0 using equations 6 and 7.

$$x(t) = x_0 + \int_{0}^{t} V(\tau) \, \cos(\theta(\tau)) \, d\tau \qquad (6)$$

$$y(t) = y_0 + \int_{0}^{t} V(\tau) \, \sin(\theta(\tau)) \, d\tau \qquad (7)$$

3 Handwriting generation

With the model described above, the generation of "human-like" handwriting can be considered as a reverse engineering problem, where the parameters used for the generation of realistic movements are based on the analysis of typical human samples. An example of "human-like" letters generated by the vectorial delta-lognormal model is presented in figure 1. A sample of 25 allographs of the alphabet letters presented were inspired from the shape and the kinematic of letters written by a human subject. The number of strokes, the kinematic properties (such as stroke overlapping time and movement time) and the general shape were respected. Some scaling and slant adjustments were, however, made to idealize the allograph models. A more general approach could be also possible if we consider the kinematic of several subjects writing the same allograph. The group of samples produced with the same spatio-temporal constraints could then be used to find out the general kinematic pattern from which we can extract, using an analysis-by-synthesis approach, the common parameters of each allograph for each letter.

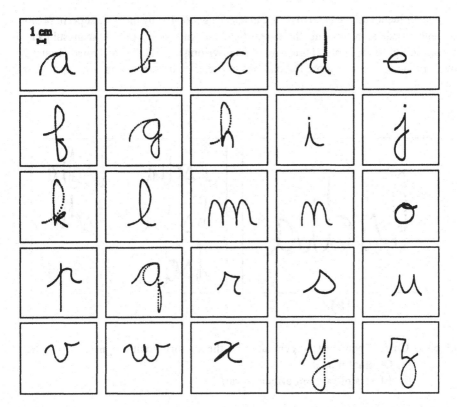

Figure 1 Models of letters generated by the vectorial delta-lognormal model.

The algorithm used for the analysis-by-synthesis of a given handwritten trace can be summarized as follow [PLA-96a][GUE-96]:

BEGIN
    - Compute the module and the direction of the curvilinear velocity from the pentip coordinate x(t), y(t),
    - Make a first level segmentation of the word into components, using the pen-paper contact information,
    FOR each component DO
        - Determine the number of strokes,
        FOR each stroke DO
            - Compute an initial estimate of the 9 parameters,
        END (for)
        - Extract the optimal group of parameters of the set of strokes by optimizing the matching of the resulting curvilinear velocity vector with equations 4 and 5,
        - Save the group of parameters that best fits each component.
    END (for)
END.

Fluent cursive handwriting can also be generated by the time superimposition of letter models, respecting the shape (and the kinematic) of a human subject, as shown in figure 2a. More "human-like" handwriting can also be generated (figure 2b) by introducing some random fluctuations of the effector parameters (μ_1, μ_2, σ_1 et σ_2).

(a)　　　　　　　　　　　　(b)

Figure 2 Cursive handwriting generated by the vectorial delta-lognormal model:
    (a) Original model
    (b) Effects of random effectors variability

4 Variations modeling

To better visualize how the fluctuations of a particular parameter can affect the shape of the generated movement, we will show the effect of a variability scheme on the particular case of the allograph of the letter "a" presented in figure 1.

Considering the role of each parameter of the vectorial delta-lognormal model (as presented in equations 1, 2 and 3), we can group the sources of possible fluctuations in three classes. The first group, includes the variations caused by the fluctuations of the kinematic command parameters (D_1, D_2 and t_0), while the second group concerns the variations caused by static or postural command parameters (C_0 and θ_0). Finally, the variations caused by the fluctuation of the effectors parameters (μ_1, μ_2, σ_1 and σ_2) constitute the third group.

The variablility of the the first group parameters affects mainly movements amplitude and timing properties. The control of the kinematic command parameters D_1 and D_2 concerns the scale and the spatial precision needed to generate a particular allograph. The values of those parameters control the total displacement D of each stroke, as measured by the difference between the agonist and the antagonist commands $D = D_1 - D_2$, and the spatial precision required for a specific movement, as reflected by the ratio $R = D_1/D_2$ [PLA-93b][PLA-95b]. Figure 3a shows the effect of the variation of the movements amplitude D with respect to the original scale (1.0), by a factor of 25, 50, 75 and 100%. Figure 3b shows the effect of varying the movement precision by changing the ratio of D_1/D_2 from 2.0 to 50.0. The larger is the ratio, the higher is the movement precision and the longer is the movement time. To have a good trade off between movement time and spatial precision, one should take an average value of the ratio D_1/D_2 between 4.0 and 10.0. The original ratio used for the letters of Figure 1 is 5.0.

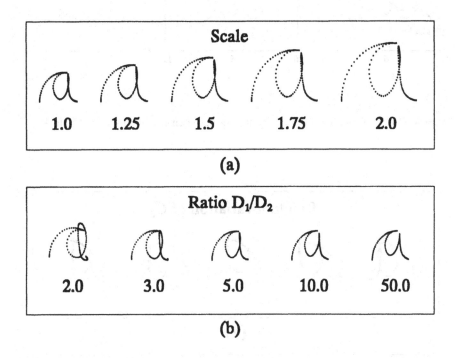

Figure 3 Effect of: (a) scaling (b) varying the spatial precision

The fluctuation of the timing parameter t_0 affects mainly the overlapping of strokes and the fluency of handwriting. The more significant is the overlapping effect, the more fluent is the handwriting. A discontinuous style of handwriting might be caused by a poor movement time overlapping between successive strokes. An excessive overlapping time can however greatly affect the resulting shape. As we can see in figure 4, the second and the third stroke of the syllable "al" are progressively overlapped until it hides part of each other. When the overlapping time becomes too high, it can even completely hide (or cancel) the effect of one or more successive commands.

The second group of parameters concerns the static or postural parameters C_0 and θ_0. Fluctuations of curvature or initial orientation have a visible effect on the smoothness and the orientation of the letters generated [GUE-96][PLA-96b]. Figures 5a and 5b show respectively the effects of a random variation of the parameters C_0 and θ_0.

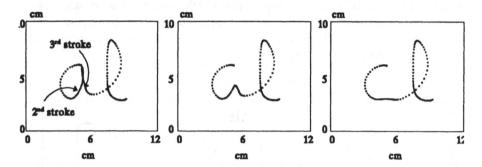

Figure 4 Effect of excessive time overlaping between the 2nd and the 3rd stroke.

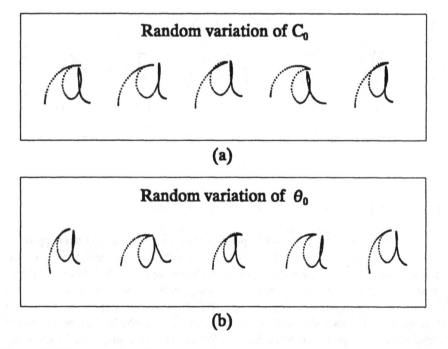

Figure 5 Effect of random fluctuation of the postural parameters C_0 and θ_0.

The third source of variability is related to the fluctuation of the effector parameters μ_1, μ_2, σ_1 and σ_2. Effectors variations have a large influence on the shape of the velocities profiles and highly affects the overlapping of strokes and the global shape of an allograph. Parameters μ_1 and μ_2, for example, are related to the movement time, usually the smaller they are, the smaller is the movement time. In a controlled experiment where we try to reproduce the same allograph under the same spatio-temporal constraints, the variations of the effectors parameters (μ_1, μ_2, σ_1 and σ_2) can be represented as a random variable distributed around an average value with a certain standard deviation [PLA-96c]. If we suppose, in our model, that the effectors parameters are distributed according to a normal distribution around average values (for example, the values used to generate the allographs presented in the figure 1), a distortion strategy can be used to generate theoretically an infinite number of different movements from a single representation of an allograph. Figure 6a and 6b show respectively the effects of a random variation of the parameters μ_1, μ_2, σ_1 and σ_2 on a letter allograph.

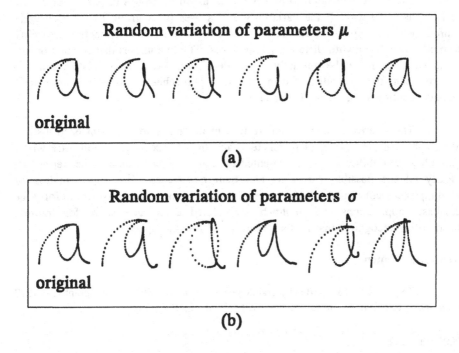

Figure 6 Effect of random fluctuations of the effector parameters on the example of the allograph of the letter "a".

5 Conclusion

In this paper, we have shown that the use of handwriting generation model can help to represent handwriting variability. Schemes of characters distortions have been proposed that could be used to generate large databases of character sets for learning or validating of handwriting recognizers.

The vectorial delta-lognormal model presented in this paper was successfully used for the generation of "human-like" handwriting. Models of letters are proposed as a sample of typical allographs. The kinematic and the shape properties for the generated letters are inspired from a reference human user. The extraction of a realistic group of parameters that model a complex movement such letters handwriting has been done using an algorithm of analysis-by synthesis. A study of the fluctuations effects of the vectorial delta-lognormal model parameters on a particular allograph shows how the generated letters are affected by some perturbations. This basic study shows the possibility to represent an allograph model at a high level and to distinguish the effect of variability from its "ideal" or average representation. A distinction is made on the fluctuations of three groups of parameters, the kinematic commands parameters, the static or postural parameters and the effectors parameters.

Even if it is difficult to presume that this three groups of parameter covers in an exhaustive way all the variability that a user (or a group of users) usually reproduce in real situations, the vectorial delta-lognomral model offers the possibility to model variability with different perspectives. The hypotheses that fluctuations of some variables might be normally distributed have also to be validated, but this already opens the possibility of using realistic approaches to model the parameter's density function for each particular user.

The interest of our method resides in its ability to represent each distinct allograph with a small set of parameters, and then to model inter and intra writer variablity. Variability is then represented by a probabilistic function that represents ideally all the possible variations and their occurrence. This representation of allographs can constitutes a high compression tool of the rough information for large databases applications and might be considered as an alternative for training handwriting recognizer that needs a large set of users samples.

Acknowledgements

This work was supported by grant OGP0915 from NSERC Canada, grant IC-4 IRIS Network of Excellence Canada and grant 2269 from FCAR Québec.

References

[BUL-93] Bullock D., Grossberg S., Mannes C., (1993) "A Neural Network Model for Cursive Script Production", Biological Cybernetics, 70, 15-28.
[GUE-95a] Guerfali W., Plamondon R., (1995) "The Delta-Lognormal Theory for the Generation and Modeling of Cursive Characters", Proc. 3rd Int. Conf. on Document Analysis and Recognition, 495-498.
[GUE-95b] Guerfali W., Plamondodn R., (1995) "Control Strategies for Handwriting Generation", Proc. 7th Biennial Conference of the International Graphonomics Society, 64-65.
[GUE-96] Guerfali W., (1996) "Modèle delta-lognormal vectoriel pour l'analyse du mouvement et la génération de l'écriture manuscrite", Ph.D. Dissertation, École Polytechnique de Montréal.

[GUY-96] Guyon I., Makhoul J., Schwartz R., Vapnik V., (1996) "What Sise Test Set Gives Good Error Rate Estimates?" Fifth International Workshop on Frontiers in handwriting Recognition, 313-316.

[MOR-94] Morasso P., Sanguineti V., Tsuji T., (19940 "A Model for the Generation of Virtual Targets in Trajectory Formation", Adevences in Handwriting and Drawing: A Multi disciplinary approach, Faure C., Keuss P., Lorette G., Vinter A. Eds., 333-348.

[PLA-89] Plamondon R., Maarse F.J., (1989) "An Evaluation of Motor Models of Handwriting", IEEE, Trans. on Systems, Man and Cybernetics, v. 19, n. 5, 1060-1072.

[PLA-93a] Plamondon R., (1993) " The Generation of Rapid Human Movements. Part I: A Delta Lognorma law", Technical Report ERM/RT-93/4, École Polytechnique de Montréal.

[PLA-93b] Plamondon R., (1993) " The Generation of Rapid Human Movements. Part II: Qudratic and power laws", Technical Report ERM/RT-93/5, École Polytechnique de Montréal.

[PLA-95a] Plamondon R., (1995) "A Kinematic Theory of Rapid Human movements. Part I: Movement Representation and Generation", Biological Cybernetics, 72:297-307.

[PLA-95b] Plamondon R., (1995) "A Kinematic Theory of Rapid Human movements. Part II: Movement time and Control", Biological Cybernetics, 72:309-320.

[PLA-95c] Plamondon R., (1995) "A Delta Lognormal Model for Handwriting Generation", Proc. 7th Biennial Conference of the International Graphonomics Society, 126-127.

[PLA-96a] Plamondon R., Guerfali W., (1996) "Why Handwriting Segmentation Can Be So Misleading?", Proc. 13th Int. Conf. on Pattern Recognition, 396-400.

[PLA-96b] Plamondon R., Guerfali W., (1996) "The Generation of Handwriting with Delta-Lognormal Synergies", Paper submitted for publication to Biological Cybernetics.

[PLA-96c] Plamondon R., (1996) "A Kinematic Theory of Rapid Human Movements. Part III: Kinetic Outcomes", Paper submitted for publication to Biological Cybernetics.

[SIN-94] Singer Y., Tishby, N., (1994) "Dynamical Encoding of Cursive Handwriting", Biological Cybernetics, 71, 227-237.

[STE-94] Stettiner O., Chazan D., (1994) "A Statistical Parametric Model for Recognition and Synthesis of Handwriting", 12th Int. Conf. on Pattern Recognition, v. II, 34-38.

Off-Line Signature Verification:
Recent Advances and Perspectives

Robert Sabourin

Laboratoire d'Imagerie, de Vision et d'Intelligence Artificielle (LIVIA)
École de technologie supérieure, Département de génie de la production automatisée
1100, rue Notre-Dame Ouest, Montréal (Québec), H3C 1K3, CANADA
e_mail: sabourin@gpa.etsmtl.ca

Abstract

This paper is a description of recent advances in off-line signature verification research performed at our laboratory. Related works pertain to structural interpretation of signature images, directional PDF used as a global shape factor, the Extended Shadow Code (ESC) and the fuzzy ESC, a cognitive approach based on the Fuzzy ARTMAP, and shape factors related to visual perception.

Keywords: Handwritten signature verification, Shape factor definition, Visual perception, Neural networks, Fuzzy inference.

1 Introduction

A lot of work in the field of off-line signature verification has been done at our laboratory since the last twelve years. The long-term goal of this research is to design a complete Automatic Handwritten Signature Verification System (AHSVS) which is able to cope with all classes of forgeries produced with or without imitation of the genuine signature. The majority of approaches reported here have been done specifically for the problem of the elimination of random forgeries defined as genuine signatures of other writers enrolled in the verification system, a prerequisite for real applications.

First, structural interpretation of signature images has been investigated in a way to implement some *graphometric features* related to the characteristics of genuine signatures used by specialists in the field of forensic science [specially from Locard, 39]. This AHSV system was design in a way to be able to cope with all classes of forgeries [2]. The complexity of this approach, especially this one related to the segmentation of the signature line in terms of arbitrarily shaped primitives, gave us the motivation to propose the Fuzzy ESC, a shape factor based on a more flexible scheme because the segmentation problem is overcome [21]. This new scheme permits the fusion of several sources of information related to the geometric shape of the signature and to the pseudo-dynamic information related to the effects of the writing dynamics like speed variations and pressure. This approach is very promising especially for real time applications.

A lot of work has been done in the context of the elimination of random forgeries. We proposed first a global shape factor, the directional PDF [15], in a way to solve this problem with a very simple approach. In fact, the results were not satisfactory (about 5% of mean total error rate with a threshold classifier and 6 reference signatures) because global shape factors like the directional PDF [see [36] for a survey of global shape factors published by others] do not take into account the spatial position of local measurements in the representation of the signature. Moreover, we observed the same level of performance with invariant moments [2].

This observation leads to the proposal of the Extended Shadow Code (ESC) as a family of shape factors related to global and to local shape factors depending on the resolution in use [see Figure 2 in [21]]. This representation showed very good results at high resolution but the best performance was obtained with the cooperation of several classifiers tailor-made for each representation [25]. In this way, several levels of perception are considered in the analysis of a test signature image.

In order to understand more deeply the underlying process related to the local analysis of signatures, a perceptual approach based on a grid of points of attention and a sliding retina has been proposed in [36-38]. Here, the pattern spectrum and elementary segments used as structural elements permit the definition of local geometrical measurements made on the signature trace or on the background area. Experimental results obtained with this representation was closed to those one obtained with the ESC.

Another approach is to put emphasis on preprocessing and to extract *perceptual features* like loops, the body of the signature, inter-words spacing, etc [35]. Here, Binary Shape Matrices are considered as a good compromise for the encoding of signature images. This approach is in fact related to the class of mixed shape factors, e.g. global shape factors where the location of measurements is embedded in the representation of the signature.

Few works have been done at the classification level. Murshed in [31] proposed a cognitive approach for the signature verification task in the context of the elimination of random forgeries. The motivation of this work is to train each classifier with genuine signatures only. In this way, a new writer can be added to the verification system without the need to retrain all other classifiers. Another interesting characteristic of this approach is the possibility to train each classifier on-line, e.g. the knowledge of the system is enhanced based on a continuous utilization of the verification system. This is a good feature for the implementation of real AHSV systems. This verification system used binary segments extracted locally by the help of an identity grid as a representation of the signature. Future works will be to integrate local shape factors like the ESC [27] and this one related to visual perception [36] in a way to enhance the discriminating power of the system.

In this paper, the related shape factors and AHSV systems are reported and experimental results are compared together. This is followed by a brief description of works in progress in this field at our laboratory.

2 Structural Interpretation of Signature Images

This study [2-14] was inspired from the work made in the field of forensic science by Locard, an expert document analyst, who proposed in [39] a survey of *graphometric techniques* used for the authentication of questioned documents. The limitations in the performance of graphometric techniques were by this time mostly attributable to the lack of accuracy of measurements taken by different expert document analysts. Locard has proposed many characteristics belonging to genuine handwritten signatures. These characteristics can be subdivided into two classes.

The first class is related to characteristics of genuine signatures imperceptible to the average forger and very difficult to imitate: the rhythmic line of the signature consisting of the positional variation of the maximum coordinate of each character composing the signature; the local variation in the width of the signature line which is closely related to the dynamics of the writing process; and the variation in aspect ratio of the whole signature followed by the local features like aspect ratio, difference in orientation, in relative position, etc., measured between pairs of characters.

The characteristics of the second class are the ones that are very easily perceived by the average forger and are consequently easier to imitate: the general design of the letter's shape; the signature's overall orientation; and the signature's position on the document.

Based on these two classes of characteristics of genuine signatures we proposed in [2-14] an image understanding system based on a signature representation defined in terms of arbitrarily shaped primitives. This approach is *text-insensitive* because no attempt is made to segment specific letters from the semantic part of the signature. Two classes of features related to those one proposed by Locard, *static* and *pseudo-dynamic* are taken into account for the representation of the signature. The former is related to the geometric shape and spatial relations between some primitives extracted from the signature trace. The latter is associated with the gray-level variation (contrast) and the texture inside the primitive. In considering these two classes of features simultaneously, and the separation from the local interpretation of the primitives followed by the global interpretation of the scene, a general purpose image-understanding system was designed for the interpretation of handwritten signature images. For a complete description of this system, see [2, 6].

Recently, an attempt was made to pursue this work related to the definition and the extraction of primitives for the structural interpretation signature images [1]. In this work, the writing trace of the signature is subdivided in terms of conics like straight lines, ellipses and hyperboles. This method permits a simplification of the signature tracing for the purpose of the elimination of random forgeries. Works are actually in progress in a way to use attributed string-matching techniques based on this representation for the design of a AHSV system for real time applications.

3 Directional PDF as a Global Shape Factor

In a system like this one described in [2], in order to take into account all classes of forgeries, the decision is made only at the end of the verification process. Consequently, this approach is a very costly solution in terms of computational resources and in terms of related algorithmic complexity. Since random and simple forgeries represent almost 95% of the cases generally encountered in fraudulent cases [40], it was decided to subdivide the verification process in such a way to rapidly eliminate gross forgeries. Thus, a two-stage AHSVS seems to be a more practical solution, where the first stage would be responsible for this rapid elimination and the second stage used only in complicated cases.

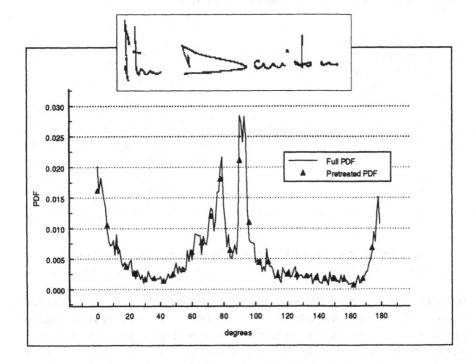

Fig. 1. A genuine signature and the related directional PDF [reprinted from 15].

The first stage of this complete AHSVS thus has two main objectives: firstly, to consider only random forgeries and, secondly, to make rapid decision. To meet the first objective, a characteristic dealing with the overall shape of the signatures has been proposed [15-20]. Accordingly, we proposed the directional Probability Density

Function (PDF) as a global shape factor [Figure 1]. Its discriminating power is not optimum because, even though it is invariant in translation and in scale, it is not invariant in rotation. To meet the second objective, we have chosen to use a BackPropagation Network as a signature classifier. The networks were trained with synthetic samples derived from original reference PDFs rotated ±6° with 1° increments. See [15] for a complete description of this system.

4 Extended Shadow Code (ESC) and Fuzzy ESC

One important characteristic that a shape factor related to the signature verification problem should have is the ability to be text-insensitive. This means that measurements taken on the signature shape do not relate to specific letters. This is especially important in the case of handwritten signatures characterized by well-written to highly personalized signatures. The main problem with global features like the directional PDF is a lack of knowledge about the location of local measurements taken on the signature considered as a pattern when they are combined together for the definition of the feature vector.

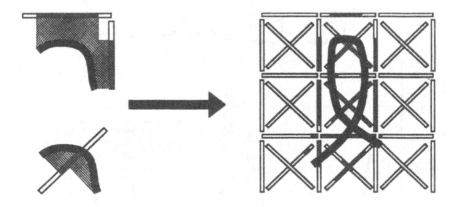

Fig. 2. Definition of the projection mechanism of the ESC [reprinted from 24].

Intrinsically, the extended shadow code is a global shape factor. The rationale behind the use of the ESC as a shape factor for the signature verification problem is that it permits the local projection of the handwriting without losing the knowledge of the location of measurements in the 2D space [Figure 2]. Thus, this shape factor seems to be a good compromise between global features related to the general aspect of the signature, and local features related to measurements taken on specific parts of the signature without requiring the low-level segmentation of the handwriting into primitives, which is a very difficult task [2]. This is achieved by the bar mask definition [21,27], where at low resolution [Figure 3a] the ESC is related to the overall proportion of the signature. In the opposite case, the values of the projections

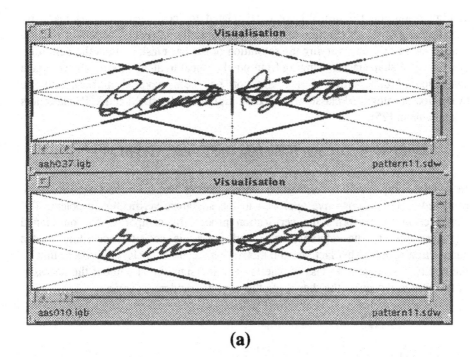

(a)

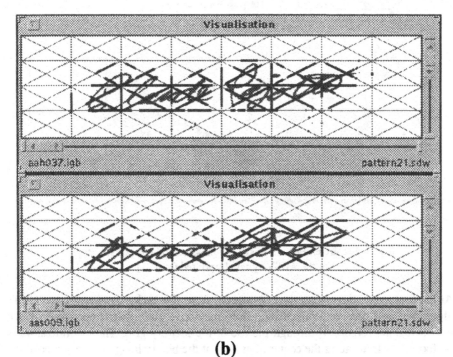

(b)

Fig. 3. Examples of the ESC representation at low resolution (a)
and at high resolution (b) [reprinted from 24].

at high resolution [Figure 3b] could be related to local measurements taken on specific parts of the signature without the segmentation of specific letters or specific primitives. In this way, varying the resolution of the bar mask permits the definition of a family of shape factors ranging from purely global to almost local. The complete evaluation of this family of representation has been reported in [27], and the evaluation of integrated classifiers based on individual threshold and kNN classifiers is available in [25].

5 A Cognitive Approach Based on the Fuzzy ARTMAP

In order to implement a verification system acting like an expert examiner, we proposed in [31] a cognitive approach to the signature verification problem. The expert examiner performs the verification process by comparing the questioned signature only to the reference ones and gives his decision according to the comparison results. This fact was the motivation for the one-class approach, that is, the ability to recognize a class of objects (genuine signatures) without the necessity of learning to recognize the shapes of other objects (random forgeries).

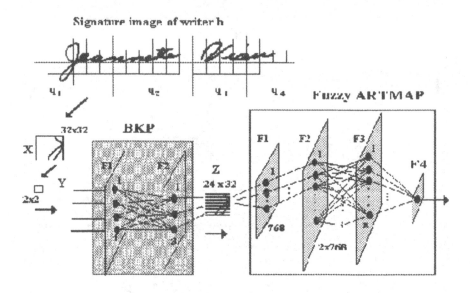

Fig. 4. Feature extraction, image compression and comparison stages
of the AHSV system [reprinted from 31].

At the first stage, the signature is segmented from the background using the Otsu's algorithm, and then centralized onto the image area such that it becomes divided into m regions [Figure 4] through the use of the identity grid. The centralization is performed by translating the center of gravity of the binary image to the center of the image area. Thereafter, graphical segments of size 16x16 pixels with 50% overlapping in the x and y directions are extracted from each region in the binary signature and applied to a Back-propagation network which reduces the size of these segments by 1/3.

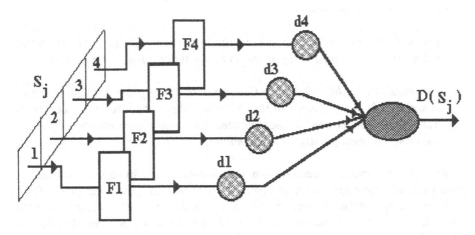

Fig. 5. Final comparison and decision stages of the AHSV system [reprinted from 31].

The reduced graphical segments are then applied to the comparison stage for learning/verification. This stage is composed of m Fuzzy ARTMAP networks, each of which is responsible for one region in the signature [Figure 5]. This structure can be viewed as having different experts examining different regions of the signature. Finally, the decision stage analyzes the results produced by each Fuzzy ARTMAP and gives the decision of the system with respect to authenticity of the unknon signature. The complete description of this AHSV system is available in [31].

The main drawback of this approach lies in the use of graphical segments related to a specific region without taking into account its local position in the image. Moreover, new local shape factor like the ESC is needed in a way to enhance the performance of the AHSV system. This approach based on cognitive concepts is very promising and more research in this area will be realized in a near future.

6 Visual Perception

We proposed in [36] a new formalism for the definition of a signature representation based on visual perception. As mentioned above, local approaches are better suited for defining signature representations than global ones. Based on the observations made earlier on the superposition of genuine signatures [Figure 6], the following assumptions can be made with respect to defining a shape descriptor tailored for the signature verification problem:

1) *The overall orientation and the overall proportions of genuine signatures written in a constrained 2D area are relatively stable for each writer, and*

2) *The local variability of the writing trace of the signature is an intrinsic characteristic of the identity of the writer and should be taken into account as well. This phenomenon is characterized by local displacements of strokes following the principal axis of the signature.*

From these assumptions, a certain invariance in rotation and in scale results; hence, their explicit requirement is not needed in the definition of a shape descriptor. Only a correction in translation remains necessary. Assumption 1 is in accordance with the opinion of expert examiners in the field of forensic science [39-41]. Finally, North American signatures are "cursive" in nature; this could partially explain the fact that their overall proportions are relatively stable over time.

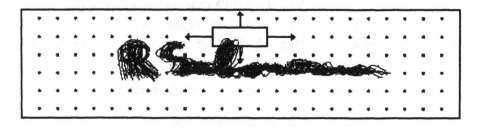

Fig. 6. A rectangular retina is shown with the field of points of attention uniformly distributed in the image space [reprinted from 36].

The originality of the proposed approach is that local measurements are not made on specific parts (primitives) of the signature, but on specific areas around foci of attention in the image plane. The identification and segmentation of feature points or primitives on the writing trace of the signature are performed by just assuming that the foci of attention are specified arbitrarily in the image space. This leads to a simplification of the training phase of the verification system. The method proposed for feature extraction is simple. A signature image of 512x128 pixels is centered on a grid of rectangular retinas which are excited by local portions of the image; each retina has only a local perception of the entire scene [Figure 6], and the measurement made on the subset of pixels related to a specific retina will reflect the local activity of the signal.

The definition grid (i.e. the position and the number of foci) together with the size of retinas have a great impact on the performance of this approach and a prototyping phase is necessary [see Section 5.1 in [36]]. An important aspect for signature representation is that the consistent absence of signal activity in specific areas of the image will be taken into account in an attempt to characterize the shape of the signature. This is achieved by Assumption 1 which stated that the overall proportions of genuine signatures are relatively stable. In other words, not only the signature itself but the background also are considered in the definition of the shape descriptor.

From Assumption 2, the local variability of the writing trace of the signature could be taken into account easily by adding a certain percentage of horizontal and/or vertical overlap between neighboring retinas. As the visual observations made on examples in our signature database showed that in general the displacement of strokes follows the principal axis of the signature, this fact suggests that horizontal overlap alone is enough.

The measurement applied to the set of pixels related to a retina should be capable of detecting the presence, or absence, of any signal activity, and it should be capable in some way of quantifying it. We think that local measurement should be capable of describing the information contained in a binary image and the manner in which it is distributed between the "fine" and "coarse" details. The morphological operations of opening and closing are useful for this task. The *pattern spectrum* is an internal information non-preserving morphological shape descriptor called the *pecstrum* [36]. The pecstrum is computed by measuring the result of successive morphological openings of the image, as the size of the structuring element increases. The sequences of openings so obtained are called granulometries. Spectra obtained from the local analysis based on elementary segments as structural elements are shown in Figure 7.

Shape matrices have been used as a representation of planar shapes like industrial parts or printed characters. In [35], we investigated the use of shape matrices as a mixed shape factor for off-line signature verification. By mixed shape factor we mean any global shape factor where the position of local measurements are take into account in the definition of a similarity measure between two representations. Several preprocessing algorithms like loop filling, smearing and morphological closing have been used for emphasize some local characteristics (*perceptual features*) of the signatures. It was demonstrated that using a good similarity measure between two shape matrices, this shape factor is relatively well suited for the global interpretation of signature images. Experimental results reported in [35] are closed to those observed with local shape factors like the ESC [27] and local measurements based on the pattern spectrum [36].

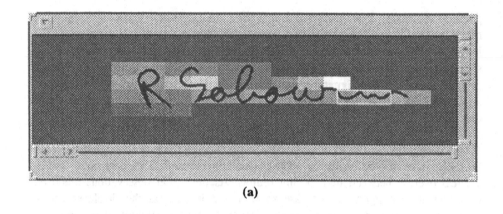

(a)

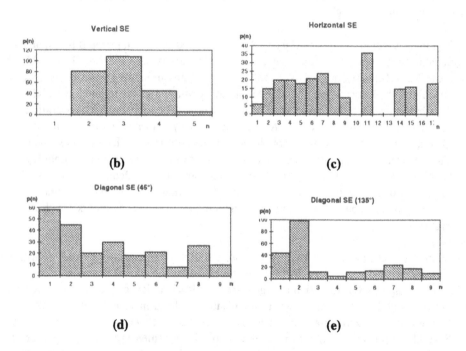

(b)   **(c)**

(d)   **(e)**

Fig. 7. A signature image (a) and the pecstrums based on structural elements in {l,—,/,\} are depicted in (b) to (e) [reprinted from 36].

7 Conclusion and Perspectives

Experimental evaluation of approaches reported in this paper has been made using a signature database of 800 genuine signatures from 20 individuals. Two types of classifiers, a Nearest Neighbor (NN) and a Minimum Distance (MD) classifier with threshold (Nref=6 reference signatures) was used for the implementation of all verification systems except this one based on the Fuzzy ARTMAP [Table 1].

Approaches	NN	MD with Nref=6	Nref=18
Structural Approach [2]		2.49%	
Directional PDF [15]	2.69%	5.61%	
ESC [27]	0.01%	0.88%	
ESC with Integrated classifiers [25]	0.0% (0.18%)	0.05% (2.88%)	
Cognitive Approach [31]			3.56%
Pattern Spectrum (PS) [36]	0.02%	0.85%	
PS with Integrated classifiers [36]	0.0% (0.05%)	0.27% (2.38%)	
Binary Shape Matrices [35]		0.84%	

Table 1. Experimental results, mean total error rate E_t in % (Rejection rate R_t)

We can see from Table 1 that local shape factors like the ESC and this one based on the pattern spectrum outperform global shape factors like the directional PDF and the structural approach. Moreover, cooperation of classifiers based on local shape factors shows very good results on this signature database [25, 36]. In the case of the AHSVS based on the Fuzzy ARTMAP, the experimental results are closed to those obtained with global shape factors and structural approaches. In this case, the performance of the verification system can be enhanced significantly if binary segments are replaced by local shape factors.

This survey of recent advances in off-line signature verification shows the importance of local approaches in the definition of a signature representation tailor-made for the elimination of random forgeries. Further work is needed to find a signature representation based on local measurements where several information sources can be taken into account, as is the case for the Fuzzy ESC [21]. In this way, several classes of forgeries can be considered without the need to segment the trace of the signature in terms of primitives, a prerequisite for real applications. Moreover, this research on signature representation should be integrated into the cognitive approach proposed by Murshed in [31]. If the merge of these two schemes succeed, it could be possible to implement a real AHSV system able to cope with all classes of forgeries, using only genuine signatures for training.

Works are in progress in a way to relate points of attention [36] to the dynamics of the writing process. In this way, the number, the position and the size of the retinas will be based on the model of the signature. This means that the representation of the signature will be adapted to the intrinsic characteristics of each writer. The main objective of this approach is the elimination of random, simple, simulated and freehand forgeries.

A lot of experiments showed the superiority of AHSV systems based on the cooperation of classifiers [25,36], as is actually the case in the field of character recognition. More research is needed on the design of integrated classifiers for signature verification because the number of data available for training is always small in real application. This situation restricts us to the use of integrated classifiers based on the voting principle because no training is required.

In conclusion, due to the confidentiality of this kind of data, no international signature database is available for research. This is one limitation of this research area. Experimental results can be compared only if the same experimental protocol and database are used for the evaluation of different verification schemes. We hope that an international standard database will be available soon.

8 Acknowledgments

This work was supported in part by grant OGP0106456 from the NSERC of Canada.

9 References

STRUCTURAL APPROACHES

[1] Bastos, L.C., Bortolozzi, F., Sabourin, R. et Kaestner, C. "Mathematical Modelation of Handwritten Signatures by Conics", to appears in Revista da Sociedade Paranaense de Matemática, october 1997.

[2] Sabourin, R., Plamondon, R. et Beaumier, L., "Structural Interpretation of Handwritten Signature Images", in the International Journal of Pattern Recognition and Artificial Intelligence, Series in Machine Perception & Artificial Intelligence, World Scientific publishing co., vol. 13, 1994, pp. 709-748.

[3] Sabourin, R., Plamondon, R. et Lorette, G., "Off-line Identification with Handwritten Signature Images: Survey and Perspectives", in "Structured Document Image Analysis", Baird, H.S., Bunke, H. and Yamamoto, K. editors, Springer-Verlag, pp. 219-234, October 1992.

[4] Sabourin, R., "Choix d'un Facteur de Forme Basé sur l'Utilisation des Matrices d'Échantillonnage Polaire (MEP)", Conf. et Expo. sur l'Automatisation Ind., Montréal, pp 20.21-20.25, June 1992.

[5] Sabourin, R. and Plamondon, R., "Observability and Similarity in Spatial Relations in the Structural Interpretation of Handwritten Signature Images", The 7th Scandinavian Conf. on Image Analysis, Aalborg, Danemark, 12-17 august, pp. 477-485, 1991.

[6] Sabourin, R., «Une Approche de Type Compréhension de Scène Appliquée au Problème de la Vérification Automatique de L'Identité par L'Image de la Signature Manuscrite», Thèse de Ph.D.,École Polytechnique de Montréal, December 1990.

[7] Sabourin, R. and Plamondon, R., "Steps Toward Efficient Processing of Handwritten Signature Images", Vision Interface 90, Halifax, Nova Scotia, May 1990, pp. 94-104.

[8] Sabourin, R. and Plamondon, R., "Progress in the Field of Automatic Handwritten Signature Verification Systems Using Gray-level Images", Proc. of the Inter. Workshop on Frontiers in Handwriting Recognition, Montréal, april 2nd and 3rd ,1990, pp. 1-12.

[9] Sabourin, R., Plamondon, R. and Lorette, G., "Off-line Identification with Handwritten Signature Images: Survey and Perspectives", IAPR workshop on Syntactic and Structural Pattern Recognition, AT&T Murray Hill, New Jersey, June 1990, pp. 377-391.

[10] Plamondon, R., Lorette, G. and Sabourin, R., "Automatic Processing of Signatures Images: Static Techniques and Methods", in Handwriting Pattern Recognition, R. Plamondon, G. Leedham eds, World Scientific Pub., Singapore, 14 pages, July 1989.

[11] Sabourin, R. and Plamondon, R., "Segmentation of Handwritten Signature Images Using The Statistics of Directional Data", Proc. of the 9th ICPR, Rome, Italy, pp. 282-285, November 1988.

[12] Sabourin, R. and Plamondon, R., "On the Implementation of Some Graphometric Techniques for Interactive Signature Verification: A Feasibility Study", Proc. of The Third Int. Symp. on Handwriting and Computer Applications", Montreal, Canada, pp 160-162, 1987.

[13] Sabourin, R. and Plamondon, R., "Preprocessing of Handwritten Signatures from image Gradient Analysis", Proc. of 8th ICPR, Paris, France, pp 576-579, 1986.

[14] Sabourin, R. and Plamondon, R., "La vérification par ordinateur des Signatures Manuscrites", ACFAS-86, Université de Montréal, may 1986.

DIRECTIONAL PDF

[15] Drouhard, J.P., Sabourin, R. and Godbout, M., "A Neural Network Approach to Off-Line Signature Verification Using Directional PDF", Pattern Recognition, Vol 29, No 3, March 1996, pp 415-424.

[16] Drouhard, J.P., Sabourin, R. and Godbout, M., "A Comparative Study of the k Neareast Neighbour, Threshold and Neural Network Classifiers for Handwritten Signature Verification Using an Enhanced Directional PDF", Third IAPR Conf. on Document Analysis and Recognition, August 14-16, 1995, Montreal, Canada, pp 807-810.

[17] Drouhard, J.P., Sabourin, R. and Godbout, M., "Evaluation of a Training Method and of Various Rejection Criteria for a Neural Network Classifier Used for Off-Line Signature Verification", IEEE International Conference on Neural Networks, Orlando, Florida, june 26 - july 2, 1994, 4294-4299.

[18] Sabourin, R. and Drouhard, J.P., "Off-Line Signature Verification Using Directional PDF and Neural Networks", Proc. of the 11th ICPR, The Hague, The Netherlands, pp 321-325, August 1992.

[19] Sabourin, R., "Évaluation d'un Classificateur de Type LVQ de Kohonen pour la Vérification Automatique de l'Identité", Actes du Colloque National sur l'Écrit et le Document, Nancy, France, 6-7 july, pp 160-167, 1992.

[20] Sabourin, R., Drouhard, J.P., Gagné, L. and Paquet N., "Automatic Handwritten Signature Verification Using Directional PDF and Neural Networks: A Feasibility Study", Conférence et Exposition sur l'Automatisation Industrielle, Montréal, 1-3 June, pp 20.17-20.20, 1992.

EXTENDED SHADOW-CODE (ESC) and FUZZY-ESC

[21] Simon, C., Levrat, E., Sabourin, R. and Bremont, J. "A Fuzzy Perception for Off-line Handwritten Signature Verification", Proc of the BSDIA'97, Curitiba, Brazil, november 1997.

[22] Simon, C., Levrat, E., Brémont, J. and Sabourin, R., "Codage d'images de signatures manuscrites pour la vérification hors-ligne", LFA'96, Nancy, 4-5 december, pp 23-30, 1996.

[23] Simon, C., Querelle, R., Levrat, E., Brémont, J. and Sabourin, R., "Codage d'images de signatures manuscrites pour la vérification hors-ligne", 4ième Colloque National sur l'écrit et le document, Nantes, 3-5 july, pp 171-177, 1996.

[24] Sabourin, R. and Genest, G., "Définition et Évaluation d'une Famille de représentations pour la vérification hors-ligne des signatures", Traitement du Signal, Vol.12, No. 6, 1995, pp 585-596.

[25] Sabourin, R. and Genest, G., "An Extended-Shadow-Code Based Approach for Off-Line Signature Verification: Part -II- Evaluation of Several Multi-Classifier Combination Strategies", Third IAPR Conf. on Document Analysis and Recognition, August 14-16, 1995, Montreal, Canada, pp 197-201.

[26] Sabourin, R. and Genest, G., "Coopération de Classificateurs pour la Vérification Automatique des Signatures", Colloque National sur l'écrit et le Document, Rouen, France, july 6-8, 1994.

[27] Sabourin, R. and Genest, G., "An Extended-Shadow-Code Based Approach for Off-Line Signature Verification: Part -I- Evaluation of The Bar Mask Definition", 12th ICPR, Jerusalem, Israel, October 9-13, 1994, pp 450-453.

[28] Sabourin, R., Cheriet, M. and Genest, G., "An Extended-Shadow-Code Based Approach for Off-Line Signature Verification", Second IAPR Conf. on Document Analysis and Recognition, Tsukuba, Japan, october 20-22, pp 1-5, 1993.

[29] Simon, C., Levrat, E., Sabourin, R. and Bombardier, V., "Vérification Automatique des Signatures Manuscrites par Classification Floue", Troisièmes Journées Nationales : Les Applications des Ensembles Flous, Nîmes, october 26-27 1993.

FUZZY ARTMAP, LOCAL TEMPLATE MATCHING

[30] Murshed, N.A., Bortolozzi, F. and Sabourin, R., "Binary Image Compression Using Back-propagation Neural Network", SPIE Congress on Electronic Imaging (EI'97), San Jose, California, Feb 1997.

[31] Murshed, N.A., Bortolozzi, F. and Sabourin, R., "A Cognitive Approach to Signature Verification", in the International Journal of Pattern Recognition and Artificial Intelligence, Special Issue on Bank Cheques Processing, to appear in 1997.

[32] Murshed, N.A., Bortolozzi, F. and Sabourin, R., "Off-Line Signature Verification Using Fuzzy ARTMAP Neural Networks", IEEE Inter. Conf. on Neural Networks, Perth, Western Australia, november 27 - december 1st 1995, pp. 2179-2184.

[33] Murshed, N.A., Bortolozzi, F. and Sabourin, R., "Off-line Signature Verification Without Requiring Random Forgeries for Training", 3rd Inter. Computer Science Conference., Hong Kong, december 11-13 1995, pp 107-115.

[34] Murshed, N.A., Bortolozzi, F. and Sabourin, R., "Off-line Signature Verification, Without a Priori Knowledge of class w_2. A New Approach", Third IAPR Conf. on Document Analysis and Recognition, August 14-16, 1995, Montreal, Canada, pp 191-196.

VISUAL PERCEPTION

[35] Sabourin, R., Drouhard, J.-P. and Sum Wah, E., "Shape Matrices as a Mixed Shape Factor for Off-line Signature Verification", Proc. of the ICDAR'97, Ulm, Germany, August 18-20, 1997.

[36] Sabourin, R., Genest, G. and Prêteux, F., "Off-Line Signature Verification by Local Granulometric Size Distributions", to appear in IEEE TPAMI (accepted for publication in june 1997).

[37] Sabourin, R., "Une approche perceptuelle pour la définition d'une représention adaptée au problème de la vérification hors-ligne des signatures manuscrites", ACFAS-96, Colloque sur l'intelligence artificielle dans les technologies de l'information : les enjeux et les techniques, may 15-16 1996, Univ. McGill, Montréal.

[38] Sabourin, R., Genest, G. and Prêteux, F., "Pattern Spectrum as a Local Shape Factor for Off-Line Signature Verification", 13th ICPR, Viena, Austria, August 1996, pp C43-C48.

RELATED REFERENCES IN THE FIELD OF FORENSIC SCIENCES

[39] Locard, E., "Traité de Criminalistique", Lyon, Payoy, 1936.

[40] Harrison, W.R., "Suspect Documents, Their Scientific Examination", Nelson-Hall Publishers, Chicago, 1981, 583 pages.

[41] Mathyer, J., "The Expert Examination of Signatures", Journal of Criminal Law, Criminology and Police Science, Vol. 5, No. 3, May-June, pp. 122-133, 1961.

Research Advances in Graphics Recognition: An Update

Rangachar Kasturi and Huizhu Luo

Computer Science and Engineering
The Pennsylvania State University
University Park, PA 16802, USA
{kasturi, hluo}@cse.psu.edu

Abstract. Localization, recognition, and interpretation of graphical objects such as line art, maps and engineering drawings in document images has been a topic of much research attention in recent years. A comprehensive overview of research problems, solution procedures and typical systems for graphics recognition was published in 1990 [KasRam90] and an update was published in 1994 [AriKas94]. In this paper we provide a further update on the state of research in graphics recognition with a particular emphasis on those systems which have been presented at recent workshops on graphics recognition [KasTom96, TomCha97].

1. Introduction

The most common sequence of steps taken for document image analysis for graphics interpretation are

1. Data Capture and Preprocessing
2. Region Segmentation
3. Vectorization
4. Feature Segmentation
5. Semantic Interpretation
6. Object Recognition.

These topics are discussed in detail in several publications [KasRam90, OgoKas97]. We very briefly describe these topics here. Data capture and preprocessing includes operations such as scanning, noise filtering, and thresholding, for converting a paper-based document to a reasonably noise-free binary image. A typical image of a document contains regions of text, graphics, and half-tone images. It is then necessary to segment these into their individual regions to facilitate application of appropriate algorithms depending upon their class. For example, text regions after separation can be used as input to an OCR system; graphics regions can be processed by a graphics interpretation system, and image regions can be input to a data compression system or an image interpretation system as appropriate. Often, it is necessary to segment a graphics image into solid objects (completely filled entities) and thin lines. Since straight lines and curves are basic components in a graphics diagram, substantial data compaction for further processing is achieved by analyzing bit-mapped graphics image to generate line attribute description file. Such a file contains not only the location of various lines but also their attributes such as line thickness, lengths of segments of dashed lines etc. Line description files are generated by fitting lines to bit-mapped data or by thinning and line following operations. Identification of graphical objects such as polygons requires detection of feature points such as corners and points of transitions from straight lines and curves. Locating such points is made difficult by the artifacts introduced during digitization and thinning. The segmented line description file is then processed by a semantic interpretation system which includes task dependent symbol recognition algorithms. For example, in an electrical schematic recognition system, the algorithms are designed to locate electrical components in the image whereas a symbol polygonal approximation may be adequate for a flow chart recognition system. For 3-D objects represented by their 2-D projections, algorithms have been developed to combine information in one or more views to generate a three dimensional object description. For example, information in orthographic projections in engineering drawings can be combined to create 3-D shape description.

Recent advances in computer hardware has further increased the interest in this area since it has now become practical to build commercial systems for converting large engineering drawings and maps. Two focussed workshops have been held since 1995 to provide a forum for exchange of research ideas [KasTom96, TomCha97]. Papers on this topic have also been published in recent document analysis conferences and workshops such as the International Conferences on Document Analysis and Recognition, Workshop on Document Analysis Systems. The objective of this paper is to provide an update to the state of research in graphics recognition since we published a similar paper in 1994 [AriKas94]. Clearly, we are not able to include all significant contributions to this field due to limited space; we have limited our update to papers which are published in well-known journals and major graphics recognition related workshops and conferences. We organize these papers into three sections: Diagram and Symbol Recognition, Engineering Drawing Understanding, and Map Processing.

2. Diagram and Symbol Recognition

A variety of diagrams such as flow charts, logic and electrical circuits, and chemical plant diagrams are used. In general, understanding a diagram consists of two phases. One is to recognize the symbols from the drawing and the other is to analyze spatial and logical relationships among symbols so as to interpret the diagram. As such, symbol recognition is generally application-specific. Overviews of symbol recognition methodologies have been presented by Blostein [Blo96] and Chhabra [Cha97]. In this section we discuss a few recent papers on this topic.

An automatic learning and recognition of graphical symbols in diagrams is proposed by Messmer and Bunke [MesBun96]. In their system, the learning procedure and the recognition procedure work alternatively. The known symbol database is generated by the learning algorithm and used for recognition by the recognition algorithm. After recognition, symbols identified are removed from the drawing and remaining line segments are again grouped into possible new symbols. Some representatives are learned and added to the symbol database. The above procedure is repeated until all symbols are identified. Hierarchical matching is applied during the recognition procedure. A symbol is decomposed into a few different sets of components, each of which is used at a different level of matching. As a consequence symbol searching is computationally efficient. The learning procedure is based on a generic, meta-level definition of the symbols which is guided by the application domain. Therefore the system can be easily modified for use with different types of documents.

Cesarini et al [CesGor96] have developed a system for locating and recognizing low level graphic items. Locating is done by morphological operations and connected component analysis. For example, erosion operation is used for breaking the lines connecting adjacent items such as diagram symbols since the lines are typically much thinner than symbols. Dilation is performed on each of the remaining connected components so as to connect broken parts within the symbol caused by previous erosion. The classification of graphic items is implemented as an autoassociator-based neural network, where the input and output layer have the same number of units. Before recognition, several instances of the items are required for training. Training is typically computation-intensive.

Chemical structure diagrams consist essentially of solid and dashed lines, characters and other symbols. An automatic system for interpretation of chemical structure drawings has been described by McDaniel et al [McDBal96]. The system consists of vectorization, dashed line detection, character recognition, post processing, display, and editing steps. Identification of dashed lines is done by exhaustive search. After removing recognized characters using an OCR module, graphic compilation is used to generate a connection table of the remaining vector data. Some errors are corrected during post processing and manual editing.

An automatic understanding system for circuit diagrams is described by Yu [Yu95]. The system is based on block adjacency graph (BAG) representation of diagrams. BAG contains shape information and topological structure information in the original drawing. This representation also requires much less memory space than original raster data and is useful for higher level processing such as the extraction of shape and structure information. Character separation from symbols and lines is first performed using the BAG representation. Vectorization and symbol identification are then performed. Fourier descriptors are used for symbol recognition because of its invariance to translation, scaling and rotation.

A number of diagram analysis algorithms use constraint grammars. Grammar-based systems are either domain-specific or domain-independent. Domain-specific systems are typically capable of analyzing more complex diagrams than domain-independent systems. Futrelle et al have described an analysis of complex diagrams using constraint-based grammar [FutNik95]. They assume the diagram is a connection of graphics primitives such as lines, polygons, circles and text. Their algorithm combines the flexibility of domain retargeting by writing alternate grammars with the efficiency of the domain-specific systems. Witterburg et al [WitWei91] have proposed a bottom-up parsing algorithm that has been applied for flow chart and mathematical expressions interpretation.

Yu et al. [YuSam97] developed a system for recognizing a large class of drawings. The class includes domains such as flowcharts, logic and electrical circuits and chemical plant diagrams. The system includes two principal stages. One segments symbols from connection lines in the image after it has been thinned, vectorized and preprocessed by a commercial software. Preprocessing here includes removal of text, artificial gaps and spurs from drawing images. The second stage classifies symbols and corrects errors automatically using a set of domain specific matchers. The following generic rules are used by the segmentation algorithm to separate connection lines from symbols:

1. Connection lines consist of horizontal and vertical lines.
2. Loops with simple geometric shapes such as circles and rectangles are part of the symbols.
3. Slanting lines and open lines are parts of symbols.
4. Symbolic loops do not contain crossing horizontal and vertical lines and hence are distinguishable from non-symbolic loops formed by crossing connection lines.

And the symbol classification process consists of the following three steps:

1. Group the set of potential symbol entities into connected components and classify the entities in each component.
2. Use the domain specific symbol library to initially recognize symbols. The library can be changed for different domains of drawings.
3. Traverse the drawing starting from the well recognized symbols. Some errors in segmentation and symbol classification are corrected during the traversal.

After the two automated stages a graphical user interface is also provided to correct residual errors interactively and to log data for reporting errors objectively.

3. Engineering Drawing Interpretation

Due to their complexity and diversity, there are no fully automatic interpretation systems which can be used for converting all types of engineering drawings. Much progress has occurred in specific application domains and a few systems which depend upon human interaction for conversion/interpretation are on the market. For an overview of the current state of literature see [Tom97]. Here we consider a few recent efforts at engineering drawing interpretation in specific domains.

Professor Dori and his group have been developing a generic machine drawing understanding system for several years. Specifically, they have developed a consistent procedure to describe both objects and their transformation by the application of processes using an *Object-Process Diagram (OPD)* [Dor95]. OPD provides a systematic way of describing a higher level object in terms of its elemental components and processes that contribute to such a composition.

A framework for an automatic CAD conversion system is described in [LiuDor96]. It includes a description of the process of understanding mechanical engineering drawings from scanning to 3-D interpretation. The system is characterized by hierarchies of object classes. In this framework, sparse pixel vectorization method is used for vectorization, which is an improved version of orthogonal zig-zag (OZZ) method. Using this vector information and a neural network based OCR algorithm characters are segmented and recognized. Arcs, arrow heads and dashed/hatched lines are then detected according to their respective syntax in the drawing. Work towards 3-D reconstruction from 2-D views is in progress [WeiDor97].

Dori et al. [DorVel96] have also described an object-process based system for segmentation and recognition of ANSI and ISO standard dimensioning text. These standards specify the features of all dimensioning text which are useful for devising techniques for recognition and association of dimensioning text. In their segmentation algorithm, text-wire candidates are first selected from the output of OZZ vectorization process. A parameter specifying the size of the characters is then used to form logical text boxes. The drawing standards are applied to guide the separation of the logical boxes into single text strings and associating the string with the corresponding arrowheads. For each string, single characters are segmented again and recognized by a neural network. After single characters are recognized, they are concatenated into strings which are verified by contextual analysis.

An important final task in drawing interpretation is to reconstruct 3-D object models from the corresponding orthographic projections (2-D drawings). For simple drawings, interpretation of lower level data is adequate to obtain consistent interpretation. Simple elements such as nodes and lines are combined to create vertices and edges and then to construct surfaces. These elements are again combined to create solid blocks, out of which objects are then assembled. However, some errors during scanning are likely to introduce ambiguities which can not be resolved from low level data. Thomas et al. [ThoPol96] have proposed a combined high and low level approach to combine low level features (nodes and lines) and high level features (textual content of drawings). They extract textual information by matching a coded phrase to a syntactic net stored in the knowledge base. Then a semantic template is associated with a particular sub-object which specifies all the information required for that type of object. With this knowledge from the textual information, many features can be identified even if they are not shown in the drawing. As a result, combined low level and high level 3-D reconstruction and error correction are achieved. Another complete 3-D interpretation system using Dempster-Shafer theory of evidence was developed by Lysak and Kasturi in 1991 [LysKas91] and was subsequently verified experimentally by Devaux et al [DevLys95].

The method of function-based vision is not new in pattern recognition and computer vision field. However, line drawing understanding systems have not incorporated many of the concepts from this approach. Capellades and Camps [CapCam96] have attempted to apply this concept into a line drawing understanding system. Their first step is to extract some functional parts from mechanical engineering drawings such as screws, hinges and gears, etc. Since they assume that the drawings are drawn to meet the ANSI standard, screw detection can be simply done by recognizing the unique structure of the schematic representation of threads drawn as parallel lines. Such function based line drawing understanding system helps to interpret complex engineering drawings effectively.

There are many applications for utility drawing interpretation. Utility drawings are typically very large in size and are often of poor quality due to their age and numerous updates over the years. Interpretation systems designed for specific classes of telephone utility drawings are described in [AriKas95, LuoKas97-1]. The drawings converted by these systems are typically 34x44 inches in size and are composed primarily of horizontal and vertical lines. At 300 DPI, a typical image has 10,200 columns by 13,200 rows resulting in a raw image of over 130 MB. Domain-specific information is incorporated to achieve high efficiency in processing such large drawings. Basically, the algorithms first take advantage of run length encoding to reduce the amount of memory and processing time. Then the primitives such as intersections for each class of drawings are extracted and combined to generate the higher level structures such as lines, boxes and tables.

One of the common tasks in processing graphics-rich documents is the segmentation of text from graphics. Many algorithms have been reported in the literature for this task for over ten years. More recently, Luo et al [LuoAga95] used directional mathematical morphology to decompose the edges of the map into eight directional edge images. Most of the touching characters and lines can be split into one of these directional edge images. It is easy to remove the short character edges and keep longer ones that belong to the line since the character edges are much shorter than line edges. As a result, the characters can be separated from lines. An efficient implementation of this algorithm using run-length-encoded image representation has also been proposed [LuoKas97-2].

4. Map Processing

Traditional maps contain information about land usage, transportation network, political boundaries, topographic and hydrographic features, and annotations. The rapid growth in the volume and the complexity of maps naturally led to the development of computerized Geographical and Land Information System (GIS and LIS) to facilitate information management. Computerized representation of maps enables efficient searching and updating of information. However the conversion of existing paper-based maps into computer-readable databases is a time consuming and tedious task. Many research groups have been developing methods to automatically extract information from scanned images of paper maps.

There are a variety of maps for use with different application realms. For example, a hydrographic map [Trier96] consists of the data about the coast and seabed including the depth values, contours of constant depth, and special symbols denoting submerged rocks. All these information is useful for navigation. Land register maps [Boa92] on the other hand are used for urban planning and contain information about parcels of land represented by polygons, buildings, streets, and annotations. Other GIS applications include map production, utility network management, market analysis, and transportation planning.

During the past few years there has been an increased research interest in automated map interpretation. Some of these efforts have been directed toward the development of overall systems [LuSak94, OgiLab93, NakYaz95, LuOku95, SmeKat96, MyeMul96, ShiTak93, ShiHor92, Boa92, SamSof96, HarKat96, EbiLau94, Har95, TriTax97] while others have emphasized algorithm development for a particular aspect of the problem [YamYam93, LuoAga95, AblFra94, EllSen94, Mow92]. In traditional bottom-up map interpretation systems, basic features such as characters and symbols are first extracted from the map and then the interpretation methods are applied to graphical entities for vectorization and recognition purposes. Alternatively, knowledge-based map interpretation approaches process and recognize all object features such as lines, characters and symbols in a unified manner. The knowledge about objects and their relationships is necessarily predefined. Many early methods process only gray level or binary maps while some recent approaches use color maps. Color segmentation [LauAnh94, Yan93, GaoHe93] then becomes a critical step.

An impressive system for automated map interpretation has been developed by Hartog [Har95, HarKat96]. This system is characterized by its knowledge base and contextual reasoning. It has been observed that maps are drawn based on some specific rules and a limited set of symbols which can be used to create the knowledge base. For several distinct types of maps, the framework of map interpretation can be the same, only difference being the specification of knowledge base and knowledge reasoning. In this system, a top-down control mechanism is integrated with a bottom-up object recognition strategy. The control mechanism is based on spatial relationships among objects. Both declarative and procedural knowledge is made available in an explicit form yielding a significant reduction in the time and effort required to adapt the system to a specific application. Top-down segmentation is integrated with the contextual reasoning framework which helps to detect inconsistencies between recognized objects and prior knowledge. One reason for inconsistency is global poor segmentation. Thus, local resegmentation is first tried. With the knowledge base, resegmentation and contextual reasoning, the map interpretation process is simplified and the classification errors are reduced. However, the system requires significant computation resources; static parameter specification of knowledge base is another limitation of the system.

Myers et al [MyeMul95] developed a hypothesize-and-verify approach to convert USGS topographic maps. Models of map features from scanned images of legends and map specifications are created in an off-line training process. Linear, point, and area features are extracted by generating and verifying hypotheses about the features in an object recognition type scheme without presegmentation of graphical entities. The system also uses the recognized features and lexicons of place names obtained from a gazetteer to hypothesize the position and content of text labels that appear on the map and then attempts to recognize the text. The system is well suited to take advantage of the rich a priori knowledge associated with maps. Another advantage of a verification-based approach is that only the data that is of interest is extracted requiring less processing time than approaches that try to interpret everything on the map.

The PROMAP system described by Ebi et al [EbiLau94] performs automatic conversion of high resolution color topographic maps. The image is split into layers of predefined map colors which are then vectorized. Neural network-based symbol and object recognition is used for the extraction of attributed structure primitives. Knowledge-directed interpretation is used in the hierarchical structuring of the map with map objects and relations and in the use of concepts (prototypes) as the basic control mechanism.

The system developed by Samet and Soffer [SamSof96] is designed to use map legends to drive the interpretation process. Information layers in the map are input to the system as individual layers. These layers are separately processed. Map legends are used to create a training set library. Symbols in the map are identified using a weighted several nearest neighbor classifier. The logical representation of physical objects is used as an index to information in the map images.

Ablameyko and Frantskevich [AblFra94] use labeling, a well known computer vision technique, for the interpretation of black and white layers of geographical maps. Elliman and Sen-Gupta [EllSen94] describe a method for the segmentation of linear features, symbols and textured areas within binary map. The paper demonstrates full segmentation of components within binary images of fairly complex maps.

A system for recognizing short character strings in maps is described by Kim et at [KimPar96]. Characters touching boundaries are recognized by template matching. Region boundaries enclosing the character strings are then vectorized using an iterative bisection method to facilitate efficient storage. Most of the region names are located inside the boundary. However when a region is small and the region name is outside the boundary, an indication line can be found to point to the corresponding region. In both cases, association of region name with region boundary can be done.

A three year project on topographic map interpretation at the National Geographic Institute of France has been recently reported [DesMar97]. In this project the goal is to extract automatically different layers such as textured areas, buildings, other symbols, text strings, and road network sequentially from the maps. As each layer is extracted, it is subtracted from the map image so that the interpretation of the remaining layers is simplified. Other papers describing specific aspects of this project may also be found in the same proceedings [TomCha97].

There is a growing interest in color map interpretation. In a color map the same line may have different colors in different regions and objects represented in different colors are more likely to overlap and intersect each other. An optimized nearest neighbor rule based method to segment color map images is described by Yan [Yan93]. In this approach, training image pixels are selected from representative areas interactively or using unsupervised learning and cluster validation algorithms. A set of color features at each training pixel are extracted to build an image classifier. A small number of prototypes are generated from the training samples and optimized using a multi-layer neural network. The optimized prototypes are then used to classify test images. The segmentation method is applied only for separation of the foreground from background instead of characters and lines.

An adaptive system for association of street names with streets on large, complex, color composite maps has been presented [NagSam97]. Color is used as a discriminating feature to separate text from other features. Subsequent processing steps such as text processing, vectorization, and name association are governed by an expert system. Information extracted from such systems can be used to guide an image analysis system operating on digital orthophotos (high resolution aerial images which are photogrametrically corrected to register with their corresponding maps) to determine changes and create an updated map.

5. Conclusions

While there is a wealth of literature with solutions proposed to solve various problems in graphics recognition and interpretation, there has been very limited effort at formally evaluating the performance of various approaches to determine their accuracy, efficiency, and sensitivity. In particular, characterization of the degradation of algorithm performance due to factors such as drawing complexity and noise level is very important. A formal effort at performance evaluation of algorithms began only recently. The first *contest* for evaluating the performance of dashed-line detection algorithms was held in 1995 [KonPhi96] and a second one on text segmentation and vectorization methods was held in 1997 [LiuDor97, PhiLia97]. These exercises have demonstrated the difficulty and complexity of the task of formally evaluating the performance of image analysis algorithms. At the same time as our field matures it is increasingly important to make such formal evaluations so that different approaches for solving the same problem can be characterized in a consistent and unbiased manner. Information gained from such efforts is very useful not only to researchers but also to system designers who must make a decision on their choice of procedures without reimplementing and evaluating every available approach.

Looking back at the progress made in engineering drawing interpretation over the past few years, it is clear that few breakthrough methods have been developed in recent years [Tom97]. Most recently published algorithms are just minor variations of old methods. This should be a serious concern to all researchers and we must pay more attention to study the reason for this stagnation and formulate plans to direct future research.

References

[AblFra94] Ablameyko, S. and O. Frantskevich,"From computer vision to document recognition or using labeling technique for map interpretation," *IAPR Workshop on Machine Vision Applications*, Kawasaki, 1994, pp. 255-258.

[AriKas94] Arias, J. and R. Kasturi., "Recognition of graphical objects for intelligent interpretation of line drawings," in *Aspects of Visual Form Processing*, eds: C. Arcelli, L.P. Cordella, and G. Sannitti di Baja, World Scientific, Singapore, 1994, pp. 11-31.

[AriKas95] Arias, J., R. Kasturi and A. Chhabra, "Efficient techniques for telephone company line drawing interpretation," *Proc. 3rd International Conference on Document Analysis and Recognition*, 1995, pp. 795-798.

[Blo96] Blostein, D., "General diagram recognition methodologies," in *Graphics Recognition: Methods and Applications*, eds. R. Kasturi and K. Tombre, [Post-Proceedings of The First International Workshop on Graphics Recognition (August 1995) Penn State], Springer-Verlag, LNCS 1072, Heidelberg, Germany, 1996, pp. 106-122.

[Boa92] Boatto L. etc. "An interpretation system for land register maps," *Computer*, July 1992, pp. 25-32.

[CapCam96] Capellades, M.A. and O.I. Camps, "Functional parts detection in engineering drawings: Looking for the screws," in *Graphics Recognition: Methods and Applications*, eds. R. Kasturi and K. Tombre, [Post-Proceedings of The First International Workshop on Graphics Recognition (August 1995) Penn State], Springer-Verlag, LNCS 1072, Heidelberg, Germany, 1996, pp. 246-259.

[CesGor96] Cesarini, F., M. Gori, S. Marinai and G. Soda, "A hybrid system for locating and recognizing low level graphic items," in *Graphics Recognition: Methods and Applications*, eds.

R. Kasturi and K. Tombre, [Post-Proceedings of The First International Workshop on Graphics Recognition (August 1995) Penn State], Springer-Verlag, LNCS 1072, Heidelberg, Germany, 1996, pp. 135-147.

[Cha97] Chhabra, A.K., "Graphic symbol recognition: An overview," *Proceedings of the Second International Workshop on Graphics Recognition*, Nancy, France, 1997, pp. 244-252.

[DesMar97] Deseilligny,M. and R. Mariani, "A three year project on topographic maps interpretation," *Proc. of 2nd IAPR Workshop on Graphics Recognition*, Nancy , France, 1997, pp. 160-167.

[DevLys95] Devaux, P.M., D.B. Lysak, C.P. Lai, and R. Kasturi, "A complete system for recovery of 3-D shapes from engineering drawings," *Proceedings of the IEEE Symposium on Computer Vision*, Coral Gables, Florida, 1995, pp. 145-150.

[Dor95] Dori, D., "Object-Process analysis: Maintaining the balance between system structure and behaviour," *Journal of Logic and Computation, Vol. 5,*1995, pp. 227-249.

[DorVel96] Dori, D, Y. Vel and W. Liu, "Object-process based segmentation and recognition of ANSI and ISO standard dimensioning texts,"in *Graphics Recognition: Methods and Applications*, eds. R. Kasturi and K. Tombre, [Post-Proceedings of The First International Workshop on Graphics Recognition (August 1995) Penn State], Springer-Verlag, LNCS 1072, Heidelberg, Germany, 1996, pp. 212-232.

[EbiLau94] Ebi, N., B. Lauterbach, and W. Anheier, "An image analysis system for automatic data acquisition from colored scanned maps," *Machine Vision and Applications, 7,* 1994, pp. 148-164.

[EllSen94] Elliman, D. G. and M. Sen-Gupta,"Automatic recognition of linear features, symbols and textured areas within maps," *IAPR Workshop on Machine Vision Applications*, Kawasaki, 1994, pp. 239-242.

[FutNik95] Futrelle, R.P. and N. Nikolakis, "Efficient analysis of complex diagrams using constraint-based parsing," in *Proc. of 3rd International Conference on Document Analysis and Recognition*, Aug. 14-16, 1995, pp. 783-790.

[GaoHe93] Gao, Q., J. He, G. Qiu, and Q. Shi, "A color map processing system PU-CMPS," in *Proc. of 2nd International Conference on Document Analysis and Recognition*, Tsukuba, Oct. 20-22, 1993, pp. 874-877.

[Har95] Hartog, J. E., *A framework for knowledge-based map interpretation*, Ph.D. thesis, 1995, Dept. of Signal Processing, TNO institute of Applied Physics, Netherlands.

[HarKat96] Hartog, J. E., T. K. ten Kate, and J. J. Gerbrands, "Knowledge-based segmentation for automatic map interpretation," in *Graphics Recognition: Methods and Applications*, eds. R. Kasturi and K. Tombre, [Post-Proceedings of The First International Workshop on Graphics Recognition (August 1995) Penn State], Springer-Verlag, LNCS 1072, Heidelberg, Germany, 1996, pp. 159-178.

[KasRam90] Kasturi R., R. Raman, C. Chennubhotla, and L. O'Gorman, "Document image analysis: An overview of techniques for graphics recognition," *Pre-Proc. IAPR workshop on Structural and Syntactic Pattern Recognition*, pp. 192-230, New Jersey, June 1990, also in *Structured Document Image Analysis*, eds: H.S. Baird, H. Bunke, and K. Yamamoto, Springer-Verlag, 1992, pp. 285-324.

[KasTom96] Kasturi, R. and K. Tombre, Editors, *Graphics Recognition: Methods and Applications*, [Post-Proceedings of The First International Workshop on Graphics Recognition (August 1995) Penn State], Springer-Verlag LNCS 1072, Heidelberg, Germany, 1996

[KimPar96] M.-K. Kim, M.-K. Park, O.-S. Kwon and Y.-B. Kwon, "Automatic region labeling of the layered map," in *Graphics Recognition: Methods and Applications*, eds. R. Kasturi and K. Tombre, [Post-Proceedings of The First International Workshop on Graphics Recognition (August 1995) Penn State], Springer-Verlag, LNCS 1072, Heidelberg, Germany, 1996, pp. 179-189.

[KonPhi96] Kong, B., I.T. Phillips, R.M. Haralick, A. Prasad, and R. Kasturi, "A benchmark: Performance evaluation of dashed-line detection algorithms," in *Graphics Recognition: Methods and Applications*, eds. R. Kasturi and K. Tombre, [Post-Proceedings of The First International Workshop on Graphics Recognition (August 1995) Penn State], Springer-Verlag, LNCS 1072, Heidelberg, Germany, 1996, pp.270-285.

[LauAnh94] Lauterach, B. and W. Anheier, "Segmentation of scanned maps in uniform color spaces," *IAPR Workshop on Machine Vision Applications*, Kawasaki, 1994, pp. 222-225.

[LiuDor96] Liu W. and D. Dori, "Automated CAD conversion with the machine drawing understanding system," *Proc. of Document Analysis Systems*, Malvern, Pennsylvania, U.S.A., October 1996, pp. 241-259.

[LiuDor97] Liu, W. and D. Dori, "A protocol for performance evaluation of algorithms for text segmentation," *Proceedings of the Second International Workshop on Graphics Recognition*, Nancy, France, 1997, pp. 317-324.

[LuoAga95] Luo, H., G. Agam, and I Dinstein,"Directional mathematical morphology approach for line thinning and extraction of character strings from maps and line drawings," *Proc. of 3rd International Conference on Document Analysis and Recognition*, Montreal, Canada, 1995, pp. 257-260.

[LuOku95] Lu, W., T. Okuhashi, and M. Sakauchi,"A proposal of efficient interactive recognition system for understanding of map drawings," *Proc. of 3rd International Conference on Document Analysis and Recognition*, Montreal, Canada, 1995, pp. 520-523.

[LuSak94] Lu, W., and M. Sakauchi,"An interactive map drawing recognition system with learning ability," *IAPR Workshop on Machine Vision Applications*, 1994, Kawasaki, pp. 235-238.

[LuoKas97-1] Luo, H., R. Kasturi, J. Arias and A. Chhabra, "Interpretation of lines in distribution frame drawings," *Proc. of 4th International Conference on Document Analysis and Recognition*, Ulm, Germany, 1997, pp. 66-70.

[LuoKas97-2] Luo, H. and R. Kasturi, "Improved morphological operations for separation of characters from maps/graphics," *Proceedings of the Second International Workshop on Graphics Recognition*, Nancy, France, August 1997, pp. 8-15.

[LysKas91]Lysak, D.B. and R. Kasturi, "Interpretation of engineering drawings of polyhedral and nonpolyhedral objects," *Proc. 1st International Conference on Document Analysis and Recognition*, St. Malo, France, 1991, pp. 79-87.

[McDBal96] McDaniel, J.R. and J.R. Balmuth, "Automatic interpretation of chemical structure diagrams," in *Graphics Recognition: Methods and Applications*, eds. R. Kasturi and K. Tombre, [Post-Proceedings of The First International Workshop on Graphics Recognition (August 1995) Penn State], Springer-Verlag, LNCS 1072, Heidelberg, Germany, 1996, pp. 148-158.

[MesBun96] Messmer, B.T. and H. Bunke, "Automatic learning and recognition of graphical symbols in engineering drawings," in *Graphics Recognition: Methods and Applications*, eds. R. Kasturi and K. Tombre, [Post-Proceedings of The First International Workshop on Graphics Recognition (August 1995) Penn State], Springer-Verlag, LNCS 1072, Heidelberg, Germany, 1996, pp. 123-134.

[Mow92] Mowforth, P., "DATA conversion for GIS," *IAPR Workshop on Machine Vision Applications*, Tokyo, Japan, 1992, pp. 403-406.

[MyeMul96] Myers, G. K., P. G. Mulgaonkar, C.-H. Chen, J. L. DeCurtins, and E. Chen, "Verification-based approach for automated text and feature extraction from raster-scanned maps," in *Graphics Recognition: Methods and Applications*, eds. R. Kasturi and K. Tombre, [Post-Proceedings of The First International Workshop on Graphics Recognition (August 1995) Penn State], Springer-Verlag, LNCS 1072, Heidelberg, Germany, 1996, pp. 190-203.

[NagSam97] Nagy, G., A. Samal, S. Seth, T. Fisher, E. Guthmann, K. Kalafala, L. Li, P. Sarkar, S. Sivasubramaniam, Y. Xu, "A prototype for adaptive association of street names with streets on maps," *Proceedings of the Second International Workshop on Graphics Recognition*, Nancy, France, August 1997, pp. 168-176.

[NakYaz95] Nakejima, C. and T. Yazawa, "Automatic recognition of facility drawings and street maps utilizing the facility management database," *Proc. of 3rd International Conference on Document Analysis and Recognition*,Montreal, Canada, 1995, pp. 516-519.

[OgiLab93] Ogier, J.M. J. Labiche, R. Mullot, and Y. Lecourtier, "Attributes extraction for french map interpretation," *Proc. 2nd International Conference on Document Analysis and Recognition*, Tsukuba Science City, Japan, 1993, pp. 672-675.

[OgoKas97] O'Gorman, L. And R. Kasturi, *Document Image Analysis: Executive Briefing*, IEEE Computer Society Press, Los Alamitos, California, 1997

[PhiLia97] Phillips, I.T. , J. Liang, and R. Haralick, "A performance evaluation protocol for engineering drawing recognition systems," *Proceedings of the Second International Workshop on Graphics Recognition*, Nancy, France, 1997, pp. 333-346.

[SamSof96] Samet, H. and A. Soffer, "MARCO: Map retrieval by content", *IEEE Trans. Pattern Analysis and Machine Intelligence, vol 18*, no. 8, 1996, pp. 783-798.

[ShiHor92] Shimotsuji, S., O. Hori, M. Asano, K. Suzuki and F. Hoshimo, "A robust recognition system for drawing superimposed on a map," *Computer, 25*, 7, pp. 56-59, 1992.

[ShiTak93] Shimada, S., Y. Takahara, H. Suenaga, and K. Tomita, "Paralleled automatic recognition of maps and drawings for constructing electric power distribution databases," *Proc. of 2nd International Conference on Document Analysis and Recognition*, Tsukuba, 1993, pp. 688-691.

[SmeKat96] Smeulders, A. W. M. and T. ten Kate, "Software system design for paper map conversion," in *Graphics Recognition: Methods and Applications*, eds. R. Kasturi and K. Tombre, [Post-Proceedings of The First International Workshop on Graphics Recognition (August 1995) Penn State], Springer-Verlag, LNCS 1072, Heidelberg, Germany, 1996, pp. 204-211

[ThoPol96] Thomas, P.D., J.E. Poliakoff, S.M. Razzaq and R.J. Whitrow, "A combined high and low level approach to interpreting scanned engineering drawings," in *Graphics Recognition: Methods and Applications*, eds. R. Kasturi and K. Tombre, [Post-Proceedings of The First International Workshop on Graphics Recognition (August 1995) Penn State], Springer-Verlag, LNCS 1072, Heidelberg, Germany, 1996, pp. 233-245.

[Tom97] K. Tombre, "Analysis of engineering drawings: State of the art and challenges," *Proc. of the Second International Workshop on Graphics Recognition*, Nancy, France, 1997, pp. 54-61.

[TomCha97] Tombre, K. And A. Chhabra, Editors, *Proceedings of the Second International Workshop on Graphics Recognition*, Nancy, France, August 1997

[TriTax97] Trier, O.D., T. Taxt and A. Jain, "Recognition of digits in hydrographic maps: Binary vs. topographic analysis," *IEEE Trans. Pattern Analysis and Machine Intelligence, vol.19*, no. 4, 1997, pp. 399--404.

[Trier96] Trier, O.D., *A Data Capture System for Hydrographic Maps*, Ph. D. thesis, Dept. of informatics, University of OSLO, Norway.

[WeiDor97] Weiss, M. and D. Dori, "A graph theoretic approach to the reconstruction of 3D objects from engineering drawings," *Proceedings of the Second International Workshop on Graphics Recognition*, Nancy, France, 1997, pp. 78-80.

[WitWei91] Wittenburg, K.L., L. Weizman and J. Talley, "Unification-based grammars and tabular parsing for graphical languages," *Journal of visual languages and computing, Vol. 2*, 1991, pp. 333-346.

[YamYam93] Yamada, H., K. Yamamoto, and K. Hosokawa, "Directional mathematical morphology and reformalized Hough transformation for the analysis of topographic maps," *IEEE Trans. Pattern Analysis and Machine Intelligence, vol. 15*, No. 4, April 1993, pp. 380-387.

[Yan93] Yan, H., "Color map image segmentation using optimized nearest neighbor classifiers," *Proc. of 2nd International Conference on Document Analysis and Recognition*, Tsukuba, Japan, 1993, pp. 111-114.

[YuSam97] Yu, Y., A. Samal and S. Seth, "A system for recognizing a large class of engineering drawings," *IEEE Trans. Pattern Analysis and Machine Intelligence, Vol. 19*, 1997, pp. 868-890.

[Yu95] Yu, Bin, "Automatic Understanding of Symbol-Connected Diagrams," *Proc. of 3rd International Conference on Document Analysis and Recognition*, Montreal, Canada, 1995, pp. 803-806.

Future Trends in Retrospective Document Conversion

A. Belaïd

CRIN-CNRS, Campus Scientifique, B.P. 239 F-54506 Vandœuvre-lès-Nancy Cedex
France

Abstract. This paper describes a framework for retrospective document conversion in the library domain. Drawing on the experience and insight gained from several projects launched over the present decade by the European Commission, it outlines the requirements for solving the problem of retroconversion and traces the main phases of associated processing.

1 Introduction

The success of library automation, resulting in user-friendly on-line catalogues[1] integrated with the WEB and other circulation-systems facilities, has created an urgent need for retroconversion of the older parts of catalogues [2,30]. As users get used to the new catalogue medium, the documents not registered in machine-readable form become "invisible" and unreadable. This has meant for many libraries the relegation of an important part of their rich stock of documents to a state of accessibility.

Such obvious waste of library collections in addition to the cost difference between manual handling and an equivalent set of automatic routines has made a strong case for the need to convert a library's entire collection of works to machine-readable records, in the interest of ensuring an efficient use of the investment in the new technology.

This has led to the search for cost-effective tools for the conversion of old catalogues into machine-readable forms. This search has not been limited to the sole problem of conversion but has been extended to embracing other objectives such as ensuring very high rates of distribution and sharing of documents between several libraries.

Drawing on [35], cataloguing methods can be classified into four main categories:

1. Retrocataloguing creating the catalogues from scratch by using new cataloguing tools and inherent standards ensuring the same structure and level of information as the library OPAC[2].

---

[1] A catalogue is a list of bibliographic descriptions of items held by a specific library, set up to give access to the items.
[2] Online Public Access Catalogue.

2. Retroconversion by direct keying from the paper catalogues and other media using automated input tools without adding or changing information in the input data to make it compatible with common standards.

3. Retroconversion by OCR and automatic formating, creating a structured and a tagged result of the machine-readable record.

4. Retroconversion by the substitution of old records with preexisting machine-readable records of the same records, provided by conversion agency or service production of national bibliographies.

The choice of method is conditioned by the state of funds available and the quality aimed at [22]. If the aim is to create a structure which is consistent with the rest of the OPAC records and respects its standards, then the first method appears as the most natural choice. Cost and time, however, constitute the two serious drawbacks of this choice.

On the other hand, if the aim is to make a maximum number of records available with the possibility of sharing with other sites, the substitution method then seems the most suitable choice. This choice also corresponds to the 1989 European Council Guidelines recommending not only inter-library cooperation in the creation of machine-readable records but also taking advantage of national bibliographical machine-readable records as well as those of important (private) collections [9–11]. In this context, appointed national agencies are to be charged with the responsibility of converting catalogues of national collections and making the results available to other libraries. The attractiveness of this solution, especially with respect to saving local efforts, is however marred by associated delays in making foreign collections available to users, a solution placing libraries in the unacceptable situation of having to wait for the new catalogues before making foreign collections accessible.

Fully aware of the limitations of these methods, most libraries are embarked on a search for optimal solutions. There is growing understanding that a combination of OCR and structure recognition may provide the basis for serious solutions. We can single out three of the several reasons for this. The first is that this approach avoids manual procedures and allows an integrated solution capable of adapting to the specific structure of the catalogue to be converted. Secondly, it allows a greater flexibility in the structuring of records without the necessity of having to reproduce a structure similar to the rest of the OPAC. Finally, the solution can easily be extended to other situations of retroconversion, requiring less rigid format structures, and where there is need to cope with free variations in formats as in scientific references.

The problem is not satisfactorily solved at the moment. Existing systems using OCR based solutions are not tangible enough to allow an assessment of the success of the technique. There is therefore room and need for more investigations to improve on the modest results obtained so far in this domain.

Drawing heavily on the experience and insight gained from FACIT[3] [35–37] and MORE[4] [4,24], and particularly on the various reports and recommendations published on the two projects, this paper outlines the main phases of retroconversion for a real production chain and states the relevant requirements of the retroconversion operation in such a chain.

2 Cataloguing Specification

2.1 Cataloguing Formats

The macrostructure organization of all bibliographical catalogues, be they on cards as in FACIT or in volumes as in MORE, can be divided into references.

References can be converted individually into machine-readable records. Each reference may itself be divided into elements which can be coded into a readable format on the target machine.

The bibliographical reference contents of the catalogues of all libraries share a number of similar features just as the bibliographical reference contents of the catalogues belonging to the same library but covering different periods. Normalised in form as they are today, all bibliographical references invariably contain information which either guides the reader in his research or advertises published documents available elsewhere, *e.g.* in national bibliographies.

The middle of the 19th century saw the birth of an international movement towards the unification of catalogue construction. This marked the beginning of a progressive development of research tools culminating in the ISBD[5] of the nineteen-seventies, an international effort to harmonize cataloguing rules [15].

ISBD is a standard that specifies for each type of document:

– the full list of information elements that a complete bibliographical reference can contain along with the associated hierarchical structure,
– the sequential conventional notation for the normalized presentation of these elements (initially on paper support). ISBD rules aim among other things at a presentation of bibliographical information which favours easy user understanding without the need for him to be familiar with the language of publication, typical names, etc.

This normalisation not only led to the emergence of machine-readable bibliographical format, but was also instrumental in the relation of the international UNIMARC[6] format in 1976. UNIMARC is an attempt to standardize the different varieties of the MARC[7] format in use in the U.S.A. and several European countries [7,16,17].

---

[3] Fast Automated Conversion with Integrated Tools, project supported by the EU under the Libraries section of the Telematics program between 1993 and 1996.
[4] Marc Optical REcogntion, supported by the same organism between 1992 and 1994.
[5] International Standard Bibliographic Description.
[6] UNIversal MAchine Readable Catalogues.
[7] MAchineReadable Cataloguing format.

The ISBD punctuation facilitates the labelling of bibliographical information during catalogue conversion into machine-readable records. The segmentation of pre-ISBD written references into bibliographical units is still possible, as the realization of catalogues almost invariably follows the same rules, rules which practice has been preserved and evolved into national and international harmonizations.

In the MORE project, the tendency was to work on pre-ISBD catalogues which were the overriding form in which European bibliographical information was kept at the time. Catalogues were printed in chronological order, thereby ensuring unity of presentation and of cataloguing rules.

The MORE project is based on year 1973 of the Belgian bibliography. The target machine-readable format is UNIMARC with the incorporation of a parametrised version of the VUBIS system used in the Belgian Royal Library.

The FACIT project on its part explores the possibilities of fast and low cost mass conversion of typed or printed card catalogues using OCR.

2.2 General Specification of the Bibliographical Information

The use of generic tools to manipulate bibliographical information almost invariably poses the same problems. These are related to the following facts. Bibliographical information is made up of text containing a large number of abbreviated words, not only in the document language but in the cataloguing language as well. It also contains numerical information, sometimes in Roman numerals, and an important quantity of names. To these must be added the multiplicity of languages used and the use of a wide range of stressed characters not in keeping with Latin writing styles. There is higher frequency of punctuation marks than in ordinary text. In addition to their natural role, punctuation marks are used as separations to delimit logical elements of information. The presence of several similar character sets such as hyphens and long dashes, parentheses and the square brackets, further increase their frequency. Printed catalogues make use of typography to differentiate between sets of elements belonging to the same logical category. Unlike card catalogues, the layout is more elaborate, including systematic justification of text, variable spacing, and at times word cutting at the end of line. Some of the word cuts belong to the very publishing language covered by the catalogue to be converted.

This diversity of information to process is beyond the processing possibilities of OCR tools, major part of the information being relevant mainly for the recognition of the structure.

3 General Approach to Conversion

The retrospective conversion of bibliographic catalogues can be technically divided into two functions:

- the transcription of textual information constituting the data itself,

– the coding of the data structure and some of the information not generally appearing in the paper catalogues [6,28,29,33].

In terms of charge, the two functions are of equal importance for the retroconversion process. For the coding, the logical structure of the visible information can be deduced from the physical structure of bibliographical references. Some of the information to be coded is provided by the content of certain areas. One can thus proceed with the coding once these areas are recognized.

The contribution expected from projects working in this area are:

– The role and the use of generic and specific dictionaries in the bibliographical information. An important point is the creation of specific dictionaries in relation with the existing readable information in the machine and the authority lists[8].
– The modelling and analysis of the catalogue structure. In particular, how the introduction of automatic methods can modify the content of the retrospective conversion specification elaborated by the library.
– Integration of OCR techniques and structure recognition tools in order to constitute a toolbox for the retrospective conversion of library catalogues. These tools should be parameterized to allow the treatment of all the catalogues and to give a high confidence score allowing the application of automatic processing.
– Assessment on the technical and economic feasibility of the application of such techniques for their integration into a real production chain [22].

4 OCR in Retroconversion

As with all the other techniques applied to manually created textual data, several attempts have been made from the very start to use OCR to solve the problem of retrospective conversion [26,34]. At first, it was used on strictly structured forms to capture catalogues. The next stage was its use to recognize part of the frame of the logical structure so as to create the mainframe of the structure or build automatic search keys for accessing the bibliographical resources.

The current market proposes a wide range of OCR packages incorporating many features to address the needs of office automation. These packages also incorporate a variety of processing approaches involving the use of dictionaries and the recognition of the language of the characters. Most of the proposed techniques are not applicable to retroconversion of catalogues owing to the specificity of the associated data. The speed and the accuracy with which individual characters are recognized and the set of characters supported by the OCR are among the main features of a good OCR package for catalogue retroconversion. To these must be added the ease with which practical scanners and feed controllers can be used.

---

[8] List of "authorized" forms of personal and corporate authors, titles of works and terms for topical subjects established by the responsible cataloguing agency to get a more homogenous search tool.

4.1 Scanning Technology

[14,26] present useful surveys of scanning technology from the point of view of library needs.

The most desirable feature of a scanner resides in the ability of its feeding mechanism to automatically handle a large volume of catalogue papers and cards in a minimum time.

Hand-held and overhead scanners must obviously be avoided for their lack of a mechanism for feeding sheets. High among the special considerations influencing the final decision on the choice of a feeder are the thickness of the card or catalogue paper the feeder is able to handle, the volume of paper to be processed, feeder speed and interface with the computer and skewness. The ability to handle double-face documents, to control brightness, to set contrast automatically and a good resolution are other important features to look for.

4.2 Choice of Recognition Method

Any method which is omni-font in its approach can work for retroconversion, as modern offices these days use a wide variety of standard fonts. However, as catalogues are printed or machine written using old and non standard fonts, this can drastically decrease the performance of the OCR packages. Moreover, references are often written separately and pasted onto the catalogue pages. This creates a mixture of fonts and variation in printing quality making it difficult to maintain the same standard of recognition for the entire set of references, even in the case of packages having a built-in training mechanism to enable them to recognize new fonts [5,25,26].

4.3 Character sets

Most OCR packages still use an 8-bit character set which limits the range to only 256 different characters, of which 100-114 represent letters. Some systems are limited to the 7-bit ASCII[9] character set, a standard code for the representation of characters (letters, digits, punctuation marks and special characters) used in computers [18,19,21,23]. The first 128 characters are fixed and include some non-printing "control characters" such us "Tab", "New Line" and "Carriage Return". The characters from 129 to 256 have variable meanings according to the code page used and compose the so-called "international characters".

Because of the large number of characters used in catalogues, including letters with diacritics and special symbols (such as Greek characters), it is expected of a retroconversion OCR package to be able to handle 16-bit UNICODE or a near subset. A number of packages are able to handle a wide range of characters. Recognita Plus is one such example. This package can code 319 characters including 200 Latin letters, 65 Greek characters, 10 digits and 44 punctuation marks. It maps characters into the 8-bit character set through a suitable assignment of codes. Most packages allow the user to go beyond this range thanks to

---

[9] American Standard Code for Information Interchange.

the combination of the 256 characters available. The diacritical character "å" will for instance be represented as "#ao".

4.4 Commercial Packages

Because of the variation in the catalogue content, font and quality, the developers abandon the idea to create a new specific OCR for their catalogues and try on the contrary to use commercially available packages, combining some of them. In the FACIT project, the packages used are Recognita Plus, Ocron Perceive and Ligature CharacterEyes. Recognita Plus was selected because of the large range of international characters supported and the possibility to export to a 16-bit UNICODE text. Ocron Perceive and Ligature CharacterEyes were selected because of their training facility to recognize non-standard typefaces. Moreover, Perceive has the ability to display the trained characters using a font provided by the user, a very useful feature for retroconversion. The combination is made either by majority vote or by dynamic programming, as in the MORE project.

Packages using trained typefaces all present a similar level of accuracy when tested on classical documents, situated around 98% of correct recognition [31,32]. For paper catalogues, this means an average of 2 to 4 errors per reference. The accuracy is much less for worn out references containing poor quality writing and old fonts. By way of conclusion, any package can be made to meet the requirements of retroconverting older multi-font, multi-lingual paper or card catalogues. What is important is the need to carefully evaluate the content of each catalogue with respect to the range of characters and the nature of the typefaces it contains.

4.5 Output Format

At the end of the coding step, the catalogue is stored in a machine-readable file in a standard format.

In the FACIT project, a plain text format is assumed ("ASCII"). In the MORE project, the input file which is the result of the segmentation is enriched with the ASCII text and with typographic and lexical information. The state of the document is modified to mark its passage through the OCR. We used SGML[10] to indicate this additional information [20].

The results of the segmentation, such as positions of columns and lines in the page, and the quality level of text recognition, are given in order to indicate to the next process in the system to decide for a manual acquisition or not. Each line found by OCR is marked by a "line tag" containing the line coordinates followed by the ASCII text. Others tags in the text inform on typographic events (beginning of typestyle "bold, italic underlined", end of typestyle, indices, exponents, etc.). Each word is followed by a tag giving the level of the recognition quality and can be followed in turn by lexical tags which give the list of lexicons containing the word and probably the list of close words according to a chosen

---

[10] Standard General Mark-up Language.

similarity distance. The following table gives an example of this type of coding.

```
<DOC VER=03.00 TY=N PROV=STRAUT IMA=/home/aragorn/brb01/
73.0/pages EG=OK NTN=56 NPN=208 5 NDN=2471 FORMAT=TIFF
COORIG=HG COUNIT=PX LARG=2592 HAUT=3508 LA=F> <DP N=1>
<TAGTXT TY=P XHG=189 YHG=1377 XBD=1374 YBD=3855> <LIG
XHG=578 YHG=1050 XBD=597 YBD=1099 YBSL=1084 ST=c>0</LIG>
<LIG XHG=431 YHG=1109 XBD=741 YBD=1158 YBSL=1143 ST=c>
<LEX L=GFR><RED F=91.66>GÉNÉRALIT ÉS.</LIG> ...
```

5   Layout Analysis

Though the physical structure of the catalogue is not so dense, some elements seem to be associated with specific positions in the text area, which can contribute to their final identification [1,8,12,13]. The reference format can change from library to library or for the same library working on different kind of catalogues and references.

Figure 1 shows some variations in the layout between different European libraries. The examples given for each library show some variations in layout and cataloguing rules from real card catalogues (see figure 1.a) and catalogues in volume for the French National Library (see figure 1.b) and for the Belgian Royal Library (see figure 1.c), highlighting what to look for.

The last two examples show that some of the bibliographical elements may be optional and may therefore be filled in the text of the reference. This means that in a context of unfilled optional entities, one must exercise caution during the analysis aimed at discovering the underlying structure. In the French National Library, the use of cross references (see third example of figure 1.b) is widespread. References have no "Cote" and are composed of two elements separated by cross formula (Voir, Classé à, etc.).

The typography can be associated with the layout description of the catalogues. In fact, for the French National Library, there are two different typographical sets for the references, with similar probabilities. This particularity complicates in a peculiar manner the set of "relevant indices" in the image, specially for the possible field separators:

- the printed references: used for French books. The typestyle (italic, bold and underlined) and the mode (capital, small capital and small letter) are "relevant features" to extract because they are able to give indication on the type of the corresponding content fragment.
- the machine-written references: they concern the foreign books and the cross references. In this case, the relevant features are poorer than in the previous case. Capitals and spaces letters are used to highlight informations.

6   Structure Analysis

The next and important step of the analysis would then be to provide all the elements that make up the bibliographical reference. These elements are set up

Format   Examples

Top Left Corner		Top Right Corner		61.642

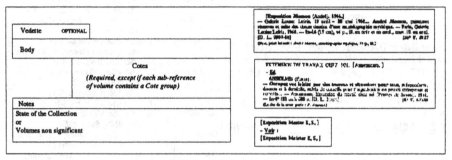

Format (left box):

Top Left Corner	Top Right Corner
Headers : Main area : Title and statement of Authorship - Edition - Imprint - Collation Supplementary Area : series - Notes - ISBN Binding - Price - Number of copies	
Bottom Left Corner	Hole Bottom Right Corner

Examples (right box):

61.642
(38.43)

Keen, Ernest
Three faces of being. Toward an existential
clinical psychology.
New York (Cop. 1970).
X11 + 367 s.
(The century psychology series)
DSH   (hole)

(a) Danish Library

Vedette	OPTIONAL	
Body		
		Cotes
		(Required, except if each sub-reference *of volume contains a Cote group)*
Notes		
State of the Collection or Volumes non significant		

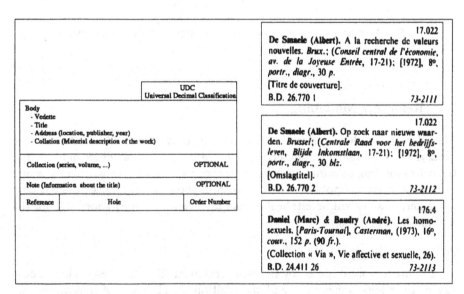

(b) French Library

	UDC Universal Decimal Classification
Body - Vedette - Title - Address (location, publisher, year) - Collation (Material description of the work)	
Collection (series, volume, ...)	OPTIONAL
Note (Information about the title)	OPTIONAL
Reference Hole	Order Number

17.022
De Smaele (Albert). A la recherche de valeurs
nouvelles. *Brux.*; (*Conseil central de l'économie,
av. de la Joyeuse Entrée,* 17-21); [1972], 8°,
portr., diagr., 30 *p.*
[Titre de couverture].
B.D. 26.770 I   *73-2111*

17.022
De Smaele (Albert). Op zoek naar nieuwe waar-
den. *Brussel*; (*Centrale Raad voor het bedrijfs-
leven, Blijde Inkomstlaan,* 17-21); [1972], 8°,
portr., diagr., 30 *blz.*
[Omslagtitel].
B.D. 26.770 2   *73-2112*

176.4
Daniel (Marc) & Baudry (André). Les homo-
sexuels. [*Paris-Tournai*], *Casterman*, (1973), 16°,
couv., 152 *p.* (90 *fr.*).
(Collection « Via », Vie affective et sexuelle, 26).
B.D. 24.411 26   *73-2113*

(c) Belgian Library

Fig. 1. General Layout of Bibliographical Information.

by grouping successive character strings in the text identified as bibliographical information.

If the target format is UNIMARC, to understand the role of the analysis step, one must look into the result (given in figure 2) of transfering information from the reference image (on the left) to a digital file (on the right) where the bibliographical elements are separated from the initial text and tagged.

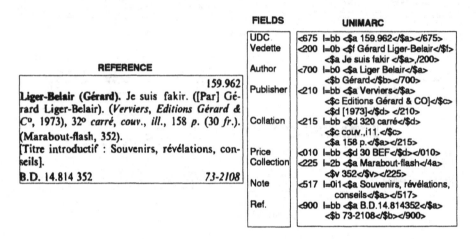

Fig. 2. Result of the structure analysis step in the MORE project.

6.1 Reference Modelling

The reference modelling is established from the cataloguing rules written by the library. This first document describes from the target format where one can find the corresponding information in the reference as well as their writing rules and conversion rules to apply on them. It is from these detailed specifications that the generic model will be established and used as an *a priori* knowledge for the automatic structuring.

Cataloguing Rules

The main problem posed by the bibliographical references resides in the density of their logical structure and the multiplicity of choice of information sequences. In fact, several cataloguing entities are optional and repetitive. These information elements are required only for the cataloguer, if the information exists in the catalogued document. Furthermore, these elements can depend on the kind of the document and of course on the kind of references, such as "monograph" or periodical publications, or as in certain catalogues, on "principal" or "secondary" reference. Finally, a practice inherited from printed catalogues is at the root of the current use of punctuation marks as a means of condensed representation of information. The ISBD normalization on the international level further reinforces this.

In the particular case of the bi-lingual Belgian Library [24], two separate groups work on the cataloguing. One group is dedicated to the French version and the other to the Dutch version. Slight differences appear in the way the two groups apply the cataloguing rules in handling details such as abbreviations or the restitution of information absent from the main reference. Such differences are easily noticed on documents published in the two languages, where contiguous references in the catalogue often differ from one another in their presentation and sometimes in their content. The two examples presented in fig. 1.c on the Belgian library help to understand these differences.

The recognition of some fields depends on the recognition of some key words in specific lexicons. In these lexicons we can find all the cataloguing vocabulary and all the words that exist in bibliographical work titles and insertions concerning the "authorship responsibility". Punctuation is currently less reliable than that of ISBD. Some words are related to the publication language (title fields, edition, address, collection) and others are related to the cataloguing language (collation and notes). Finally, all the words have to be taken into account in a complete form and also in an abbreviated form, knowing that they were not normalized at the time of the tests.

Modelling Methodology

Modelling is a work jointly carried out by the cataloguer and the developer. The expertise of each one in his domain is therefore very helpful for the model construction. But as the main sources of the analysis of catalogues are the references themselves, and knowing the variability of the content involving exceptions and particular cases, we must obviously attach the greatest importance to the content at the modelling stage. As the approach is not quite straightforward, we outline below some useful recommendations concerning the content and the structuring.

- *Classify the element set*: before modelling, it is important to establish the types of bibliographical elements to look for. These types are: main entry references, added entry references with secondary authors, titles or subject terms, follow-on references, references for author names or subject names. It is important for all these types to note the possible changes in the cataloguing rules. For example, in the Italian bibliographic tradition, the sequence of the physical description used to be:<size>cm. <pages> p. [or pp.]. These conventions have changed after the introduction of the RICA rules in 1979. The new sequence is: <pages> pp. <size>cm. Another example: in the Belgian bibliography the writing rules can change for each language such as the use of abbreviations and sometimes of punctuation.
- *List the conventional sequence of elements*: the order and location of the reference elements can change from library to library and between periods in the same library. It is crucial to identify these elements and their change of position in the reference area.

The main elements existing in almost all references or cards are: location mark (Italy : "segnatura", Danish: "signatur", Belgian: "CDU", located on the first line, often at the right corner), Heading (Author, beginning of title or subject terms), Main area (Title area, Area of authorship, statement or responsibility and Edition area), Imprint area (Place of publication, Publisher and year of publication), Physical Description area (Pagination illustration, Size and additional material), Notes area (Author, title, etc.) and collection area (Multivolume statement), Tracings area and Other information (Holdings statements, accessions number, class marks, subject terms, extra locations marks).

- *Identify typographic signs* and the context of each element such as the typestyle, the surrounding punctuation, and try to understand the specificity of their use in the identification of bibliographical information. Another point concerns the standard sequence of constituent elements which is of capital importance for the structure identification (i.e. the position of first names in relation to last names, call names, family names, titles, etc.).

- *Highlight sources of ambiguity* in the delimiters. Although delimiters are important for the separation between bibliographic elements, some may be difficult to recognize by OCR or totally absent. So the greatest care must be taken in the definition of the model and the description of the element separators. It seems judicious to double the separators or to reinforce them by additional information on the mode. For example:
  - the use of a capital letter or an article like "the", "a" in brackets or parentheses may indicate the presence of a title,
  - an indentation of the line following the heading followed by a line or a dash, "-", may also indicate the beginning of a title,
  - words and abbreviations usually found in names may indicate an authorship statement, etc.

- *Insert the target format* of each element in its description. At this level of the description, it is important to know which elements are needed for the conversion and into which format. A full knowledge of the target format is necessary as is the format of each bibliographical elements. For example, the conversion of the author names is treated as the "heading" in UNIMARC 700, which means that "**Last name (First name).**" becomes "First name Last name", by inversion of the first and last names, suppression of parentheses, full point and writing in a regular typestyle. Another example concerns the price, where the string "fr." indicating an amount in Belgian francs will be converted in the string "BEF".

6.2 Structural Description

This part is a formal description of the cataloguing rules to be used as an input in the structural analysis phase. This can be made in different ways: by formal grammars [4,35–37], by concept nets [28,29] or by constraint graphs [6].

Without entering into all the details, we can just outline the main information that such a formal description should contain in general [3].

- **Terms**: the formal description can identify the set of terminals and non-terminals representing all the bibliographic elements at different levels.
- **Constructors and qualifiers**: they are generic elements used to generalize the description to a class. The constructor specifies the order of appearance of subordinate objects such as "SEQUENCE" and "AGGREGATE" or gives a "CHOICE" between several descriptions. The qualifier expresses the object occurrence in the term; each object may be accompanied by a qualifier such as *optional* (?), *repetitive* (+) or *optional-repetitive* (*).
- **Separators**: since the structuring is based on ISBD, punctuation is an important clue to the bibliographical structure [4]. In general, spaces and empty lines, punctuation marks and special typographic vocabulary of bibliographies will be used in the automatic structuring. This is why these features will have to figure prominently in the formal analysis of bibliographical references. In the MORE project, different kinds of delimiters are explicitly introduced in the formal description.

  The following example describes the term "TITLE" as a logical sequence of two objects: "PROPER-TITLE" and "REST-OF-TITLE", where the style is not italic (may be bold or standard). It is located at the beginning of the line with a comma as a separator.

TITLE	::= *seq* PROPER-TITLE REST-OF-TITLE
Style	-Italic
Position	Begline
Sep	Comma

- **Attributes**: because of the weakness of the physical structure and the multiplicity of choices represented in the model, some attributes given by the library specification can be added to the formal description to better specify the description of the reference components. Several kinds of attributes can be defined, among them, *Type* (string, line, word, char, etc.), *Mode* (capital, numeric, alphabetic, punctuation, etc.), *Style* (bold, italic, standard, etc.), *Position* (beginning of line, inside, end).
- **Weights**: given an uncertain OCR flow to be able to evaluate the solution retained, weights can also be used. In this manner, the user can specify the importance that he attaches to each subordinate object.
- **Dictionaries**: for verification of words, a set of dictionaries may be set by the user. The formal language includes orders to call dictionaries for word validation. Typical dictionaries may include many lists of: publishing places (with country codes), important first names, last names and family names, publisher's names, typical words and expressions used in cataloguing such as "Edited by", "Herausgegeben von", "S.L.", "S.A.", etc.

6.3 Structure Analysis

A general objective of the analysis is to introduce qualitative "reasoning" as a function of the recognition evaluation. This evaluation allows:

- the reduction of errors and ambiguities dues to faulty data (OCR errors, data not fitting the model specification, etc.),
- taking into account what is important to recognize,
- the qualitative evaluation of the obtained solutions,
- the isolation and separation of doubtful areas.

To achieve this objective in the MORE project, weights are first assigned by the user to the terms and attributes according to the degree of importance he attaches to them. Objects for which weights are not specified are automatically assigned weights as a function of the number of objects present for this term as well as the symbolic weights already assigned to the other objects by the user.

The analysis structure is a tree of nodes corresponding to the terms analyzed. At each step, the system further decomposes terms stacked in an agenda. This stack is sorted by decreasing order of *a priori* scores. Since the agenda is always sorted in decreasing order of hypothesis scores, the analyzer is said to work in an opportunistic mode. That is, it always first selects the term that looks most promising. Thus, it can move from one branch of the hypotheses tree to another in no "apparent" order. Terms that are no longer decomposable (i.e. leaves) are directly verified and as such either pass or fail. If the analysis fails and a recovery function is present, it is first executed, before the next item on the agenda is selected and processed. This process continues until the agenda is empty. The result and evaluation scores can be obtained for the different nodes in the hypotheses tree explored to produce the output data for the fields that need to be restituted. For example, in the references we treated, names are transformed from (Surname First name(s)) to (First name(s) Surname) while addresses are not restored.

The analysis task is repeated as long as there are hypotheses to verify. The constraint imposed on the system is to find all solutions (not just the first, even if this sometimes leads to ambiguities).

In the event of errors, the system generates a fictive UNIMARC code 903 which it uses to demarcate the zone it should have recognized for a field but which does not quite fit the characteristics as specified by the user. This helps in modifying the model to take care of exceptional cases or to really determine that the reference was badly formed as a result of OCR errors, the printers outright bad transcription of the reference.

When the system finds more than one solution for a given zone, it equally generates a fictive UNIMARC code 902, that it puts around each of the possible solutions, which are then presented to an operator who has to make a choice.

7 Retroconversion Evaluation

Beyond the fashion phenomena, the quality of a retrospective operation is obviously very important because it conditions all the future access to the documentary funds during all the production step. The quality can be evaluated in the course of production on some document samples selected randomly but sufficiently representative of the document being studied. So it is necessary to define the quality level as well as the criteria on which the evaluation can be based.

In the FACIT and the MORE projects, the criteria are measured in terms of erroneous characters, structure faults and bad coding of bibliographical information.

For textual information, the evaluation is made by counting erroneous characters. In the MORE project, we used about 200 references and more than 60,000 characters. The percentage of correct recognition was about 99.97% and errors are essentially due to the Dutch dictionary used.

For structure recognition, the evaluation is carried out on the number of bibliographical elements and on their content according to the specification of the target format (UNIMARC). The evaluation depends on the importance of fields and the sub-fields in the fields. For this reason, we introduced in the MORE project an automatic evaluation of the recognition quality made individually on each reference. It shows what can be considered as an error and what can be considered as an ambiguity. In both cases, this information is taken into account during a manual phase. The total evaluation can be limited to the counting of bad recognition cases. In a test made on 4548 references, 22.3% errors were found on the structure recognition, mostly on the generation of the country codes and languages. 5.4% of the references were rejected.

The quality level and the document sampling for the evaluation control can be determined with respect to the standard "AFNOR NF X 06-022", which is equivalent to the "ISO 2859-1" standard. So, before the beginning of the retroconversion, we can establish a quality table that can be applied on each set of samples according to their dimensions. Other standards can also be used for scanning, such as "AFNOR Z 42-012", which proposes 13 types of different information and characteristics measured for each one of them.

8 Conclusion

The aim of this paper is to enhance understanding of the issues involved in the retroconversion process and to show the advances in the field of character recognition and structure interpretation and their usefulness in the development of solutions to the retroconversion problems.

The prototypes produced in FACIT and MORE projects constitute important results of the European approach, and the different syntheses produced by them, corresponding to different retroconversion steps and analysis, are very precious and should provide a broader basis for further work in the field of library automation.

The cooperation in the field of libraries has produced valuable insights into practices for the retroconversion of old catalogues. This has contributed to a better understanding of the problems in establishing common standards and shared resources.

A number of problems specific to bibliographical information remain to be tackled. These relate to:

- Character acquisition: the processing of an important proportion of non textual characters, such as punctuation (marks), does not take advantage of intelligent processing. There is a need to make different OCRs cooperate.
- Structure recognition: an important proportion of microstructure codes depends on ambiguous characters or on the interpretation of very short textual contexts, the structure being very rich and dense. Finally, the majority of content fragments are optional or repetitive and the number of cases is very high. The modelling of such a structure is very difficult to realise. Specific tools are not available for doing so.
- The creation of coded information: the creation of the publication language code and the country publication code is made by the analysis of the content of some areas and comparison with dictionaries. The results of automatic processing are consequently function of the characteristics of the collections presented in the catalogue, i.e. function of the number and the relative proportion of languages and publication countries and their relative ambiguities.

In the two mentioned projects, dictionaries are very important. They are used at the same time for character recognition, structure recognition and in creating coded information. These dictionaries are composed of specific and general tools for the definition of bibliographical information, and also for the country where references are written. They therefore include the cataloguing rules and the tools of the local library.

The experimentation conducted in the framework of the two projects highlights how difficult a task it is to generalize the retroconversion procedure to different libraries, because of the specificity of bibliographical information and catalogues.

Acknowledgements

We are very grateful to the European Commission for making available to us the necessary documents for this paper.

References

1. André J., Furuta R., Quint V.: Structured Documents, The Cambridge Series on Electronic Publishing, 2, Cambridge University Press, 1989.
2. Beaumont J., Cox J. P.: Retrospective Conversion. A practical Guide for Libraries. Meckler, Westport/London. 1989. 198 p.
3. Belaïd A., Chenevoy Y., Brault J. J.: Knowledge-Based System for Structured Document Recognition. MVA'90 IAPR Workshop on Machine Vision Applications, Tokyo, Japan, 1990.
4. Belaïd A., Chenevoy Y., Anigbogu J. C.: Qualitative Analysis of Low-Level Logical Structures. In *Electronic Publishing EP'94*, volume 6, pages 435–446, Darmstadt, Germany, April 1994.

5. Belaïd A.: OCR Print - An Overview, R. A. Cole and J. Mariani and H. Uszkoreit and A. Zaenen and V. Zue Edts., Chapter 2, Center for Spoken Language Understanding, Oregon Graduate Institute of Science & Technology, 1996. http://www.cse.ogi.edu/CSLU/

6. Belaïd A., Chenevoy Y.: Document Analysis for Retrospective Conversion of Library Reference Catalogues, ICDAR'97, ULM, Germany, August 1997.

7. Bokos G.: UNIMARC, CDS/ISIS and Conversion of Records in the National Library of Greece. In Program, (4)2, pp. 135–148, 1993.

8. Brown B.: Standards for Structured Documents. In the Computer Journal, 32(6), 1989.

9. CEC, DG XIII B: Report of the Workshop on Retrospective Conversion of Catalogues. Problems, Priorities and Projects under the Library Plan, Commission of the European Community, Directorate General XIII, B, Luxembourg, Printed as Draft, 1990.

10. CEC, DG XIII B: Libraries Programme, Telematics Systems in areas interest 1990-1994: Libraries, Synopses of Projects. http://www2.echo.lu/libraries/en/libraries.html

11. Council of Europe: Guidelines for Retroconversion Projects prepared by the LIBER Library Automation Group, Council of Europe, Council for Cultural Co-operation, Working Party on Retrospective Cataloguing, 1989.

12. Crawford R. G., Lee S.: A prototype for fully Automated Entry of Structured Documents. In The Canadian Journal of Information Science, (15)4, pp. 39–50, 1990.

13. Harrison M.: Retrospective Conversion of Card Catalogues into Full Marc Format Using Sophisticated Computer-Controlled Visual Imaging Techniques. In Program, (19), pp. 213–230, 1989.

14. Hein M.: Optical Scanning for Retrospective Conversion of Information. In The Electronic Journal, (4)6, 1986.

15. ISBD (G): General International Standard Bibliographic Description: Annotated Text. Prepared by the Working Group on the General International Standard Bibliographic Description set up by the ILFA Committee on Cataloguing. London, 1977. 24 p.

16. UNIMARC Manual: Edited by Brian P. Holt with the assistance of Sally H. MacCallum & A.B. Long. IFLA Universal Bibliographic Control and International MARC Programme/ British Library Bibliographic Service, London. 1987, 482 p.

17. UNIMARC/Authorities: Universal Format for Authorities, recommended by the IFLA Steering Group on a UNIMARC Format for Authorities. K. G. München 1991, 80 p.

18. ISO 5426: Extension of the Latin Alphabet Coded Character Set for Bibliographic Information Interchange. Second Edition. International Standards Organization. 1983.

19. ISO 6937: Information Technology - Coded Graphic Character Sets for Text Communication - Latin Alphabet. Second Edition. International Standards Organization. 1993.

20. International Standard Organization: Information processing, text and office systems, standard generalized markup language (sgml). Draft International Standard ISO/DIS 8879, International Standard Organization, 1986.

21. ISO 8859-1 to 7: Information Processing - 8-bit single-byte Coded Graphic Character Sets - Part 1-7: Latin Alphabet No. 1 to 7. International Standards Organisation. 1987.

22. Lupovici C.: The retrospective Conversion of Documents. In Document Numérique, (1)2, June 1997.

23. Mackenzie C. E.: Coded Character Sets, History and Development. Addison-Wesley Publish. Co., Reading (Mass) / London, 1980. 513 p.

24. Lib More: Marc Optical Recognition (MORE), Proposal No. 1047, Directorate General XIII, Action Line IV: Simulation of a European Market in Telematic Products and Services Specific for Libraries, 1992.

25. OCR: Ocr software reviews and background material. Technical report, Emap computing Labs, 1995.

26. Ogg H. C., Ogg M. H.: Optical Character Recognition: A Librarians Guide. Meckler, London. 1992. 171 p. ISBN 0-88736-778-X

27. Ortiz-Repiso V., Rios Y.: Automated Cataloguing and Retrospective Conversion in the University Libraries of Spain. In Online & CD-ROM Review, (18)3, pp. 157–167, 1994.

28. Parmentier F., Belaïd A.: Bibliography References Validation Using Emergent Architecture. In *Third International Conference on Document Analysis and Recognition (ICDAR'95)*, volume II, pages 532–535, Montréal, Canada, Aug. 1995.

29. Parmentier F., Belaïd A.: Logical Structure Recognition of Scientific Bibliographic References, ICDAR'97, Ulm, Germany, August 1997.

30. Schottlaender B.: Retrospective Conversion: History, Approaches, Considerations. Haworth Press, NY. (1992).

31. Simon B.: Recognita Plus: OCR with Strength in Hardware. In PC Magazine (10)50, April 1991.

32. Smith J. W. T., Merali Z.: Optical Character Recognition: the Technology and its Application in Information Units and Libraries. Library and Information Research Report 33. British Library, London 1985, 125 p.

33. Süle G.: Bibliographic Standards for Retrospective Conversion. In IFLA Journal (16)1, pp. 58–63, 1990.

34. Wayner P.: Optimal Character Recognition. In Byte, December 1993, pp. 203–210.

35. Wille N. E.: Optical Character Recognition for Retroconversion of Catalogue Cards: Hardware, Software and Character Representation. FACIT Technical Report no 1). Statens Bibliotekstjeneste, Copenhagen. October 1996.

36. Wille N. E., Valitutto V.: A Framework for the Analysis of Catalogue Cards. (FACIT Technical Report no 2). Statens Bibliotekstjeneste, Copenhagen. Revised version, October 1996.

37. Wille N. E.: Retroconversion of Older Card Catalogues using OCR and Automatic Formatting. Project Overview and Final Report. (FACIT Technical Report no 5). Statens Bibliotekstjeneste, Copenhagen. October 1996.

Using Binary Pyramids to Create Multiresolution Shape Descriptors

Gunilla Borgefors[+] Giuliana Ramella[*] Gabriella Sanniti di Baja[*]

[+]Centre for Image Analysis, SLU, Lägerhyddvägen 17, S-752 37 Uppsala, Sweden
[*]Istituto di Cibernetica, CNR, Via Toiano 6, I-80072 Arco Felice (Napoli), Italy

Abstract

The analysis of a 2D graphical document can be accomplished by using a suitable linear representation, e.g. the skeleton, of the pattern included in the document. Multiresolution representation and description are desirable for pattern recognition applications, as it reduces the complexity of the matching phase. In this paper, multiresolution shape descriptors of 2D graphical documents are obtained by using binary AND-pyramids. A multiscale representation is first obtained by simply extracting the skeleton of the pattern at all resolution levels of the pyramid. The so obtained skeletons are then transformed into multiresolution structures by suitably ranking skeleton subsets, based on their permanence at the various scales. The two different types of hierarchy built in this way both contribute to facilitate recognition. In fact, the skeleton is available at various scales, so one could initially match roughly using only skeletons at lower scales, where only the most significant parts of the pattern are represented. In turn, at each scale, skeleton subsets are furthermore ranked according to their permanence along the pyramid levels, thus reducing the number of prototypes for which a more detailed comparison is necessary.

1 Introduction

The analysis of 2D graphical documents often implies conversion of a document into suitable linear representation. The labelled skeleton, e.g., [1], is an example of linear representation that can be profitably employed for graphical documents like maps or engineering drawings. It accounts for topological and geometrical properties of the pattern, such as connectedness, symmetry, elongation, and width. Each pixel of the skeleton is labelled with its distance from the complement of the pattern and can be interpreted as the centre of a disc which fits the shape. The shape of the disc depends on the adopted distance function, and the radius is proportional to the label of the pixel. A correspondence can be established between any subset of the skeleton and the region of the pattern, union of the discs associated with the pixels constituting the skeleton subset. Due to the information on pattern thickness carried on by the skeletal pixels, the labelled skeleton can be used as a shape descriptor also in the case of patterns having variable thickness.

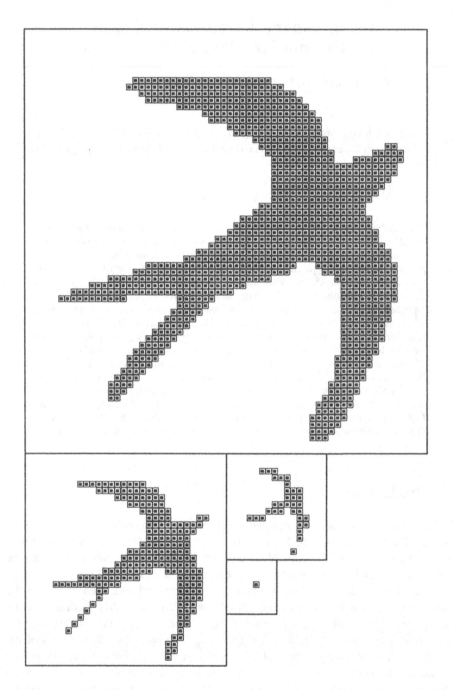

Fig. 1. The AND-pyramid of a test pattern. Only the (four) significant levels are shown. The first level has size $2^6 \times 2^6$.

Multiresolution representation is regarded as useful to reduce the complexity of recognition, when this is accomplished by matching (the representation of) the pattern at hand with (that of) a set of prototypes. In fact, matching can be preliminary accomplished at low resolution, so as to reduce the number of prototypes to be taken into account for a successive matching at higher resolution. If skeleton subsets are suitably ranked, a multiresolution skeleton can be obtained, e.g., [2-3]. Alternatively, one could first record the pattern on a resolution pyramid (so as to transform it from a single-scale data set into a multiscale data set), and then extract the skeleton at all pyramid levels. The so obtained set of skeletons, each of which representing the pattern at a different scale, provides a multiscale representation structure, [4].

In this paper, a multiresolution skeleton is originated at each level of a binary pyramid. Ranking of skeleton subsets is done by exploiting the information on the structure of the skeleton at lower resolution levels. The resulting multiscale set of multiresolution skeletons could be employed to reduce significantly computation complexity in a recognition process involving matching. In fact, one could initially match roughly using multiresolution skeletons at lower scales, where only the most significant parts of the pattern are represented. Moreover, at any selected scale, one could consider first only the most significant subsets of the relevant multiresolution skeleton.

2 Pyramid and Skeleton Computation

Let B and W be the sets of black and white pixels of a binary picture P having size $2^n \times 2^n$. (If the original picture is not $2^n \times 2^n$, white rows and columns are added as needed.) For the sake of simplicity, we suppose that the pattern B consists of a single 8-connected component. The background W may include any number of 4-connected components. We assume that all the pixels on the border of P are white.

The picture P is stored in the highest resolution level (also called the first level) of an AND-pyramid. The next, lower, resolution level is built by partitioning the bitmap in 2×2 blocks of pixels, the *children*, and associating a single pixel, the *parent*, to each block. The colour of the new pixel (black or white) is determined by the logical AND operation. The process is repeated for the previously computed lower resolution representation, and then further iterated to build all the resolution levels regarded as significant in the problem domain. Resolution levels sized $2^2 \times 2^2$ or less are too small to be meaningful for shape analysis purposes. Therefore, we will consider the $2^3 \times 2^3$ image as the last pyramid level, giving n-2 levels in the pyramid. As an example, see Fig. 1 where the n-2 significant levels of the AND-pyramid of a test pattern are shown. The first level of the pyramid has size $2^6 \times 2^6$. We note that when the resolution decreases the pattern becomes narrower and narrower so that some of the initial regions either completely vanish or become disconnected. Thicker regions appear at all resolution levels and constitute the most significant pattern components.

At each level i of the pyramid, the distance map of B_i with respect to W_i is a replica of B_i, where the pixels are labelled with their distance from W_i. Different distance functions might be used to compute the distance map, depending on the desired properties, [5]. In this paper, we use the city-block distance even though we are aware that this provides only a rough approximation of the Euclidean distance. However, the city-block distance is suited to an image tessellated by squares and to its segmentation into square blocks. Moreover, the computation of the city-block distance map is the most convenient one from a computational point of view.

Distance map based skeletonization is accomplished simultaneously at all resolution levels. First, the city-block distance maps are computed efficiently at all pyramid levels by a parallel algorithm that does not use arithmetic, [6]. Then, the sets of the skeletal pixels are identified on the distance maps by using an iterative process that, at each iteration, is active only on pixels whose distance label is equal to the iteration number. At iteration k , pixels with label k remain black, if they are detected as skeletal pixels, otherwise they are changed to white. A pixel p labelled k is identified as a skeletal pixel if at least one of the following conditions holds:

1.   No pair of horizontal or vertical neighbours of p exists, such that one neighbour is white and the other is black and is labelled more than k;

2.   A 2×2 block of pixels, including p, exists, such that the diagonal neighbour of p is black and is labelled at most k , while the other two neighbours are white.

After a number of iterations, equal to the maximal label in the distance map at the highest resolution level, the sets of skeletal pixels S_i's are obtained at all levels. These sets are at most two-pixel wide. They include the centres of the maximal discs in the city-block distance maps. Thus, recovery of the corresponding B_i's is guaranteed. At each pyramid level, S_i and the corresponding B_i have the same number of components and holes (for the relevant proofs on topological correctness see [7], where a city-block based skeletonization algorithm is illustrated from which the algorithm used in this paper is derived).

A final step would be necessary to compress the sets of the skeletal pixels to the unit wide skeletons. This process might be accomplished easily, at all resolution levels, by applying standard topology preserving 3×3 removal operations to the sets of the skeletal pixels. However, this process would unavoidably prevent complete pattern recovery as also some centres of maximal discs would be removed. In this paper, we do not accomplish final thinning, as we favour complete reversibility. To avoid lengthy periphrases, we will refer in the following to the sets of skeletal pixels as to the skeletons.

3 Multiresolution Skeletons

In the AND-pyramid, every pixel of the B_i at level $2^3 \times 2^3$ is parent of four black pixels at level $2^4 \times 2^4$, grand-parent of sixteen black pixels at level $2^5 \times 2^5$, and so on. A similar *parent-child relationship* is expected to hold also among the pixels of the skeletons S_i's, as well as among suitable components into which the skeletal pixels can be grouped, according to some common feature. For instance, if the feature is adjacency, the skeletal pixels could be grouped into connected components. Alternatively, if the common feature is collinearity (in geometry or distance label), the skeletal pixels could be grouped by means of polygonal approximations.

Once the desired components of skeletal pixels have been singled out on the skeleton at the lowest resolution level ($2^3 \times 2^3$), for each of them the corresponding child component and all the descendants, could be found at the successive higher resolution levels by establishing the parent-child relationships. Skeletal pixels at level $2^4 \times 2^4$, that do not belong to any child component, are in turn grouped into new parent components, directly originating at level $2^4 \times 2^4$. Their children and grandchildren could be found in the higher levels. Similarly, new parent components and their descendants could be found at all levels.

Skeleton components at different scales could be assigned a *permanence number*, by counting the number of pyramid levels we should pass through to find the most remote corresponding ancestor component. All skeleton components found at level $2^3 \times 2^3$, directly originate at that level and are accordingly assigned permanence 1; components at the successive level $2^4 \times 2^4$ will be ascribed permanence 2 or 1, depending on whether they are generated from components at level $2^3 \times 2^3$ or originate directly at level $2^4 \times 2^4$; generally, components at level $2^n \times 2^n$ will be assigned permanence n-2, n-1,..., 2, 1, depending on whether their most remote ancestor component are found at level $2^3 \times 2^3$, $2^4 \times 2^4$,..., $2^{n-1} \times 2^{n-1}$, or they originate directly at level $2^n \times 2^n$.

The parent-child relationship is expected to maintain connectedness and shape when components are detected from lower resolution levels to higher resolution levels. The first process we perform to establish the parent-child relationship is termed *projection*.

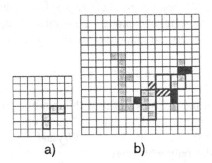

a)   b)

Fig. 2. Pixels (grey squares) in a parent skeleton component, a), and their skeletal children (grey squares in the 2×2 sets enclosed by thick frames) identified by the projection process, b). Dashed squares and black square respectively denote pixels identified by the expansion process and the propagation process.

Projection. Every skeletal pixel p in a (parent) skeleton component projects over a 2×2 set of children at the immediately higher resolution level (*quadruplet*). A quadruplet generally includes both white pixels and skeletal pixels. The latter pixels are assigned to the corresponding child skeleton component.

The projection process can be followed with reference to the example shown in Fig. 2. The pixels in the parent skeleton component are shown as grey pixels in Fig. 2a, and their 2×2 sets of children are enclosed by thick frames in Fig. 2b; among the children, only the (grey) skeletal pixels are interpreted as belonging to the child component.

Unfortunately, due to the discrete scheme used to build the binary pyramid, a child component might have some of its pixels slightly shifted compared to their expected positions. As a consequence, the set of skeletal pixels identified via the projection process in correspondence of a connected parent component might not be connected. This is evident in Fig. 2b, where all pixels in one of the quadruplets enclosed by thick frames are white.

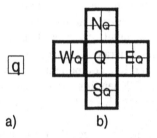

Fig. 3. A skeletal pixel q at level $2^k \times 2^k$, a); the quadruplet Q corresponding to q and the neighbouring quadruplets at level $2^{k+1} \times 2^{k+1}$, b).

To limit the shifting effect, one should interpret as belonging to the current child component also a suitable number of skeletal pixels, neighbours of the projected pixels. These extra pixels should be enough to guarantee maintenance of topological properties and should not be too numerous so as to guarantee also preservation of component's shape.

To identify the extra pixels, we use two processes, respectively termed *expansion* and *propagation*.

Expansion. Let Q be the 2×2 quadruplet of children of a given skeletal pixel q, and N_Q, S_Q, W_Q, E_Q be the four neighbouring 2×2 quadruplets. See Fig. 3. During the projection process, an expansion onto N_Q, S_Q, W_Q and E_Q is performed whenever Q is empty, i.e., Q includes only white pixels.

Expansion interprets as belonging to the current child component the skeletal pixels of the quadruplets that are neighbours of an empty quadruplet, provided that these pixels have not already been marked as belonging to any other child component.

The pixels identified by the expansion process are shown as dashed squares in Fig. 2b.

Propagation. This process is accomplished after all pixels of the parent components have been projected (and possibly expanded) on the next higher resolution level.

Propagation restores connectedness of child components when, though the projected quadruplets, are not empty, the distribution of skeletal pixels inside them does not result in connected components.

As an example, refer to Fig. 4, showing two pixels connected at level $2^k \times 2^k$ and their non-connected children at level $2^{k+1} \times 2^{k+1}$.

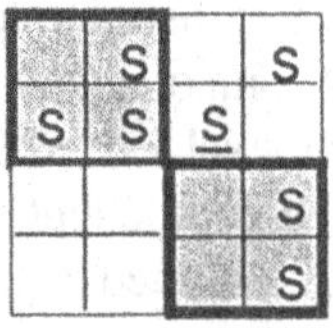

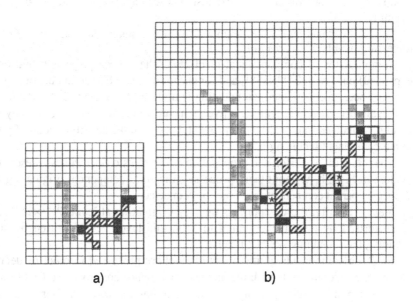

Fig. 4. The two grey squares with thick frames denote two skeletal pixels at level $2^k \times 2^k$. Letter "s" denotes skeletal pixels at level $2^{k+1} \times 2^{k+1}$: skeletal pixels in the grey quadruplets are not connected.

For every pixel p in a child component, propagation ascribes to the same component also the skeletal neighbours of p, provided that they have not yet been assigned to other components. With reference to Fig. 4, propagation would ascribe also the pixel denoted by "$\underline{s}$" to the child component including the pixels denoted by "s" in the grey quadruplets.

The pixels identified by the propagation are shown as black squares in Fig. 2b and Fig. 5.

Fig. 5. Dashed and black squares denote pixels identified by projection-and-expansion and by propagation, respectively. In the child component, b), starred squares are pixels identified by suitably projecting the pixels denoted by black squares in the parent component.

Pixels assigned to a component by propagation have to be distinguished (and, then, differently treated) from those assigned by projection and expansion. This is necessary to prevent an excessive spreading of child components over the skeleton, and hence to preserve components shape, during the successive projections onto higher resolution levels. In our case, distinction is done by marking with a negative sign the pixels ascribed to a child component by the propagation process.

Detection of all components at level $2^{k+1} \times 2^{k+1}$ requires five steps.

Step 1. Pixels ascribed to each component at level $2^k \times 2^k$ by the projection and expansion from $2^{k-1} \times 2^{k-1}$ (i.e., dashed squares in Fig. 5a) are the first to be projected, and possibly expanded, on the next level $2^{k+1} \times 2^{k+1}$, so detecting the pixels denoted by dashed squares in Fig. 5b. The permanence of each component at level $2^{k+1} \times 2^{k+1}$ is equal to the permanence of its parent component at level $2^k \times 2^k$ plus 1.

Step 2. The pixels remaining in any current component at level $2^k \times 2^k$ (i.e., black squares in Fig. 5a) undergo a slightly different projection on level $2^{k+1} \times 2^{k+1}$.

Let q be any of these pixels and let Q the corresponding quadruplet on the next higher resolution level. Among the skeletal pixels in the quadruplet Q that have not yet been ascribed to any other component, only those having at least one neighbour already assigned to the current component are interpreted as belonging to the same child component. With reference to the example of Fig. 5, and to the black squares of Fig. 5a, these pixels are the squares starred in Fig. 5b. Note that not all skeletal pixels present in a quadruplet are necessarily ascribed to the child component by this modified projection. Moreover, no expansion is accomplished if a projected quadruplet is empty.

Step 3. Propagation on the neighbouring pixels takes place at level $2^{k+1} \times 2^{k+1}$ (detecting the black squares in Fig. 5b).

Step 4. After all components at level $2^k \times 2^k$ have been projected (and possibly expanded) onto level $2^{k+1} \times 2^{k+1}$, and for each projected child component propagation has also been accomplished, an updating is done on level $2^{k+1} \times 2^{k+1}$. Pixels due to propagation (i.e., pixels with negative sign) are examined. The sign of any such a pixel p is changed into positive if all skeletal pixels in the neighbourhood of p belong to the component including p.

The effect of expansion and propagation as concerns connectedness preservation is evident in Fig. 5b. Without these processes, the child component originating from a connected parent component would not be connected.

Step 5. Finally, the components directly originating at level $2^{k+1} \times 2^{k+1}$ (hence characterised by permanence equal to 1 on this level) are identified by suitably grouping the remaining skeletal pixels.

Skeleton components at any pyramid level are processed in order of decreasing permanence. Components with the highest permanence are processed first to identify the corresponding child components. Components with smaller and smaller permanence assign to their child components only the children that have not yet been included in other child components with larger permanence.

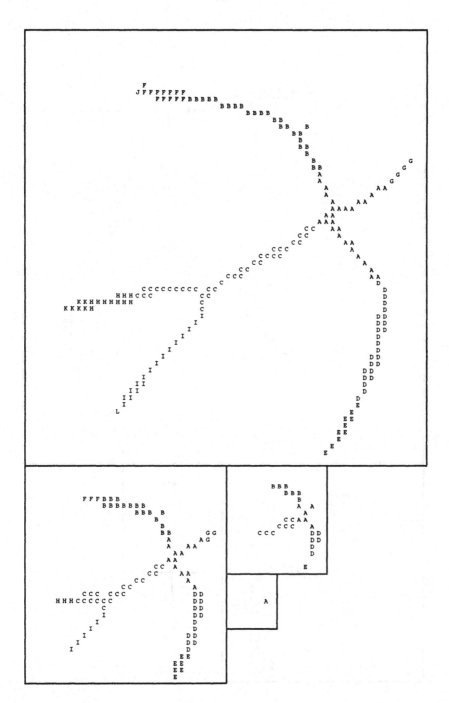

Fig. 6. Multiresolution skeletons. The same letter denotes components found by the parent-child relation.

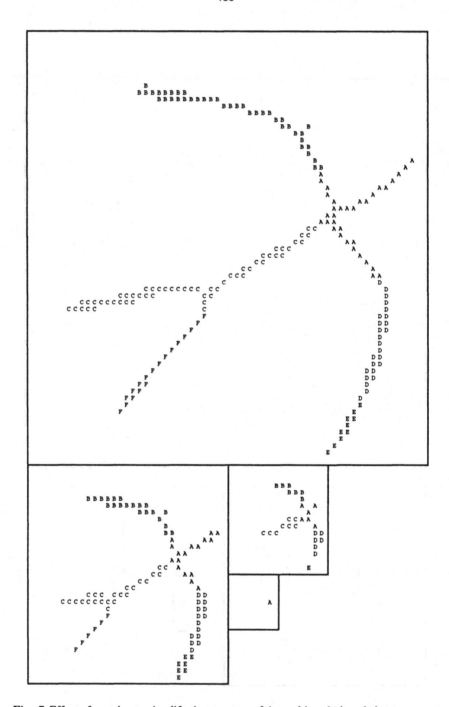

Fig. 7. Effect of merging to simplify the structure of the multiresolution skeletons.

The multiresolution skeletons obtained in this way can be seen in Fig. 6. At each pyramid level, letter A denotes the components with the highest permanence, while letters B-E, F-H, and J-L respectively denote components with smaller and smaller permanence. The same letter is used to denote components that are found by the parent-child relation.

In general, not all components with permanence equal to 1 should be regarded as equally meaningful. Indeed, small components can directly be originated at level $2^{k+1} \times 2^{k+1}$, for instance in between components whose parents were adjacent at level $2^k \times 2^k$. These small components should be merged to adjacent components to simplify the structure of the skeleton. To this aim, a *merging* process is performed.

Merging. The process is effective, at each level of the pyramid, only on components with permanence number equal to 1. These components can be merged to adjacent components (with larger permanence number), depending on component *significance*. Significance is measured in terms of *length* and *average distance label*. Both features are computed while the components are singled out; in particular component length is half the number of moves necessary to trace the component completely.

A component C with permanence equal to 1 and length L is merged to the adjacent component, say C_1, with larger permanence and length L_1, provided that L/L_1 is smaller than a fixed threshold q. When all components adjacent to C have the same permanence, C is merged to the component having the largest average distance label.

The effect of merging can be appreciated in Fig. 7, where the merging threshold is q=0.15. As before, letters A denotes the components with the highest permanence, while letters B-E, and F respectively denote components with smaller and smaller permanence.

The complete algorithm can be summarised as follows:
1. Store a binary picture of size $2^n \times 2^n$ in the first level of an AND-pyramid and build the successive n-2 (significant) levels, ending at level $2^3 \times 2^3$.
2. Extract the (at most) two-pixel wide sets of skeletal pixels at all resolution levels of the AND-pyramid. Set k=3
3. Assign permanence = 1 to all skeletal pixels not yet labelled on level $2^k \times 2^k$; group these pixels into components.
4. Suitably project (and expand) from level $2^k \times 2^k$ onto level $2^{k+1} \times 2^{k+1}$ the obtained skeleton components, in order of decreasing labels. Perform propagation on level $2^{k+1} \times 2^{k+1}$. Set k=k+1.
5. Merge meaningless components found at level $2^{k+1} \times 2^{k+1}$.
6. If k<n, go to 3.
7. Display the multiresolution skeletons and stop.

4 Conclusion

A method to create multiresolution skeletons at all levels of an AND-pyramid has been suggested, which exploits the information on skeleton structure derived from lower resolution levels. Skeleton components are hierarchically ranked, based on their permanence at the various scales. Components at any pyramid level are processed in order of decreasing permanence; a merging process is introduced to simplify the structure of the obtained multiresolution skeletons.

The criteria used to establish the parent-child correspondence favour skeleton components that are more stable, that is present at many pyramid levels.

The multiresolution skeleton representation obtained in this way includes two different types of hierarchy which could both contribute to facilitate object recognition. The first type of hierarchy is intrinsic in the fact that the pyramid includes different resolution levels. It allows one to match initially only the parts of the pattern having large size. The prototypes for which matching is favourable, are then compared with the pattern to be identified at the next higher resolution level. The second type of hierarchy facilitates matching at a given resolution level, since the components of the corresponding skeleton can be ranked according to some relevance criteria (their permanence in the skeleton along the resolution levels, in our case). Distance labelled skeletons are used, so that a set of multiresolution shape descriptors are available that can be used to analyse patterns non necessarily characterised by constant thickness.

Though not intended for a specific application, the proposed method could be employed in areas such as recognition of engineering drawings, maps and logic circuit diagrams.

References

1   G. Sanniti di Baja, "Well-shaped, stable and reversible skeletons from the (3,4)-distance transform", *Visual Communication and Image Representation*, 5, 107-115, 1994.

2   A.R. Dill, M.D. Levine, P.B. Noble, "Multiple resolution skeletons," *IEEE Trans. Patt. Anal. Mach. Intell.*, 9, 495-504, 1987.

3   S.M. Pizer, W.R. Oliver, S.H. Bloomberg, "Hierarchical shape description via the multiresolution symmetric axis transform," *IEEE Trans. Patt. Anal. Mach. Intell.*, 9, 505-511, 1987.

4   G. Borgefors, G. Sanniti di Baja, "Multiresolution skeletonization in binary pyramids", *Proc. 13 International Conference on Pattern Recognition*, Vienna, Vol. D, 570-574, 1996

5   G. Borgefors, "Distance Transformation in Digital Images," *Comput. Vision Graphics Image Process.* 34, 344-371, 1986.

6   G. Borgefors, T. Hartmann, and S.L. Tanimoto, "Parallel Distance Transforms on Pyramid Machines: Theory and Implementation," *Signal Processing* 21, 61-86, 1990.

7   C. Arcelli, G. Sanniti di Baja, "A one-pass two-operations process to detect the skeletal pixels on the 4-distance transform," *IEEE Trans. Patt. Anal. Mach. Intell.* 11, 411-414, 1989.

Contour Pixel Classification
for Character Skeletonization

Maria Frucci° and Angelo Marcelli*

° Istituto di Cibernetica, CNR
Via Toiano, 6
80072 Arco Felice (NA), ITALY

* Dipartimento di Informatica e Sistemistica
Universita' di Napoli "Federico II"
Via Claudio, 21 - 80125 Napoli, ITALY

Abstract

In this paper it is proposed a mechanism for implementing the isotropic propagation of the figure border to obtain the skeleton of elongated shapes. The mechanism allows for detecting, classifying and labelling the contour pixels depending on the characteristics of the wavefronts which interact during the propagation. The skeleton provided by the algorithm is not affected by the distortions which arise in correspondence of regions where the parts of the figure interact. Moreover, it is given in terms of a set of digital lines, each one corresponding to one of the figure parts, rather than by a connected set of pixels.

1. Introduction

Printed and handwritten characters, as well as line-drawings, chromosomes and maps are examples of stick-like figures, i.e. figures whose length is appreciably greater than width. However, once scanned by a digitizer to be converted in a form suitable for computer processing, all of them appear as ribbons of finite thickness evolving into a bidimensional space. In these cases, processes such as feature extraction, decomposition and description may benefit by the analysis of the skeleton of the figure. A skeleton of a binary figure is a linear subset of the figure that preserves aspects of the shape and topology of the original set. The notion of skeleton of a continuous binary image is elegant and consistent; the skeleton of a set X, under the names of medial axis or symmetric axis, is defined as the locus of the maximal discs fitting in X [1]. Although several people have developed a theory for the skeletons of digital binary images [2], there is a discrepancy between the properties of skeletons in continuous and in digital spaces. Problems mainly regard preservation of homotopy, and stability (invariance or continuity) with respect to image transformations such as scaling and rotation.

Authors have often given direct definitions for skeletal pixels in the discrete plane and have designed algorithms based on four criteria: thinness, position, connectivity and stability for skeleton pixels. According to these criteria, the methods to obtain the skeleton of a binary figure can be distinguished in three main classes [3,4]: analytical methods, thinning methods and distance transformation based methods. Whichever the approach used, one basic concept is shared by these methods: the skeleton should be spatially placed along the medial region of the figure. Thus, in the discrete plane the detection of the medial axis of a figure has been understood as equivalent to the detection of the connected set of the most internal pixels of the figure, according to the metric used and under the constraint of the preservation of the figure topology. Accordingly, the distance information assumes more importance than other structural features of the figure, such as the curvature of the border. Only the methods of the first class try to propagate toward the interior of the figure some information related to the curvature of the border, but this information is successively lost, because the linking of the branches is essentially based on topology preserving operations. As a consequence, some problems regarding both the structure of the skeleton and its description arise, which greatly reduces the appeal of the skeleton for shape analysis and description, as described below:

a) The skeleton should be a stylized version of the figure. Thus, each simple region of the figure should be adequately represented by its medial line. This requirement might not be verified whenever a ribbon self-interacts or more ribbons join. A region thicker than each single ribbon piece of the figure may be generated, so causing displacements of the medial axis and originating different types of spurious inflections of the skeleton. Many algorithms attempt to correct these distortions [5-7], but either they do not deal with all the different types of distortions, or the results are satisfactory only in a limited number of cases.

b) The skeletal structure should reflect the structure of the figure. Even though the skeleton does not present spurious inflections, such an interpretation might not be simple to achieve if only distance information is associated to the skeletal pixels.

c) Depending on the order in which the figure is processed, the structure of the skeleton may be different under figure rotation. Thus, the resulting skeletons are drammatically different from each other, and their use in a recognition process is inhibited.

d) Spurious branches of the skeleton may also be generated in corrispondence with border protrusions which are regarded as not significant in the problem domain. Most of the skeletonization algorithms proposed in the literature perform a pruning process on the basis of the length of the branches or the trend of the distance values associated to the pixels of the branch. Pruning based on these criteria may fail when two branches, having the same length or an equivalent trend of the distance values of the pixels, correspond to a not significant and to a significant protrusion, respectively.

e) A final thinning of the set of skeletal pixels is needed if the set has not unitary thickness. Generally, topology preserving removal operations are applied to obtain a unit wide skeleton. This process might too introduce distorsions on the structure of the skeleton.

To overcome these drawbacks, we have proposed a method based on the propagation of the contour of the figure and on a set of criteria to detect the changes of the contour shape during propagation [8,9]. The key idea exploited by our method is that the shape of the contour does not change while the contour points propagate towards the inner part of the figure, except for the case in which the propagation witnesses an interaction among contour parts (or contour components) associated with different ribbon sides or with different ribbons, as in Fig. 1. Such shape changes are used to stop the propagation of the contour and to preserve the contour components representing the ribbons within the region of interaction. We have shown that the method avoids the introduction of spurious inflections in the skeleton, thus providing as skeleton a set of digital lines placed along the medial axis of each simple ribbon constituting the figure, and that it allows a better and easier interpretation of the skeletal structure.

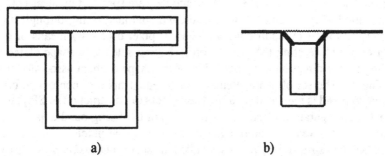

a)   b)

Fig. 1. The results of contour propagation after two successive steps. Thick lines in a) represent the interaction between the contour parts associated with opposite ribbon sides. These lines cannot be further propagated and will represent parts of the final representation of the relative ribbon. The dashed line, part of the contour component representing a ribbon side, interact with the contour components associated with a different ribbon. This line could be further propagated toward the region of interaction as shown in b), but the link between the thick lines and the propagation of the dashed line shows a change in the shape of the contour component with respect to the previous propagation step.

In this paper we describe in great detail the fundamental tool we have used to design the algorithm, namely the classification of the pixels belonging to the current contour. This classification has been obtained by considering both the distance and directional information, i.e. information about the evolution of propagation of the contour, associated to the pixels. Such a classification, performed during the propagation of the contour, is used twice: to decide which pixels of the current contour should be included in the final skeleton, and to drive the further propagation toward the interior of the figure. Note that, for the sake of simplicity, in the remaining of the paper we will assume that the figure at hand has no holes, so as the figure contour is constituted by a single digital lines. If this is not the case, the method must be applied to all the digital lines representing the current contour.

The paper is organized as follows: Section 2 introduces the pixel classification scheme and its key features. Section 3 introduces some basic definitions and notations useful for describing the classification algorithm, presented in Section 4. Discussion and conclusions are left to Section 5.

2. The Method

Three basic concepts are shared by the previous remarks:
1) a pixel p of the border of the figure cannot be considered as an element independent of the contour to which it belongs;

2) the propagation of the information on the contour of the figure must be performed isotropically;

3) no information can be lost or altered during contour propagation.

According to the first point, each pixel p of the border of the figure should be related to the adjacent pixels detected during a contour following. The adjacent pixels of p are said to be *linked together* with p, and they are specified by means of a set of pointers called *shape elements* of p. Thus, the contour passing through p can be completely followed by addressing an adjacent pixel q of p by means of a shape element of p and iterating the process on q, having selected a shape element of q depending on the direction and orientation of the shape element associated to q and addressing p. The set of shape elements of p represents the contour information associated to p and is indicated by SP_p. The process of construction of SP_p takes into account the configuration of the pixels $p_1, ..., p_n$ of the background adjacent to p; in this sense, p receives the contour information from the pixels $p_1, ..., p_n$. The set AD_p of pointers specifying $p_1, ..., p_n$ is called the set of *addressing pointers* of p.

On the basis of the second point, all the pixels of the current contour of the figure must propagate to the adjacent inner pixels their contour information at the same time. To this purpose, a set of pointers specifying the pixels candidates to receive the contour information SP_p is associated to each pixel p of the border of the figure. These pointers, called *propagation pointers*, address the adjacent pixels belonging to the figure and not linked together with p. The set of the propagation pointers of p, thus, represents the propagation information associated to p and it is indicated by PP_p. In Fig. 2, the information associated to the pixels of the border of a figure are shown.

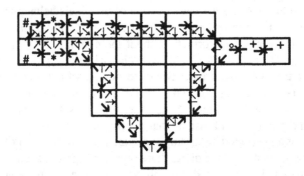

Fig. 2. The thick and the thin arrows represent, respectively, the shape elements and the propagating pointers associated to the pixels of the border of the figure. The addressing pointers, not showed for semplicity, are pointers addressing the pixels of the background. Special simbols are assigned to the pixels where opposite wave fronts collide.

Eventually, the third point requires to implement the propagation as an information preserving process. The subset APp of PPp constituted by the propagation pointers addressing only inner pixels represents the set of *actual propagation pointers* of p. When AP_p is not empty, p is said to be a *propagating pixel*. To preserve the direction and the orientation of the propagation of the contour, also the propagation information of p must be transferred toward the interior of the current figure. For this purpose, all the pixels belonging to the contour of the current figure propagate at the same time their information to the inner pixels by means of the associated actual propagation pointers. More precisely, suppose that an inner pixel q receives the contour and propagation information from different pixels $p_1,...,p_n$. Then, the set AD_q will be contituted by the pointers specifying the pixels propagating on q, the set PP_q will be the intersection of the sets $PP_{p_1},...,PP_{p_n}$, and the set SP_q will results from the union of the sets $SP_{p_1},...,SP_{p_n}$, from which each pointer corresponding to a pointer of AD_q has been deleted.

The analysis of the sets PPp, APp and SPp allows to obtain information about the evolution of the contour, namely they specify how the contour has reached p and how it will move towards inner pixels of the figure. Different wavefronts, originated from opposite sides of the figure, interact in p when one of the following configurations occurs:

a) PP_p is empty. Thus, p cannot propagate its information because there are no inner pixels in the neighbourhood of p. The contour information associated to p originates from pixels belonging to different sides of the figure (see the pixels marked "+" in Fig. 2).

b) a pointer of PP_p addresses a 4-adjacent pixel q belonging to the current contour. Then, the wavefront passing through p is opposed to the one passing through q. Thus, there is not a direct interaction of wavefronts in p, because p is a pixel belonging to a region having double thickness. Therefore, its contour information is the final result of the propagation of only one side of the figure. Two cases may occur. If APp is not empty, a part of the contour information associated to p can be further propagated (see the pixels marked "^" in Fig. 2). Otherwise, p cannot propagate its information because there are no inner pixels in the neighbourhood of p (see the pixels marked "*" in Fig. 2).

c) SP_p is constituted by more than two shape elements. Thus, starting from p, the current contour can be followed along different directions. This configuration represents the resulting union of contour information propagated from pixels belonging to sides of the figure having different curvature (see the pixels marked "o" in Fig. 2).

d) PP_p is not empty and all the addressed pixels belong to the current contour (i.e. APp is empty). In this case, the same considerations reported under point b), when APp is not empty, apply (see the pixels marked "#" and "*" in Fig. 2).

When one of the previous conditions is verified , p is called a *multiple pixel* and represents a skeletal pixel. In fact, if p is a non propagating pixel (i.e. APp is empty), its contour information cannot be further propagated and SPp is a part of the final representation of the figure. On the other hand, a propagating multiple pixel p delimits a part of the current figure, which is one or two pixel wide, from a thicker part; thus, p represents a neck of the figure and it belongs to a part of the contour (passing through p) which cannot be further propagated.

According to remarks 1 and 3), as soon as a pixel p is declared a skeletal pixel, all the pixels $q_1,...,q_n$ linked together with p should be declared as belonging to the final representation of the figure. However, if some of the pixels linked together with p are propagating pixels, such a concatenation can be broken, on condition that the shape of the contour does not change before the further propagation. In such a case, a further analysis is needed to decide whether the concatenation of p with the propagating pixels must be preserved or not.

3. Notions and Definitions

Before explaining how the information associated to p is computed, propagated and analyzed, it is necessary to introduce some definitions.

1) *Neighbours and Pointers.* A pixel q is a j-neighbour (d-neighbour) of p if q is a horizontal/vertical (diagonal) adjacent pixel of p. A pointer associated to p and addressing a j-neighbour (d-neighbour) q of p will be indicated as a j-pointer (d-pointer).

2) *Contour label.* The pixels of the figure border will be labelled i=2, while the pixels adjacent to the initial background and not belonging to the 8-contour of the figure will be labelled i+1. Successively, any pixel of the figure that receives the contour information from the pixels labelled i (i+1) will be labelled i+2 (i+3) . Thus, as soon as a pixel p is reached from the contour propagation originated by the pixels labelled i-2, p is labelled i. The value i>1 associated to p is called contour label of p. The value I=int(i/2) counts the steps of the propagation needed to reach p from the original border of the figure. The pixel p will belongs to the I-th contour if i is an even integer; otherwise some criteria are adopted to decide if p belongs either to I-th or to (I+1)-th contour. The pixels of the figure not yet reached by the propagation and the pixels of the background are labeled 1 and 0, respectively.

3) *Linked pixels.* Two pixel p and q labelled i, with i an even integer, are said to be linked together if they are detected consecutively during a tracing of the I-th contour. To avoid of tracing the contour passing through p, a set of criteria are adopted to associate to p its shape elements, specifying the pixels linked together with p. The criteria are based on the consideration that p and q are linked together if the following conditions holds:

- p and q have the same contour label i;
- it exists at least a neighbour r of both p and q labelled k, with $0 \leq k < i$ and $k \neq 1$;
- p and q are both j-neighbours of the same pixel labelled k, or p and q are adjacents to two pixels r and s both labelled k, with r a j-neighbour of s.

Thus, two linked together pixels p and q have associated respectively the shape elements h and k such that h addresses q and k addresses p; thus, h has the same direction and opposite orientation of k.

4) *Propagating pixel.* A pixel p labelled i>1 is called a propagating pixel if it has at least one adjacent pixel q such that q is labelled k, with k=1 or k>i; that is, q is a inner pixel than p. In this case, PP_p will contain a pointers addressing q. When a propagating pixel p is a multiple pixel, it is called *neck*.

5) *Contour element of p.* A *contour element* of p can be either a composed contour element or a starting contour element. A composed contour element is the merging of two shape elements associated to p such that they form an angle $\alpha \geq 90°$, where the

equality sign holds if p is a propagating pixel. A starting contour element is a shape element h of p, such that h doesn't belong to an composed contour element.

6) *Contour*. Let CE_p and CE_q two contour elements associated respectively to pixels p and q labelled i, such that CE_p and CE_q are constituted by shape elements linking together p and q. CE_p and CE_q are said linking contour elements. The merging of couples of linking contour elements associated to the pixels labelled i is a *contour component* of the I-th contour. If only one contour component exists then it represents the I-th contour; otherwise the I-th contour is the merging of all contour component.

4. Pixel Classification

An initial step is performed to associate the contour label, the contour and propagation information and the addressing pointers to the pixels of the figure border. The index i is set to 2, and a raster of the image is used to detect the pixels of the background adjacent to the figure. Each detected pixel p is stored in a list L(0) together with the set PP_p constituted by the pointers to the adjacent pixels of the figure. The list L(0) is scanned two times and a set of masks is used to define the contour of the figure. During the first analysis of L(0), only the j-pointers of PP_p are used to address the pixels adjacents to the background. Let q be a pixel of the figure addressed by means of the pointers h_k, with k=1,...,n, associated to the pixels p_k of L(0). Then, q is labelled i and it is stored in the list L(i) together with the sets AD_q, SP_q, PP_q, AP_q, with:

- $AD_q \equiv \{g_k\}$,

where g_k is a pointer having the same direction and opposite orientation of h_k;

- $SP_q \equiv \{h \in \cup_k MaskSP(h_k) : h \notin \{g_k\}\}$,

where $MaskSP(h_k)$ is the set of shape elements defined by the masks of Fig. 3, according to the direction and orientation of the pointer h_k;

- $PP_q \equiv \{h \in \cap_k MaskPP(h_k)\}$,

where $MaskPP(h_k)$ is the set of propagation pointers defined by the masks of Fig. 3, according to the direction and orientation of the pointer h_k;

- $AP_q \equiv \{\}$.

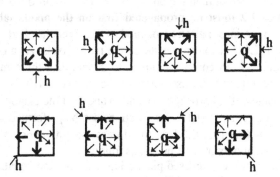

Fig. 3. The masks represent the shape elements (the thick arrows) and the propagating pointers (the thin arrows) to be associate to a pixel q according to the direction and the orientation of the pointer h addressing q.

The list $L(0)$ is analyzed again to address the pixels of the figure by means of the d-pointers of PP_p. For each addressed pixel q, if it is labelled 1, q is labelled i+1, stored in the list $L(i+1)$ and the previous procedure is applied. On the contrary, when q is labelled i, only the updating of the sets AD_q, SP_q, PP_q in $L(i)$ is required, according to the following rules:

- $AD_q \equiv \{h \in AD_q \cup \{g_k\}\}$;
- $SP_q \equiv \{h \in SP_q \cup MaskSP(h_s) \cup ... \cup MaskSP(h_l) :$
    $h \notin AD_q \cup \{g_k\}\}$,

with $h_s,..,h_l$ the d-pointers addressing q and not opposite to pointers of PP_q;

- $PP_q \equiv \{h \in PP_q \cap MaskPP(h_s) \cap ... \cap MaskPP(h_l): h \notin \{g_k\}\}$.

To explain the restriction on the set of pointers used to update SP_q and PP_q, let us recall that two pixels q and p can be linked together when the condition listed under point 3) in the previous section holds. Thus, to avoid of associating to q a shape element addressing an adjacent pixel p that cannot be linked together with q, the updating of SP_q and PP_q during the second analysis of $L(i-2)$, must be performed taking into account only the particular pointers $h_s,..,h_l$.

At the end of this initial step, we have two sets of pixels, those labelled i and those labelled i+1. The first ones constitutes the current contour of the figure, while nothing can be said, at this point, about the second ones. In particular, it can be observed that they have been addressed by the pixels of the background, but that they are d-neighbours of the background, not j-neighbours as in the case of the pixels labelled i. As a consequence, they are more internal with respect to the pixels labelled i, but not so internal to be ascribed to the next contour. Somehow, they are intermediate, and therefore any decision about their inclusion in the current contour cannot be reliably taken until the propagation and the analysis of the current contour has been performed. Such a way of dealing differently with j-neighbours and d-neighbours is one of the key features of our pixel classification scheme, in that it allows to approximate better the isotropic propagation in the digital plane. Note that similar grouping is obtained at the end of every step of the propagation.

Now, let i=i+2. To obtain the I-th contour, the propagation of the (I-1)-th contour is performed by means of the AP_p associated to each pixel p labelled i-2 or i-1. However, since the pixels belonging to $L(i-1)$ are more external than the ones labelled 1 and more internal than the ones labelled i-2, the information associated to the pixels labelled i-2 must be propagated first on the pixels labelled i-1; only successively, when the information associated to the pixels labelled i-1 is completely defined, the pixels $L(i-1)$ can propagate on the inner pixels. As a consequence, the propagation of the (I-1)-th contour requires to accomplish three main actions: the propagation of the information asociated to the pixels in $L(i-2)$ on the pixels in $L(i-1)$, the horizontal/vertical propagation and in and then the diagonal propagation of the pixels in both $L(i-2)$ and $L(i-1)$. These actions can be performed at the same time, during two scans of the lists, the first to address the j-neighbours, the second for addressing the d-neighbours of their pixels, as follows.

During the first scan, if the addressed pixel q has label i-1, the updating of AD_q, SP_q and PP_q is required, according the following rules:

- $AD_q \equiv \{h \in AD_q \cup \{g_k\}\}$;

- $SP_q \equiv \{h \in SP_q \cup SP_{p1} .... \cup SP_{pk}:$
      $h \notin AD_q \cup \{g_k\}\};$
- $PP_q \equiv \{h \in PP_q \cap PP_{p1} .... \cap PP_{pk}\};$

On the other hand, any pixel q labelled 1 may receive the information associated to the pixels labelled i-2. The same process applied during the propagation of L(0) is performed; in this case, as in the previous one, the sets MaskSP(h_k) and MaskPP(h_k) are respectively substituted by the sets SP_{pk} and PP_{pk}. In fact, in both cases, the addressing pixels have their own information to transfer on q, while during the initialization step, such information must be built starting from the initial contour of the figure.

During the second scan, for any pixel q labelled i-1, the same procedure applied during the second scan of L(0) for updating the sets ADq, PPq and SPq when q is labelled i is perfomed. Vice versa, the propagation of the contour cannot be performed on pixels labelled 1 or i, because, as already mentioned, to achieve an isotropic propagation, these pixels must receive the contour and propagation information by their j-neighbours labelled i-1; for this, the addressing of the pixels labelled i-1 must be completed.

At this point, each pixel p belonging to L(i-1) has been addressed and its information updated. It is linked together with pixels labelled i-1 or i and therefore, p is candidate to belong to I-th contour. On the other hand, if PP_p is not empty, p is candidate to propagate on the pixels labelled i. Thus, p carries conflicting information, being at the same time candidate to belong to the I-th contour and candidate to propagate on a pixel of the same contour. To solve this conflict, it suffices to note that p belongs to the I-th contour if different wavefronts collide in it, because, in this case, there are no inner contour to which p could belong. Since such a condition is very similar to the one corresponding to multiple pixels, it can be checked by means of the same configurations reported in Section 2, provided that the configuration d) is substituted by the following:

d') PP_p contains only d-pointers and p is not a j-neighbour of two pixels q and r labelled i.

The substitution of the configuration d) with d') can be easily understood by observing the patterns of Fig. 4a and Fig. 4b, where any pixel p labelled 3 must belong to the 2-th contour, indipendently on whether AP_p is empty or not.

A pixels satisfying one of the conditions a)-c), d') is called a *singular* pixel. A singular pixel is considered as belonging to the I-th contour, it is labelled i and the associated information are moved from L(i-1) to L(i).

On the contrary, in the case of a non singular pixel p, the concatenation of p with its adjacent pixels must be removed, since the propagation information is more reliable than the contour one. This remotion is achieved by deleting any shape element of p in SP_p, and coping into PP_p the ones addressing pixels labelled i. Therefore, p propagates in the same manner as a background pixel (see Fig. 4c). Note that when PPp is not empty and SPp is constituted at the most of two shape elements, to detect whether p is a singular pixel it is sufficient to address only to j-neighbours of p by means of the pointers of PP_p (or SP_p) to verify if the configuration c) (or d')) occurs. Thus, the detection of the singular pixels in L(i-1), can be performed during

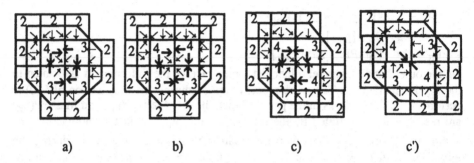

Fig. 4. Detection of singular pixels. The pixels labelled 3 in a) and b) are singular pixels according to the configuration d'). On the contrary, the pixels labelled 3 in c) are not singular pixels, and then they propagate on the inner pixels in the same manner of background pixels (c').

the examination of the list $L(i-1)$ to propagate horizontally/vertically the information associated to its pixels. For each addressed inner pixel q with label 1 or i, the same process applied during the first examination of the list $L(i-2)$ is performed; but in this case, the set SP_{pk} is substituted by the set $MaskSP(h_k)$ because no contour information is associated to the addressing pixels.

During the second scan of $L(i-1)$ also the list $L(i-2)$ is examined, taking into account only the propagation pointers addressing d-neighbours of the pixels stored into the lists. The same procedure applied during the second scanning of $L(0)$ is performed. Moreover, the information associated to each pixel p of $L(i-2)$ are analyzed to decide if p can be declared as multiple pixel. In Fig. 5a, it is shown the propagation and the analysis of the pixels adjacent to the background of a given figure.

At the end of the propagation process of the pixels labelled i-2 or i-1, if no multiple pixels have been detected, all the information about the (I-1)-th contour is completely transferred in the interior of the figure. The pixels labeled i-2 and i-1 are labelled 0 and deleted from the respective lists and will represent pixels of the background of the current figure. Otherwise, the analysis of the current contour will be performed by analyzing the multiple pixels.

During this analysis, the concatenation of a neck with a propagation pixel may be broken and substituted with a concatenation with one or more inner pixels. This concatenation is obtained by transforming into shape elements, a subset $h_1...h_n$ of AP_p and the addressing pointers, opposed to $h_1...h_n$, associated to pixels $q_1...q_n$ addressed by p. Then, the concatenations between $q_1...q_n$ are deleted and the removed shape elements are trasformed in propagating pointers of the relative pixels. In particular, p might be linked together with a pixel q of $L(i-1)$. In this case, q is declared a *connecting pixel* because it represents a pixel of the concatenation between a part of the (I-1)-th contour and the inner parts of the figure. Moreover, selected pointers of AP_q addressing pixels $s_1...s_n$ labelled i are used to concatenate q with $s_1...s_n$ (see the pixels marked ° in Fig. 5b). Also in this case the concatenations between $s_1...s_n$ are deleted and the removed shape elements are trasformed in propagating pointers of the relative pixels.

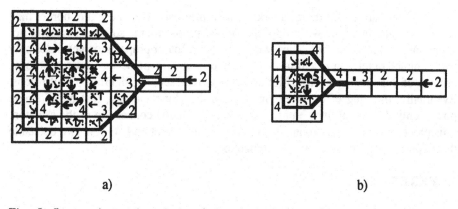

a)   b)

Fig. 5. Propagation and analysis of the pixels labelled 2 and 3. a) No countour information is associated to the pixels labelled 3 since they are not singular pixels. b) The concatenation of a neck pixel p of the figure with the propagating pixels labelled 2 is removed. The pixel marked ° represents then a connecting pixel and its propagation pointer is used to concatenate the current contour of the figure.

At the end of the analysis of the (I-1)-th contour, any pixel q of L(i-1) having SPq empty is labelled 0 and deleted from L(i-1) and it will represent a pixel of the background of the current figure. Thus, the current contour is completely defined after the analysis of the pixels with label i-1. That is, depending on the propagation and contour information associated to a pixel p of L(i-1), it is possible to decide to which contour p belongs to.

The process of the propagation and analisys of the current contour is iterated until no inner contour can be detected.

5. Concluding Remarks

In the framework of a novel approach to computation of line representation of elongated objects, we have exploit the concept of "isotropic propagation" of the figure border towards the interior of the figure. Such a concept has led us to develop a model of the propagation according the which the interactions between different wavefronts originated by different sides of the figure border result in a suitable labelling of the pixels of the figure where the interactions take place. In the paper, it has been shown that the directional information associated to the figure pixels allows us to perform such a labelling by addressing only a subset of the pixel neighborhood. Nevertheless, the directional information conveys global information about the shape of the figure, so that the isotropic propagation can be implemented. Such a combination of local operations on global information makes the approach feasible and effective in achieving the purpose, thus representing a suitable tool for computing faithful line representations.

Once the set of contours has been computed, the final skeleton is obtained by combining the pixels classification provided by the algorithm with shape preserving criteria described elsewhere [8,9].

There are two further remarks. Small protrusions on the figure border may favour the interaction between opposite frontal propagations, so giving raise to not significant skeletal branches. Thus, when a step of contour propagation generates only few skeletal pixels representing a line unnecessary to preserve figure topology, pruning of this line is performed for avoiding to represent not significant protrusions. Eventually, thinning of the skeleton is performed by removing, instead of single pixels, only the digital lines unnecessary to preserve the connection of the skeleton. This process takes into account also the better representation between two deletable lines, preserving the lines with no inflection.

REFERENCES

[1] H. Blum, "A transformation for extracting new descriptors of shape", in: *Models for the Perception of Speech and Visual Form*, (W. Watjen-Dunn, ed.), MIT Press, Cambridge:MA, 1967, pp.362-380.

[2] C. Arcelli and G.Sanniti di Baja, " Skeletons of planar patterns", in *Topological Alforithms for Digital Image Processing*, T.Y.Kong and A. Rosenfeld (Editors), 1996 Elsevier Science.

[3] F. Leymarie and M.D. Levine, "Simulating the grassfire transform using an active contour model ", *IEEE Trans. Patt. Anal. Mach. Intell.*, 14, 56-75, 1992.

[4] L. Lam, S.W. Lee and C.Y. Suen, "Thinning methodologies - A comprehensive survey", *IEEE Trans. on Patt. Anal and Mach. Intell.*, vol. PAMI-14, no.9, 1992, pp.869-887.

[5] G. Boccignone, A. Chianese, L. P. Cordella and A. Marcelli, "Using Skeletons for OCR", in *Progress in Image Analysis and Processing* (L.P. Cordella et al., Eds.), pp. 275-282, World Publisher, SINGAPORE, 1989 .

[6] S. Lee and J. C. Pan, "Offline tracing and representation of signatures", *IEEE Trans. on Syst., Man and Cybern.*, SMC-22, 1992, pp. 755-771.

[7] S. W. LU and H. Xe, "False stroke detection and elimination for character recognition", *Pattern Recognition Letters*, 13, 1992, pp. 745-755.

[8] M. Frucci and A. Marcelli, "Line representation of elongated shapes", in *Lecture Notes in Computer Science*, V.Hlavàc, R. Sàra Eds., Springer-Verlag, vol. 970, pp. 643-648.

[9] M. Frucci and A. Marcelli, "Computing line representations of ribbonlike objects", *Proc. ACCV'95 Second Asian Conf. on Computer Vision*, Singapore, December 5-8, 1995, vol. 3, pp.548-553.

Constraint Propagation vs Syntactical Analysis for the Logical Structure Recognition of Library References

A. Belaïd[1] and Y. Chenevoy[2]

[1] CRIN-CNRS, Campus Scientifique, B.P. 239 F-54506 Vandœuvre-lès-Nancy Cedex
France
[2] CRID, Faculté des Sciences Mirande, 9 rue Alain Savary - BP 400 21011 Dijon
Cedex - France

Abstract. This paper describes a constraint propagation method for logical structure extraction of Library references without the use of OCR. The accent is put on the search of anchor points from visual indices extraction. A mixed strategy is performed. For each anchor points. the system proposes in a bottom-up manner the most probable model hypothesis and tries to verify in a top-down manner its left and right contexts.

1 Introduction

As part of the European project LIB-MORE[1] [Mor 92], we were interested on the retrospective conversion of pre-ISBD library catalogues. Libraries are faced with this problem to convert their old paper catalogues into a data processing format in which they are more readily accessible to the readers. Since 1976, bibliographic references are normalized by the ISBD[2] according to a common formalism called UNIMARC[3]. However, the contents of catalogues written before this date does not totally obey this standard. Ambiguities and exceptions remain embedded in the text and are difficult to resolve.

In this study, the idea was to explore the feasibility of a recognition approach avoiding the use of OCR and basing the strategy mainly on visual indices. The challenge was to recognize a logical structure from its physical aspects. We describe the method applied for the French Library. First, visual indices are computed from the original image (particular characters, style of words, numbers, etc.). From these indices, corresponding labels are searched in the generic model of the references so that a visual characteristic correspond to an attribute of a label in the model. A list of possible labels is then attached to each word of the analysed reference. On this chain, a neighborhood constraint propagation method is applied to prune the list of possible labels. A mixed analysis finishes the recognition process and gives the hierarchical recognized structure.

---

[1] LIBrary Marc Optical REcognition
[2] International Standard for Book Documentation
[3] UNIversal MAchine Readable Cataloguing

We present in the following the modelization of the references with a particular accent on the model compilation stage. A system overview is then presented, focusing on the neighborhood constraint propagation. Finally, we present our results and conclude on the opportunity of this method compared with the full top-down syntactical analysis used for the Belgian Library [Bel 97,Che 96,Bel 94].

2 Generic Reference model

2.1 Structural aspects

The French Library catalogue is composed of 23 volumes with about 1000 pages printed in recto-verso, giving a total of 550 000 references corresponding to the year 1973. References are pasted individually on two columns per page creating some alignment defaults and a big difference in the text typesetting [Che 92]. References are separated in the columns by white spaces and are composed of few lines. There are five reference classes: *Principal, Secondary, Analytic, Collection,* and *Link.* The reference text is composed of a sequence of different parts: "*Vedette*", "*Body*", "*Cote*", and "*Note*". For example, the *Vedette* part can contain either an *author* name or the beginning of a *title*. All these entities can overlap the lines which leads the analysis to search specific field separators (such as dashes, parentheses, points, etc.). Some printed references adopt the bold or the italic style in order to point out fields. Other typed references use more capital letters to distinguish the fields. Figure 1 gives an example of a French reference.

[Exposition Marty (Édouard). 1960.]
— Exposition rétrospective des œuvres de Édouard Marty, portraitiste de la Haute-Auvergne, 1851-1913. Catalogue [par Isabelle Marty]. Musée H. de Parieu [Aurillac], 1er août-15 septembre 1960. — Aurillac, Musée H. de Parieu (Impr. moderne), 1960. — In-16 (18 cm), 32 p., portrait. [D. L. 6844-64]
[16° V. Pièce. 1462

Fig. 1. Example of a French National Library Reference.

Finally, the structure is very abundant and variable. It contains several fields and sub-fields (sub-titles, secondary authors, etc.) and several descriptions of the same field (many types of notes, of titles, etc.). A wide range of field separators is prescribed by the standards ([], (), =, /, etc.) indicating the limits between fields and sub-fields among which some are dedicated to specific fields.

2.2 Specific problems

References are open to many kinds of error sources which constitute a great handicap for a good recognition. This makes a challenge for the automatic analysis of bibliographic references. We summarize in the following some of these difficulty sources.

Typographic imperfections: This deals with style and font mixing (which can be a failure source for OCR), non-respect of the standard punctuation, and separation due to the use of old references (pre-ISBD). This, in addition to frequent typing errors, makes a straightforward recognition of the logical structure difficult.

Linguistic variabilities: Linguistic expressions must be identified in order to locate author names (e.g. "with a preface of ..."). These names have a lot of different representation: name and first name inversion, initials or not for the first name, use of acronyms, pseudonyms, etc. A finer analysis of the content is required to recognize such variabilities. Furthermore, there is no verbal structure within the fields for the use of common natural language analysis.

Structure variabilities: There is no unicity in the structure description because references concern different types of publications. Some of them need more information while some others are just links to existing references. The description can be incomplete because some fields are optional or repetitive (e.g. several authors or notes) without *a priori* knowing their occurrence. As a consequence, the references have to be separated into several classes which leads to the definitions of different sub-models.

2.3 Structure Modeling

The model is given from the examination of different reference samples. It describes the generic structure in terms of objects and relationships between them. Attributes reinforce this description in order to illustrate the content. We distinguish high-level and low-level objects. Each high-level object is given by a constructor (such as *sequence, aggregate* or *choice*), the list of its subordinate objects, and one or more separators. The latter describe the content fragments specially by the use of attributes. The model is written as a context-free grammar (in EBNF)[4]. We used a mechanism similar to SGML to represent generic constructions as object sequences or choices and object occurrences (*required, optional*, and *repetitive*) [Org 86].

2.4 Model compilation

This step allows to transform the model into a more directly usable structure. The model is first parsed in order to extract the visual indices to be searched in the references (separators, styles, etc.). Then, for each object o of the model, three sets are built: the set of *initials* (I_o^*), the set of finals (F_o^*) and the set of compatible neightborhood. These sets will be used during the constraint propagation and mixed analysis stages.

---

[4] Extended Backus-Nauer Form

Indices to be searched: Each generic object of the model (content fragment) is observed so that a list of specific attributes is summarized. These attributes correspond to pertinent visual indices to be searched in the image, such as the style (italic, bold, underlined, spaced), the mode (capitals, numbers) or the size of the words. Furthermore, special characters correspond to separators between fields and subfields. The list of these separators is generated during the compilation stage and is added to pertinent visual indices to be searched.

Initials and finals: The model can be viewed in a formal point of view as a grammar $G = (V_n, V_t, P, S)$ where P is the set of production rules of the model, V_n is the set of non terminal objects, V_t is the set of terminal objects (which can not be decomposed any more) and S is the starting axiom (the reference).

Let S_a, the set of subordinate objects of a (the right part of a rule where a forms the left part). The set of initials of a (I_a) is defined as the subset of S_a where each element can appear in first position according to the construction (choice, sequence, aggregate). By extension, I_a^*, the transitive closing of I_a, can be recursively defined as follow:
$$\text{if } a \in V_t \text{ then } I_a^* = \{a\} \text{ else } I_a^* = I_a \cup (\cup_{i \in I_a} I_i^*).$$
The set of finals of a (F_a) and its transitive closing F_a^* are defined in a same manner but correspond to elements that can appear in last position. I_a^* and F_a^* are extracted for each object a, during the compilation stage.

Compatible neighborhood: Let $N_{l_{a,p}}$, the set of the possible neighbors at the left of a in the rule:
$$p \to \lambda a \delta, \text{ with } \lambda \in (V_t \cup V_n)^*, \text{ and } \delta \in ((V_t \cup V_n) - \{a\})^*.$$
A fictitious rule $\Lambda \to \lambda$ is imagined so that $N_{l_{a,p}}$ can easily be recursively defined as follow:
$$\text{if } a \notin S_\Lambda \text{ then } N_{l_{a,p}} = F_\Lambda \text{ else } N_{l_{a,p}} = F_\Lambda \cup N_{l_{a,\Lambda}}.$$
$N_{r_{a,p}}$, the set of the possible neightbors at the right of a in the rule:
$$p \to \lambda a \delta, \text{ with } \lambda \in ((V_t \cup V_n) - \{a\})^*, \text{ and } \delta \in (V_t \cup V_n)^*.$$
is defined in a same manner. A particular case conserns repetitive objects. In this case the result is $N_{l_{a,p}} \cup \{a\}$ and $N_{r_{a,p}} \cup \{a\}$ because repetitive objects are their own neighbor.

Let $nl_{a,p}^*$, the transitive closing of the left neighborhood finals and $nr_{a,p}^*$, the transitive closing of the right neighborhood initials of a in p.
$$nl_{a,p}^* = \cup_{l \in N_{l_{a,p}}} F_l^* \qquad nr_{a,p}^* = \cup_{r \in N_{r_{a,p}}} I_r^*.$$
These neightborhoods have to be generalized to each rule where a appears in the right. NL_a^* and NR_a^* are the left and right neighborhood of a in the model and are defined as follow:
$$NL_a^* = \cup_{p \in E(a)} nl_{a,p}^* \qquad NR_a^* = \cup_{p \in E(a)} nr_{a,p}^*.$$
where $E(a)$ represent the set of possible father for a, that is the set of the left part of the rules where a appears in the right. Finally, the neightborhood compatibilities of two objects A and B (see figure 2) are defined as follow :

- A is right compatible with B if $B \in NR_A^*$ (case (a) and (c)) or $A \in NL_B^*$ (case (a) and (d)) or $\exists P_A \in E(A)$ and $\exists P_B \in E(B)$ / P_A is right compatible with P_B (case (b)).

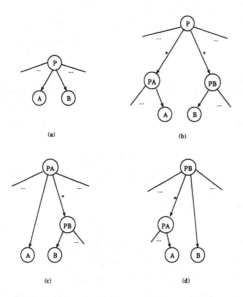

Fig. 2. Neighborhood compatibility between two content fragment A and B.

- B is left compatible with A if $B \in NL_A^*$ or $A \in NR_B^*$ or $\exists P_A \in E(A)$ and $\exists P_B \in E(B)$ / P_A is left compatible with P_B.
- furthermore, A is right compatible with B $\Longleftrightarrow$ B is left compatible with A.

During the compilation stage, the set of couples $((i, p), (j, p'))$ are extracted from the model. they represent for each two objects i and j a father (non-terminal) p (i.e. $p \to \lambda i \lambda'$) of i compatible with a father p' of j. This last set contains about 3700 pairs for the model of French references.

3 System overview

Figure 3 shows the principal components of the recognition system. As presented in the schema, the indices extraction plays a central role in the running of the process. In the following, we briefly describe these different components.

3.1 Image Pre-processing

This basic step concerns the low-level segmentation of the image (extraction of connected components, words and lines) and different measurements useful for the rest of the analysis (baselines, central bands, density, thickness, etc.). Being individually pasted into pages, the reference images are altered (skew angle, font changing, cut or connected characters, etc.). Specific algorithms had to be developed to take into account these particularities [Che 92].

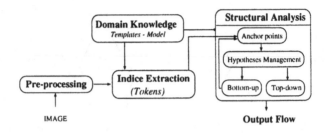

Fig. 3. System Overview.

3.2 Visual Indices Extraction

Visual indices extracted during the compilation of the model are observed on each data component (token). As no OCR had been performed for this method, specific tools had to be realized to recognize typographic styles (italic, bold, underline, spaced letters), the font family (printed or typed), the mode (number, capital, etc.), separator characters (punctuation, parentheses, arrows, particular symbols, etc.) and on particular words such as cross-references, or responsibility mentions (`Adapt.`, `Collab.`, `Comment.`, etc.). A pattern matching approach is performed for the extraction of all these characters and words using specific templates trained from samples. In the second study, conserning the Belgian Library, different commercial OCR systems where used and better results where achieved concerning indices extraction.

3.3 Analysis Strategy

The strategy is driven at the same time by the model and by anchor points extracted from the visual indices of the current reference. It operates in a bottom-up / top-down scheme. For each anchor point, the system proposes in a bottom-up manner the most probable model hypothesis and tries to verify in a top-down manner its left and right contexts. This strategy is adapted to an item where the beginning is noisy and does not favour a top-down scheme. However the strategy efficiency is proved only when the number of anchor points and hypotheses (in the grammar) is limited; which is not the case of such references. In fact, visual indices are not sufficient to generate relevant anchor points, this is due to different reasons: finer degree of some indices generating many ambiguities from the grammar (such as the punctuation which is present for about 30% of the time), non relevant answers coming from the style, and poorness of the physical representation as for typing references where some content fragments have no physical characteristics. Consequently, we use as anchor points tokens that minimize the hypothesis number.

Anchor Point Extraction: For each content fragments in the analysed reference, the sets of their possible fathers are searched. If the symbol value of a fragment A is known (for exemple a particular recognized character), the set of its possible fathers is initialized with P_A as defined in the model compilation. The problem is that in most cases, the symbol value of the fragments are not known in advance. However, some of its physical characteristics are given by the visual indices extraction. These characteristics will allow to initialize the possible fathers with the set of objects that verifies them. For each fragment, the set of its possible fathers is pruned in order to keep only those that are compatible with at least one of the possible fathers of each of its neighbors. Each suppression of a possible father constitute an information to be processed by the neighbors (propagation). The compatibility constraints are extracted during the model compilation step. This method is based on the arc consistency algorithm AC4 [Moh 86].

The basic idea is to associate a counter C_{ijp} to each arc-label couple. This counter contains the number of possible father of j ($E(j)$) compatible with the possible father p of i. Furtermore, the set $S(i, p)$ is memorized. It represent all the node-label couples (k, p') compatible with the father p of i. This set allows to find the counters $C_{(}kip')$ to be decreased when p is suppressed from $E(i)$. The cost is $O(ae^2)$ where a is the number of fragments is the chain and e is the maximal number of fathers per fragment. The algorithm contains two steps. The first one consists to initialize the counters $C_{(}ijp)$ and to build the sets $S(i, p)$. The node-label couples to be suppressed will be added to a set L, previously void. The second step allows to prune the sets of labels associated to the fragments. When a counter C_{ijp} becomes null, it means that no element of $E(j)$ is compatible any more with the father p of i. p has to be suppressed from $E(i)$ and the information has to be propagated to all the nodes that have a compatible label with p, i.e.:
$$\forall (j, p') \in S(i, p), C_{jip'} \leftarrow C_{jip'} - 1.$$
This operation is repeated each time a counter becomes null. If a set $E(i)$ becomes void, the inconsistancy of the chain is proved (see figure 4).

The example in figure 5 shows the incidence of the constraint propagation for anchor points extraction.

Bottom-up / Top-down Analysis: The analysis algorithm proposed by [Moh 88] starts from a well chosen anchor point o_k. It searches in the model the rule $(A \rightarrow \lambda o_k \lambda')$. Then, the left and right contexts (λ and λ') are verified in a top-down manner, from right to left for λ and from left to right for λ'. This procedure is then repeated on A by searching another reducing rule ($B \rightarrow \mu A \mu'$) and so on until reaching the grammar axiom. In order to optimize the bottom-up step, we choose the objects that minimize a criterion which depends on the hypothesis number and on an *a priori* probability for each of them. This probability is given at the model construction step as an additional attribute to help the analysis.

For the left and right context verification, the method is illustrated in figure 6. Considering an already verified object o_k (which can be an anchor point at the beginning), a rule $A \rightarrow \lambda o_k \lambda'$ is proposed from the model where λ and λ' are the left and right context to be verified. We describe in the following the left verification which is processed from right to left. The right verification is performed in the opposite direction in an identical manner. The left context λ can be rewritten into: $\lambda = \lambda_1 \ldots \lambda_i$. λ_i is first verified in a top-down manner,

Initialization step	discrete propagation
$L \leftarrow \phi$	consistancy $\leftarrow$ **true**
for each $(i,j) \in A$ **do**	**while** $L \neq \phi$ **and** consistancy **do**
for each $p \in E(i)$ **do**	choose (i,p) in L
for each $p' \in E(j)$ **do**	$L \leftarrow L - \{(i,p)\}$
if $P_{ij}(p,p')$ **then**	$E(i) \leftarrow E(i) - \{x\}$
$C_{ijp} \leftarrow C_{ijp} + 1$	**if** $E(i) = \phi$ **then**
$S(j,p') \leftarrow S(j,p') \cup \{(i,p)\}$	consistancy $\leftarrow$ **false**
end if	**else**
end for	**for each** $(j,p') \in S(i,p)$ **do**
if $C_{ijp} = 0$ **then**	**if** $p' \in E(j)$ **then**
$L \leftarrow L \cup \{(i,p)\}$	$C_{jip'} \leftarrow C_{ijp'} - 1$
end if	**if** $C_{ijp'} = \phi$ **then**
end for	$E(j) \leftarrow E(j) - \{p'\}$
end for	$L \leftarrow L \cup \{(j,p')\}$
	end if
	end if
	end for
	end if
	end while

Fig. 4. AC4 algorithms.

then λ_{i-1} and so on. λ_i is first checked to avoid a useless top-down analysis when it is not necessary.

Considering the input chain to be recognized $(a_1\ a_2\ \ldots\ a_n)$, different cases can be found:

- λ_i is repetitive. It is transformed into the following sequence:
$$\lambda_i \rightarrow \text{sequence}(\lambda_i \text{ (optional repetitive)}, \lambda_i).$$
- λ_i is a terminal. In this case, we observe a_f (see fig. 6.(a)), the left neighbor of o_k. λ_i is validated if it is a possible label for a_f (obtained during the propagation step).
- λ_i is a non-terminal. Here, we check if there is an intersection between the possible labels of a_f and $F^*_{\lambda_i}$, the set of finals for λ_i (see fig. 6.(b)). The syntactic analysis of λ_i is performed only in this case.
- If λ_i is verified or if it is optional, the verification is continued on λ_{i-1}. The problem is recurrent and a possible final for λ_{i-1} (a_x) has to be found in the input chain, as a left neighbor for a possible initial of λ_i (a_y) (see fig. 6.(c))
- If λ_i is required but not verified then the current rule is resigned. The failure is propagated to the upper level (A) so that other alternatives can be tested. If all the alternatives end in failure for object A, then another possible father is chosen from object O_k.
- If λ and λ' could be verified for the rule $A \rightarrow \lambda o_k \lambda'$, the set of possible fathers for A is pruned according to the neighborhood compatibilities between A and its left and right context in the chain.

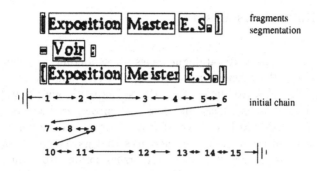

fragments
segmentation

initial chain

	before syntactic propagation		after syntactic propagation	
fragment	symbol	fathers	symbol	fathers
1	ϕ	{ "(", "[" }	"["	ZTitreF
2	ϕ	23 labels	TypeF	TitreF
3	ϕ	23 labels	ϕ	{ TypeF, NomF }
4	ϕ	23 labels	ϕ	{ Pren1, TypeF, NomF }
5	ϕ	{ ".", "," }	"."	TitreF
6	"]"	7 labels	"]"	ZTitreF
7	"-"	10 labels	"-"	NoticeT
8	ϕ	7 labels	Renvoi-T	NoticeT
9	":"	3 labels	":"	NoticeT
10	ϕ	{ "(", "[" }	"["	ZTitreF
11	ϕ	23 labels	TypeF	TitreF
12	ϕ	23 labels	ϕ	{ TypeF, NomF }
13	ϕ	23 labels	ϕ	{ Pren1, TypeF, NomF }
14	"."	24 labels	"."	TitreF
15	ϕ	{ ")", "]" }	"]"	ZTitreF

Fig. 5. Incidence of the constraint propagation.

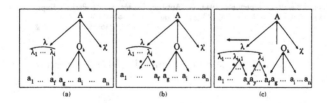

Fig. 6. Different Left-to-Right Top-Down Context Verification.

4 Results and discussion

A test has been made on 10 catalogue pages corresponding to about thirty references per page (300 references in total). Figure 7 shows an example of structure recognition. Results have shown that the structure is always well recognized when the initial data is consistent. The few encountered errors coming from bad choices (for Human) have been at last verified with the help of the model. We can estimate at about half, the references that gave a consistent chain after syntactic propagation. For the other half, errors coming from the extraction of visual indices can be recovered at 50% by redefining the model in a less strict manner, or by enhancing the low-level tools. This constitutes the major weakness of this global method. This weakness results from the difficulties to extract visual indices without OCR.

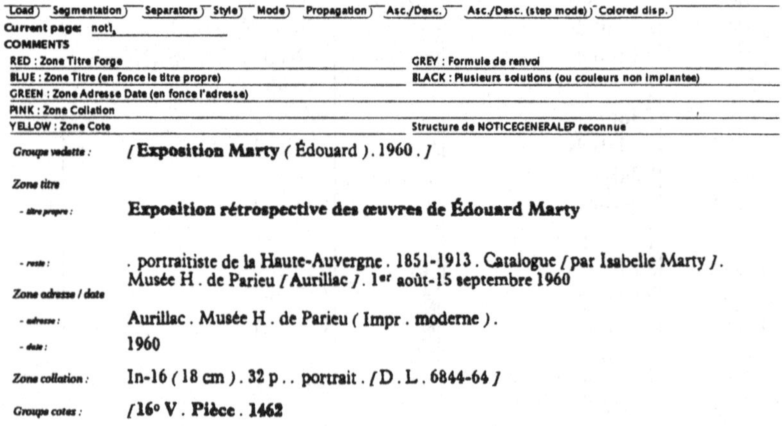

| Load | Segmentation | Separators | Style | Mode | Propagation | Asc./Desc. | Asc./Desc. (step mode) | Colored disp. |

Current page: not1,

COMMENTS

RED : Zone Titre Forge	GREY : Formule de renvoi
BLUE : Zone Titre (en fonce le titre propre)	BLACK : Plusieurs solutions (ou couleurs non implantee)
GREEN : Zone Adresse Date (en fonce l'adresse)	
PINK : Zone Collation	
YELLOW : Zone Cote	Structure de NOTICEGENERALE? reconnue

Groupe vedette : **/ Exposition Marty** (Édouard) . 1960 . /

Zone titre

- *titre propre :* **Exposition rétrospective des œuvres de Édouard Marty**

- *reste :* . portraitiste de la Haute-Auvergne . 1851-1913 . Catalogue / par Isabelle Marty / .
Musée H . de Parieu / Aurillac / . 1er août-15 septembre 1960

Zone adresse / date

- *adresse :* Aurillac . Musée H . de Parieu (Impr . moderne) .
- *date :* 1960

Zone collation : In-16 (18 cm) . 32 p .. portrait . / D . L . 6844-64 /

Groupe cotes : / 16° V . Pièce . 1462

[**Exposition Marty** (Édouard). 1960.]
— **Exposition rétrospective des œuvres de Édouard Marty,** portraitiste de la Haute-Auvergne, 1851-1913. Catalogue [par Isabelle Marty]. Musée H. de Parieu [Aurillac], 1er août-15 septembre 1960. — Aurillac, Musée H. de Parieu (Impr. moderne), 1960. — In-16 (18 cm), 32 p., portrait. [D. L. 6844-64]
[16° V. Pièce. 1462

Fig. 7. Example of a recognized reference structure.

The studied approach have revealed the possibility to recognize the structure of content fragments without the use of OCR. A mixed (bottom-up and top-down) analysis is performed on the token chain extracted from the reference image. Anchor points are computed to start the analysis from visual indices extracted from the image. These indices allow to prune the number of generated hypotheses according to neighborhood compatibility constraints between the tokens. Compared with the full top-down syntactical analysis used for the Belgian Library, several observations can be made.

Weekness of the method: the neighborhood constraint propagation needs robust tools to extract visual indices from the image. At this stage, if the structure recognition could be achieved without OCR, we have to admit that the use of one or several combined commercial OCR would have improved the recognition rate. We had to develop specific tools to recognize particular characters and words, and to recognize the styles and the modes of the tokens. This task has proved to be difficult because of the noisy multifont environment of the catalogues. For example, the punctiation is often connected with the previous word and is not identified. This lack is propagated along the chain which often leads to an inconsistancy. This weekness has been corrected in the study on the Belgian catalogues by introducing OCR and additional knowledge sources such as author and subject indexes, city and country names

Strength of the method: A bottom-up/top-down analysis allows to come back up in the reference hierarchy from interresting fragment called *anchor points*. the other fragment are just verified in a top down manner on their left and right context. This mixed analysis is much faster than a full top-down analysis because the input chain is filtered according to the extracted indices and the neiborhood constraint propagation. Fancy references are rejected by the propagation stage where they would have been rejected after the analysis for the top-down method.

Concerning both methods: The two studied approches have revealled the importance of a model for structure recognition. They also showed that the model is insufficient if we do not dispose of tools to extract pertinent indices (OCR or visual characteristics of the image). This is particularly true if the content of the document is rich and complex. Several references still remain unrecognized or ambiguous because of the complexity of such a task. The reasons of these failures are various.

First, as the model is built from non-normalized references (pre-ISBD), the knowledge is incomplete and uncertain. Furthermore, a great number of subclasses are represented in only one model which leads to difficulties during hypotheses generation and leaves ambiguities in the final structure. The principal encountered difficulty in these projects was obviously the model construction. In fact, the fine degree of specification, essentially concerning weights for objects or attributes, makes difficult the model conception when the models are ambiguous or complex. An interesting prospect of this project concerns an automatic help for model learning. Such a system was elaborated in our team for scientific paper models by [Aki 95]. It shows the interest to extend the learning to the micro-structure.

Second, the structure recognition for this kind of document is a real problem of text understanding. The interpretation is not only based on character or word recognition, but also on the recognition of specific expressions and even complete sentences. Furthermore the linguistic style of these sentences is not always fixed and known in advance (not regular). The problem here is not only syntactic recognition but also semantic understanding. For this, a perfect domain knowledge is necessary to understand the different specific expressions. Adapted tools and heuristics have to be investigated. This investigation can be based first on existing natural language processing techniques and linguistic models. Our recent work [Par 95] is oriented in this direction.

References

[Aki 95] O. T. Akindele and A. Belaïd. Construction of Generic Models of Document Structures Using Inference of Tree Grammars. In *Third International Conference on Document Analysis and Recognition (ICDAR'95)*, volume I, pages 206–209, Montréal, Canada, Aug. 1995.

[Bel 94] A. Belaïd, Y. Chenevoy, and J. C. Anigbogu. Qualitative Analysis of Low-Level Logical Structures. In *Electronic Publishing EP'94*, volume 6, pages 435–446, Darmstadt, Germany, April 1994.

[Bel 97] A. Belaïd and Y. Chenevoy. Document Analysis for Retrospective Conversion of Library Reference Catalogues. In *Fourth International Conference on Document Analysis and Recognition (ICDAR'97)*, Ulm, Germany, Aug. 1997.

[Che 92] Y. Chenevoy. *Reconnaissance structurelle de documents imprimés : études et réalisations*. Thèse de doctorat, INPL, Centre de Recherche en Informatique de Nancy, décembre 1992.

[Che 96] Y. Chenevoy and A. Belaïd. Une approche structurelle pour la reconnaissance de notices bibliographiques. *Traitement du signal*, 12(6), 1996.

[Moh 86] R. Mohr and T. C. Henderson. Arc and Path Consistency Revisited. *Artificial Intelligence*, 28: 225–233, 1986.

[Moh 88] R. Mohr and G. Masini. Good Old Discrete Relaxation. In *ECAI88*, pages 651–656, Munich, Germany, 1988.

[Mor 92] Lib More. Marc Optical Recognition (MORE), Proposal No. 1047, Directorate General XIII, Action Line IV: Simulation of a European Market in Telematic Products and Services Specific for Libraries, 1992.

[Org 86] International Standard Organization. Information processing, text and office systems, standard generalized markup language (sgml). Draft International Standard ISO/DIS 8879, International Standard Organization, 1986.

[Par 95] F. Parmentier and A. Belaïd. Bibliography References Validation Using Emergent Architecture. In *Third International Conference on Document Analysis and Recognition (ICDAR'95)*, volume II, pages 532–535, Montréal, Canada, Aug. 1995.

Grammatical Formalism for Document Understanding System : From Document towards HTML Text

S. TAYEB-BEY* & A. S. SAIDI**

* Reconnaissance de Forme et Vision Bât 403
20 Avenue Albert Einstein 69621 Villeurbanne Cedex
Email : tayebbey@rfv.insa-lyon.fr
** Ecole Centrale de Lyon Département Math-Info-Système BP 163 69131 Ecully

Abstract

This paper deals with the use of grammatical formalisms to recognize the physical and the logical structures of a composite document. We propose a new system for document recognition and analysis. The goal of this system is to identify particularly the «summaries», and as an application, to convert them into machine readable form. We translate a summary paper into a HTML (HyperText Markup Language) text.

Key Words

document analysis, logical structure, physical structure, two level grammar, HTML

1 Introduction

The goal of a document image understanding system is to extract paper document's structures and content in a hierarchical representation [Li96]. The extracted representation allows document interchange, indexing and other tasks.

Our study is centered on the document analysis and identification (text, graph, image,...). The analysis steps consist among others, of the elaboration of three complementary information types for a given document :

➢ *its physical structure* : it describes the document organization, in terms of objects (typographically homogeneous regions) and relationships between these objects (hierarchical decomposition , absolute and relative positions in the page) ;

➢ *its logical structure* : it decomposes the document in information elements, characterized by the role they play in the document (chapter titles, chapters, paragraphs...), and specifies the relationships (syntactic and semantic) between these elements;

➢ *its content* : which can be text, graphic, mathematical equations, tables...

The system uses also a learning base issued from a prior learning process. This base contains several identified summary models. Given the results of the above analysis steps, the document is then identified by matching with the models of the learning base, by using the whole information stemmed from the analysis steps.

The goal of our works [Ta96] is the construction of a system able to identify a document. As an application, we convert summaries into a readable format accessible by navigation softwares (Netscape, Mosaic, ...). In our process based on a syntactic approach, we use *two level grammars* that proved to bring original solutions to document analysis. A two level grammar, also called W-grammar, is a formal system well suited for the language definition. We use an operational version of W-grammar [ASa92] to perform document analysis and structure inference.

2 Related Works

A lot of works [Na93,Ak93] has been done in document structuring, showing the growing interest in this field in recent years. Consequently, several methods have been proposed for the structure representation problem.

Ingold [In89,In91] proposed a document description language similar to an attributed grammar. The aim in designing this language has been to top-down analyse several document classes.

Chenevoy [Be90,Ch92] proposed a general-purpose system for document structure recognition called GRAPHIEN. The system has a multi-agent architecture based on the blackboard model. Its role is to organize and control diverse document recognition processes.

Peden [Pe90] proposed a system centered on the analysis of the physical and logical organization of different text entities of a scientific publication. The system deals with an intern representation of analyzed document near to the ODA [Hor85, Mau87] norm and on an objects modelisation.

However, the construction of structure models of documents has turned out to be a difficult task, and is often carried out manually because of document diversities.

Sanfeliu [Sa92] showed that the grammatical formalism used in syntactic pattern recognition is quite suitable for document analysis. Grammatical formalism has also proved to be a powerful tool in describing a document.

3. The System Architecture

The application field of the system may be more general, though, firstly, we focus on a particular type of documents : **the summaries of reviews.**

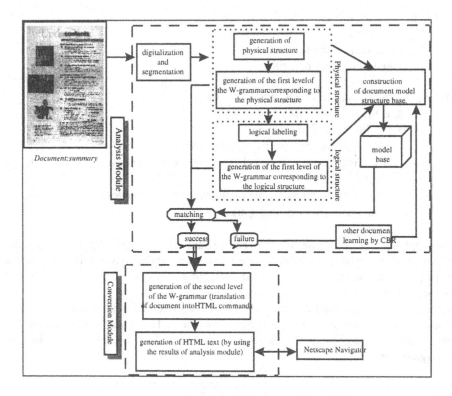

Fig. 1. : The system Architecture

Following the grammatical approach, we built a system composed of two modules (see figure 1). The first one achieves the analysis and the identification of a given summary. The second translates this summary into a HTML (HyperText Markup Language) text.

Before describing the architecture of the system, we briefly define two levels grammars.

3.1 Two Level Grammars

A two level grammar (see figure 2), called W-grammar [Wi65] is a formal system well adapted to the language definition. It is composed of two grammars called metagrammar and hypergrammar. The metagrammar defines the possible domain of values for (meta)variables. These metavariables appear in the rules of the hypergrammar.

Generally, the metagrammar is a context-free grammar describing the language structures of the hypergrammar and the domains of variables.

The hypergrammar holds the definition of a contextual grammar, describing the specification of the calculus (e.g. translation operation) expressed in semantic actions.

☞ The metagrammar $MG = (M_n, M_t, M_r)$ where

  M_n corresponds to the nonterminals,

  M_t corresponds to the terminals,

  M_r is the finite set of metarules.

☞ The hypergrammar $HG = (H_n, H_r, T, S)$ where

  H_n is the set of nonterminals,

  H_r is a finite set of hyperrules,

  $T = (TT \cup TM)$ is the terminal alphabet where

    TT : finite set of terminal symbols which appear at right side of hyperrules,

    TM : set of terminals same as the domain of variables.

  $S \in Hn$, special nonterminal (strat symbol).

By applying the principle of uniform replacement in the hyperrules (similar metavariables are replaced by the same value), we obtain a ground instance of the hypergrammar named the protogrammar containing only context-free rules. Note that the set of protogrammars may potentially be infinite.

Hence, the hypergrammar defines a « relation » between an input language (e.g. the document structures) and a close instance of its axiom. The syntax of this relation and thus, the domains of the parameters (e.g. variables) being defined by the metagrammar.

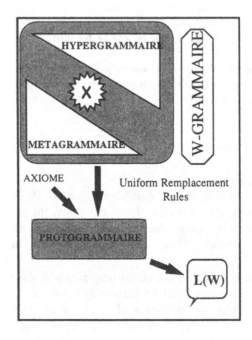

fig. 2. : A two level grammar

The aim of the metagrammar in this system is to describe the physical structure of the document, while the hypergrammar describes the logical structure of the document and the processing to be done on the document.

3.2 System Description
The system is composed of two modules :
The analysis module : its goal is to treat, to describe and to identify a given summary.
The conversion module : its goal is to translate the summary into HTML language.

3.2.1 Analysis and Identification Module

The analysis step achieves several classic operations on the document images : acquisition and preprocessing, physical segmentation, logical labeling and recognition of the structures. Generally, there are ascending [Hi93,Sa94] and descending [Wo82,Na86] methods for the segmentation. In our system, we use a mixed method (ascending descending) [Le96].
This operation yields several classes of properties of the document which, according to the physical and logical structures, will be used for the identification of the summaries and for their conversion.

At the identification level, the module operates in two modes:
"learning mode": starting from a set of samples of summaries, we infer the physical structure grammar (called physical grammar), as well as the logical structure grammar (called logical grammar). The result is a model of the learning base.
The general algorithm which allows us to build this base is the following :

```
START
        -- at this step, we construct the model base with some samples of documents --
While not end of samples
    DO
    Begin
            read the document scanned
            display the picture of document
            preprocess and segment the picture
            extract the typographic and disposition parameters
            extract the physical structure
            label logically the blocks (text, graphic, ...)
            extract the logical structure
            generate the specific metagrammar (MG) and the specific hypergrammar
            (HG) of the sample
            -- MG et HG describe respectively the physical end the logical structures of the
            document --
            insert these two grammars in the base
End While
END
```

Basic model algorithm

"identification mode": the system matches the description of one specific document with the patterns of the learning base and determines its type.

Each of the grammars inferred by this module correspond to the first level of a W-grammar (the metagrammar).

The general algorithm which allows to identify a document is the following:

```
Start
               -- in this step, we identify a document by using the base of models --

While  not end
     Do
               read a document scanned
               display the document picture
               preprocess and segment the image
               extract the typographic and layout parameters
               extract the physical structure
               label logically the blocks
               extract the logical structure
               analyse the structures by the generic W-Grammar  (MG et HG)
               If succes
                 Then document accepted and identified
                 Else Reject
               EndIf
End While
End.
```

Identification algorithm

In the reject case, if the document is a representative sample, we can update the generic W grammar.

3.2.2 Conversion Module

A HTML document is composed of text and commands. These commands (title, bold character, italic, picture, links, etc.) determine the layout of its content.

The layout of a document in the HTML format is based on the construction of the second level of the W-grammars (translation of the parameters of the first level in HTML commands), and its content is recovered from a file filled during the segmentation step.

Example of the production rules corresponding to the second level of the W-Grammar.

page_principal	→ <HTML> <TITLE> Document Title</TITRE> page_1(parameters... </HTML>
page_1(parameters....)	→ <BODY> section1, section2,... </BODY>

The results of the segmentation is used to generate the set of rules used for the conversion into HTML and to restore the content. The generated grammar recovers also the physical and the logical structures.

4 Preliminary Results

We have numerised some summaries in gray levels, and applied the segmentation to these images.

fig. 3. : Character segmentation

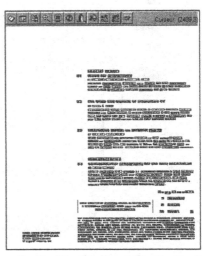

fig. 5. : Word Segmentation (general vision of the documentt)

fig. 4. : Word Segmentation (detailed vision of the document)

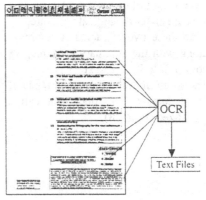

Fig. 6. : section Segmentation

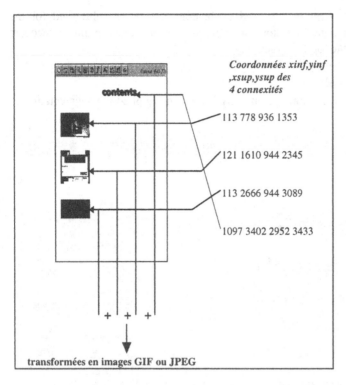

Coordonnées xinf,yinf
,xsup,ysup des
4 connexités

113 778 936 1353

121 1610 944 2345

113 2666 944 3089

1097 3402 2952 3433

transformées en images GIF ou JPEG

Fig. 7. : Détection des images

The figures (3, 4, 5, 6, 7) represent the segmentation results. The location of paragraphs, lines, words and chars allow to extract the physical structure. Important information obtained after the segmentation such as the coordinates of paragraphs, the style, height of chars and other parameters are used to achieve the logical labeling and thus to extract the logical structure.

With these two extracted structures, and by using the grammar of the second level along with the summary contents, we identify the class of the summary and finally convert it to HTML text.

The result of the grammatical parse of the hypergrammar produces the HTML text of the document. This document will be corrected by navigation software, and then visualized (figure 8).

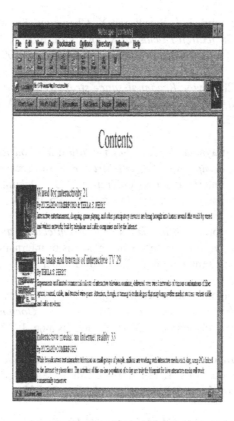

Fig. 8. HTML text

5 Conclusion and Future Work

In this paper, we presented a Document Analysis and Conversion system which is being developed.

The step of identification could reject a dismissal document (noised summary image or summary whose structures is not already learned). We currently study the use of Case Based Reasoning framework to achieve a continuous learning [Pet95].

Besides, we will improve our approach by applying it to other types of documents like the scientific publications.

References

[Ak93]   O. T. Akindele and A. Belaid. *Page Segmentation by Segment Tracing*. Second International Conference of Document Analysis and Recognition(ICDAR 93). 1993. pp.341-344.

[ASa92]   A. S. Saidi. *Extensions Grammaticales de la Programmation (en) Logique : Application à la Validation des Grammaires Affixes*. Ph. D Thesis. Ecole Centrale de Lyon. 1992.

[Be90]   A. Belaid, J. J. Brault and Y. Chenevoy. *Knowledge-Based System for Structured Document Recognition.* In MVA'90 IAPR Workshop on Machine Vision Applications, November 1990.

[Ch92]   Y. Chenevoy. *Reconnaissance structurelle de documents imprimés : Etudes et Réalisations.* Ph.D. Thesis. INRIA-Lorraine. December 1992.

[Hi93]   Y. Hirayama. *A Block Segmentation Method for Document Images with Complicated Column Structures.* In Proceedings of ICDAR'93 : 2nd International Conference on Document Analysis and Recognition. Tsukuba, Japan. 1993. Pp. 91-94.

[Hor85]   W. Horak. *Office Document Architecture and Office Document Interchange Formats.* Current status of international standardization. IEEE Computer. Vol. 18. N°10. October 1985. pp.50-57.

[In89]   R. Ingold. *Une nouvelle Approche de la Lecture Optique Integrant la Reconnaissance des Structures de Documents.* Ph.D. Thesis. Ecole Polytechnique Federale de lausanne. 1989.

[In91]   R. Ingold. *A Document Description Language to Drive Document Analysis.* First International Conference of Document Analysis and Recognition(ICDAR 91). Vol 1. pp. 294-301.

[Le96]   F. Lebourgeois. *Localisation de Textes dans une Image à Niveaux de Gris.* CNED'96. France. 1996. pp. 207-214.

[Li96]   J. LIANG, J. HA, R. ROGERS, I.T. PHILLIPS, R.M. HARALICK, B. CHANDA. *The Prototype of a Complete Document Image Understanding System* , DAS'96. Malvern, October 1996. pp. 131-154,

[Mau87]   P. Maurice. *L'Architecture d'un Document électronique : concepts et applications.* L'écho des Recherches. N°130. 4st term 1987. pp. 15-24.

[Na86]   G. Nagy, S. C. Seth and S. D. Stoddard. *Document Analysis with an Expert System.* Pattern Recognition in Practice II (E. S. Gelsema and C. N. Kanal, Eds.). 1986. Pp. 147-159.

[Na92]   G. Nagy. *A Prototype Document Image Analysis System for Technical Journals.* IEEE Computer Magazine. July 1992.

[Pe90]   D. Peden-Derrien. *Analyse des structures de documents : une approche objet.* Ph. D. Thesis Université de Rennes 1. 1990.

[Pet95]   J. Petrak. *An Object-Oriented Case-Based Learning System.* Ph. D. Thesis. 1995

[Sa92]   A. Sanfeliu. *Syntactic and Structural Methods in Document Image Analysis.* In Structured Document Image Analysis. H.S Baird, H. Bunke& K. Yamamoto (Eds.). 1992. pp-479-499.

[Sa94]   T. Saitoh, T. Yamaai and M. Tachikawa. *Document Image Segmentation and Layout Analysis.* IEICE Transactions in Information and Systems. Vol. E77-D. N° 7. July 1994. pp. 778-784.

[Ta96]   S. TAYEB-BEY, S. SAIDI, H. EMPTOZ *Grammatical Approach for the Physical and the Logical Structure of Documents Analysis : Application to Summary Documents.* MVA'96. IAPR Workshop on Machine Vision Applications, November 1996, Tokyo. pp. 341-343.

[Wi65]   A. Van Wijngaarden. *Orthogonal Design and Description of Formal Languages.* Mathematish Centrum Amsterdam, MR 76, 1965.

[Wo82]   K. Y. Wong, R. G. Casey and F. M. Wahl. *Document Analysis System.* IBM Journal of Research and Development 26. 1982. pp. 647-655.

Document Modeling for Form Class Identification

Sébastien Diana[1], Eric Trupin[1], Yves Lecourtier[1] and Jacques Labiche[2]

[1] Université de Rouen, Laboratoire PSI / La3I, 76821 Mont Saint Aignan Cédex, France
[2] Université de Caen, Laboratoire ISMRA / LACP, 14050 Caen Cédex France

Abstract. This article deals with the description of a document system analysis based on document modeling. This system is applied to forms which are used by the CAF, the French national family allowance Department -*Caisse d'Allocations Familiales*. The system is composed by three different modules which deals with the different form processes. The first module - *low-level processing* - is divided into three stages : acquisition, binarisation and skew correction. These stages allow the transformation of a paper form into an image with correct qualities. The second module - *document structuration* - processes this image to extract the information contained in the form. The information is arranged to obtain a tree. This tree shows the organisation of the form content into a hierarchical way. In addition to the tree extraction, the document structuration module allows the creation of a form model base. The last module - *form class identification* - uses the tree and the form model base. It is composed with two pre-classifiers to extract possible lists of forms and a structural classifier. The two pre-classifiers filter the form classes among the 250 classes in order to reduce the treatment of the classifier. This classifier is based on graph matching to compare the tree of the particular form and the possible list of form extracted during the two pre-classifiers.

1. Introduction

Today, Administrations and Companies (i.e., mail-order business ...) expect a system to automatically treat their documents. Thanks to their project, they want to improve the quality of the services offered to their clients. It corresponds to management of the storage and retrieval of information contain in the extensive flows of documents.

The CAF -*Caisse d'Allocations Familiales*- manages family benefits and has to cope with these problems. It has therefore decided to solve this problem thanks to a project called ADI -*Archivage et Documentation Intelligents*- which means Intelligent Documentation and Storage. The CAF wants to be able to treat automatically all the documents received. The treatment is first applied to printed forms. Figure 8 (Appendices) shows an example of the 250 form classes.

To manage all their documents or forms, Administrations and Companies need a system which is able to treat documents automatically. A such system exists but it can only deal with one kind of documents at a time. We propose a complementary method of modeling based on document analysis aiming at identify the kind of forms before dealing with them.

Generally, document analysis systems are articulated around two main modules

(Fig.1). The first module, called low-level processing, transform the paper document into an image (i.e., bitmap ...) with the use of image treatments like thresholding [1] or contrast enhancement [2]. The second module, called document structuration and text recognition, extracts a set of information represented by physical and logical structures [3, 4, 5]. We propose here a third module that uses these structures for document (form) class identification.

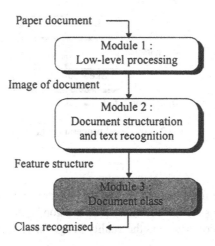

Fig 1 . Document analysis system

Based on this principle and as part as the ADI project, we are developing a form analysis system aiming first at recognising the class of forms and then treating them for the file actualisation.

This paper deals with this form analysis system which is composed as follows : Module 1 concerns low-level processing applied to the image (i.e., binarisation, skew correction ...). Module 2 concerns only the document structuration with the extraction of information. Module 3 concerns the identification of the form class.

The paper is based on the description of these modules. Section 2 describes the first stage of the module (Module 1, Fig.1) which concerns low-level processing applied to scanned images. Section 3 describes the second stage (Module 2, Fig.1) which concerns the document structuration with the extraction of features contained in the form. Section 4 presents the results concerning these two different stages. Section 5 presents the prospects for our work based on the form class identification. This module will be divided into three main stages. Finally, the last section of the paper includes discussions for our work.

2. Low-level Processing

The aim of the low-level processing is to get an image of the paper form. The different stage are represented Fig.2.

The start point is the paper form. A grey scale image is obtained after acquisition. A binarisation is applied on this image in order to reduce the data. The last stage corrects the potential skew and localises the contents in the image.

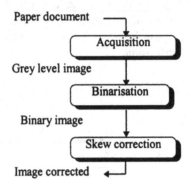

Fig.2. Low-level processing stages

2.1 Acquisition

The acquisition of the form image is made at a 300 dpi (dots per inch) resolution and in 256 grey scale levels. This resolution value is a good compromise between the size of the resulting scanned image and the quality image for an OCR processing.

2.2 Binarisation

A scanned image at 300 dpi with 256 grey scale levels takes 8 Mo without compression. To limit this important quantity of data, a binarisation is applied to the grey level image to reduce the space to 1 Mo without compression.

The binarisation techniques are based on a thresholding method. The contents and background pixels can be distinguished thanks to their grey level values. The grey level image can be transformed into a binary one by chosen a threshold between characteristics values of the content and background intensities.

Thresholding approaches can be divided into global or local techniques and they can be separated into entropic thresholding [6] and maximum likelihood thresholding [7]. Our method is based on a global approach by detection of a threshold with maximum likelihood [8]. The maximum likelihood criterion (MLC) is applied on the histogram of the grey level value $H_{glv}(x)$ to find the optimum binarisation threshold T_b (Fig.9 - Appendices). Figure 10 (Appendices) shows the correct result gives by this binarisation for these sort of forms (with two classes for black text and background). The method algorithm can be adapted to multi-thresholding if forms have text with several colours or if background is coloured

2.3 Skew Correction

In a similar way, skew correction methods are divided up into local and global approaches. The local techniques determine the skew angle of document with the straight line directions [9]. The global techniques extract the main orientation of a document [10].

Our method comes from this second approach. The technique is based on the detection of a reference. The correction of the form orientation is made by a rotation processing where the centre is the origin of the reference and the angle corresponds to the skew angle between the X-axis and the horizontal (Fig.11 - Appendices).

Processing is based on the detection of this reference with the upper straight line (X-axis) and the left straight line (Y-axis). The skew of this upper straight line is obtained with the detection of local upper contours called « modes » on the upper contour of the form. The upper straight line is adjusted on the higher pixel in the upper straight line direction. The left straight line is obtained by calculating the perpendicular straight line of the upper straight line and by adjusting it to the most left positioned pixel in the direction of the perpendicular straight line.

The global approach of this technique allows the application on other sorts of documents (i.e., scientific articles, mail-order business forms ...).

3. Document Structuration

The aim of this module is to extract, from the binary image, information about the content of the form. The information is organised in a hierarchical structure which shows the dependencies between the different parts of the form (Fig.3).

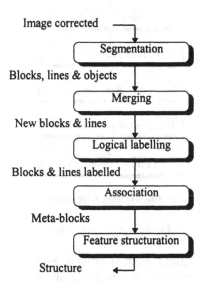

Fig.3. Document structuration stages

First, the segmentation extracts the information defined by blocks, lines and words. Secondly, the merging, logical labeling and association complete this information. The merging creates new blocks and lines. The logical labeling adds new information. The association creates new blocks, called « meta »-blocks. Finally the feature structuration organised all the information in a hierarchical structure represented by a tree.

3.1 Segmentation

Three different approaches exist for the segmentation. The top-down method cuts the document in always smaller elements. The bottom-up method extracts small elements from the document and merges them together to create larger elements. The mixed method is based on a combination of the two previous methods.

Our segmentation process is based on mixed method. First, blocks are detected with a modified contour following algorithm. In each detected block, objects, called connected objects, are extracted with a usual contour following algorithm. Blocks and connected objects are characterised by text/graphic discrimination. This part corresponds to a top-down method. Secondly, text lines are located in each block with histograms of horizontal projections. This part corresponds to a bottom-up method, and finally, the complete chaining corresponds to a mixed segmentation method (Fig.12 - Appendices).

The method is represented Fig.4 and, for more information, is detailed in the paper [11].

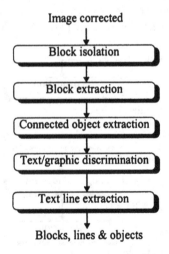

Fig.4. Segmentation stage

Other methods like RLSA (Run Length Smoothing Algorithm) [12] or mathematical morphology enable the obtaining of similar localisation for blocks entities, but computer-time consuming is disadvantageous.

3.2 Merging

The previous segmentation processing extracts the different elements with an effect of over-splitting. This effect allows the extraction of small elements .

The aim of the merging processing is to reduce the number of blocks and text lines obtained by this over-splitting effect. Merging is made in the set of blocks and in the set of text lines. For text lines, the process is applied in each block.

Merging process is driven by the application of six criteria. These criteria are established by the reading notion.

First, block b2 must be at the right of block b1. The y-positions of the blocks must overlap themselves (Fig.5). The blocks must be at a minimum distance defined by an euclidian distance. Finally, candidates must have same properties (i.e., height of text lines ...).

The merging process gives the result as presented Fig.13 (Appendices).

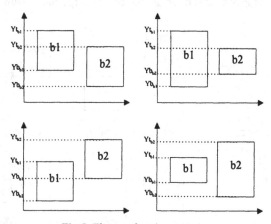

Fig.5. The overlapping criteria

3.3 Logical Labeling

This process is composed of two stages. The first stage corresponds to the labeling of blocks and text lines with general document knowledge. The second stage is based on some particularities which are met regularly in the 250 form classes. Forms are composed by a headband divided into a logo and a title, by a form number, by adress block corners which enclosed the recipient's adress and with some straight lines which can be filled up to designate some information.

First, text lines are labelled with « left indentation », « right indentation », « justified line » and « centered line ». It therefore enables the labeling of the blocks with « paragraph begin », « paragraph body », « paragraph end » and « complete paragraph ».

Secondly, other blocks are localised and labelled with the form particularities. The labels are « title part » and « address block corner ».

3.4 Association

The aim of this association process is to add new hierarchical information for feature structuration. The principle is the creation of new lists of blocks called « meta »-

blocks. Like merging process, association is driven by several criteria (Fig.6).

Criteria are established thanks to disposition notions, label notion and notions about lines. The distance criterion (Fig.6a) tests the block proximity notion with euclidian distance « vert dist ». Association takes into account labels (Fig.6b). For example, a « title part» block will only be associated with another « title part » block. Blocks must have similar text line heights (Fig.6c). The inter-line space criterion (Fig.6d) shows the line space notion in a block (i.e., « paragraph » ...) in opposition with the notion of space between blocks. The fifth criterion (Fig.6e) tests the overlapping between blocks. Two blocks must have enough common part to be associated. Finally, if there is complete overlapping, the last criterion (Fig.6f) tests the re-covering between blocks to avoid the association of large block with small block.

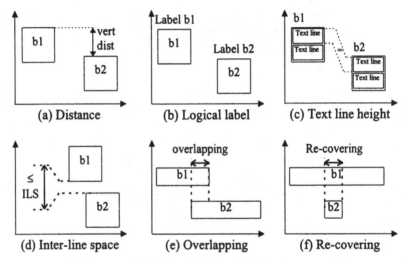

Fig.6. The association criteria

3.5 Feature Structuration

The goal of this structuration process is to organise in a hierarchical way (tree) all these lists of features extracted during the previous processes (« meta »-blocks, blocks, text lines, objects).

The tree is organised into n levels (n ≥ 4) with level 1 represented by the root. There are also three levels (block, text line and object). Between the root and the block level, several « meta »-block level may be inserted.

Figure 15 (Appendices) shows a representation of a form structure composed with five levels and only a « meta »-block level.

The structuration process allows to organise the complete set of elements in order to keep all the information extracted and the dependence between these elements.

4. Experimental Results

Results are obtained by testing each process of module 1 (low-level processing) and by testing module 2 (document structuration) like an entity.

Binarisation is evaluated with visual criteria (broken line structures ; broken symbols or text; blurring of lines, symbols and text ; loss of complete objects and noise in homogeneous areas) defined in the paper [13]. Its score is 13,6/25. This is an average performance result but with great robustness (i.e., forms printed on colour paper ...). The execution time is about 0.1s for an 634x876 image on a Sun Sparcstation20.

The low-level processing is tested on a base composed by synthetic images of forms. Forms are synthetically created in order to eliminate potential skew due to scanning. These forms are rotated with theoretical skew (TS) from -15° to +15° with a step of 1° to obtain a base of skewed forms. 97% of the calculated skews (CS) are found in [TS-0,5°;TS+0,5°]. The process is limited by the appearance of a distortion in the extreme skew (-15 or +15), by the loss of quality and the loss of information on the border of the form.

The document structuration module extracts necessary information for the form class identification module. This involves to test the stability of the document structuration module. The stability result is evaluated with the study of structures extracted during the document structuration module. Evaluation is established on the number of lists obtained and on the « meta »-blocks created during the association process.

First, the results are established on a particular base. This base is created with one kind of form scanned several times with different skews (about 50 images). We apply the low-level processing module and the document structuration module on this base and we count the number of lists obtained are counted for each form image and the number of « meta »-blocks created. The results give a great stability for the module. Systematically, we find the same number of lists and 98% of created « meta »-blocks are identical.

Secondly, other results are established on another classic base (about 120 images). This base is composed by different form classes. The results give by the treatment of this classic base (Table 1) show that the creation of structures is a stability process.

Table 1. Percentage of created lists

Number of created lists	Percentage
3	22.22
4	71.11
5	5.56
6	1.11

The created lists correspond to the « meta »-block, block, text lines and object lists. We find a minimum of three lists (blocks, text lines and objects) and the number of created « meta »-block lists is included between 0 and 3.

5. Form Class Identification Strategy

This section describes our strategy to identify form class based on a structural classification with graph matching [14, 15]. The module will be composed by two pre-classifiers and a structural classifier (Fig.7). We will use the result of document structuration module.

The aim of the pre-classifier is to reduce the research of the classifier because of the important number of classes (250 form classes). The pre-classifier will extract two lists of possible form classes. These lists will be input data for the structural classifier with a reduce number of form classes.

Pre-classifier 1 is based on the use of the form identifier number. First, a tree is created thanks to a hierarchical organisation of the list of all the identifier numbers. Secondly, the content of the form to identify is recognised with an OCR software. The principle is based on the run of the tree with the correct or incorrect OCR-ed form number (accuracy problem of OCR software).

Pre-classifier 2 is based on the quad-tree notion. The aim corresponds to extraction of information of the initial image thanks to a multiscale resolution technique and the identification with a kppv-classifier.

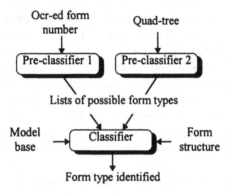

Fig.7. Description of identification module

The classifier uses the result lists emitted by the pre-classifiers (list of possible form classes). It is based on graph matching with the comparison of the particular tree (form structure) and the model trees (model base). The principle corresponds to the computation of edition distance. This distance is based on the addition of the cost of all the elementary modifications between two trees.

6. Conclusion

We have presented the treatments of forms with a document analysis system based on a document structuration module and a form class identification module. These two modules are preceded by a low-level processing module

The low-level processing module (Module 1, Fig.1) is composed by image treatments to obtain correct binary image (acquisition, binarisation and skew correction).

The document structuration module (Module 2, Fig.1) extracts the information of document and organises it in a tree (hierarchical notion). This module has two aims. First, it is used to extract feature for form class identification module. Secondly, it used to create a data base composed by the model set (information structure) of the 250 form classes.

The form class identification module (Module 3, Fig.1) will be used to recognise a particular filled up form. The module will contain two pre-classifiers based on OCR-ed number form and on a multi-resolution notion (quad-tree) applied to image. After these two pre-classifiers, the module will contain a structural classifier based on the comparison of two form structures (trees) - (One for the particular form to be recognised and one for each model).

The system will be completed by a model base organisation module. In order to accelerate the treatment, models will be organised with a resemblance notion [16].

References

[1]   Lamouche, I., Bellissant, C. : Séparation recto/verso d'images de manuscrits anciens. Colloque National sur l'Ecrit et le Document Nantes France (1996) 199-206.

[2]   Chatterjee, C., Raychowdhury, V.P. : Models and algorithms for real-time hybrid image enhancement methodology. Pattern Recognition 29-9 (1996) 1531-1542.

[3]   Sauvola, J., Pietikäinen, M. : Page segmentation and classification using fast feature extraction and connectivity analysis. International Conference on Document Analysis and Recognition Montreal Canada 2 (1995) 1127-1131.

[4]   Esposito, F., Malerba, D., Semeraro, G. : Automated acquisition of rules for document understanding. International Conference on Document Analysis and Recognition Tsukuba Science City Japan (1993) 650-654.

[5]   Tang, Y.Y., Suen, C.Y. : Document structures : A survey. International Conference on Document Analysis and Recognition Tsukuba Science City Japan (1993) 99-102.

[6]   Brink, A.D., Pendock, N.E. : Minimum cross-entropy threshold selection. Pattern Recognition 29-1 (1996) 179-188.

[7]   Kittler, J., Illingworth, J. : Minimum error thresholding. Pattern Recognition 19-1 (1986) 41-47.

[8]   Kurita, T., Otsu, N., Abdelmalek, N. : Maximum likelihood thresholding based on population mixture models. Pattern Recognition 25-10 (1992) 1231-1240.

[9]   Le, D.S., Thoma, G.R., Wechsler, H. : Automated page orientation and skew angle detection for binary document images. Pattern Recognition 27-10 (1994) 1325-1344.

[10]   Leroux, M. : P.A.B.L.O, Procédure de saisie de bordereaux par lecture optique. Colloque National sur l'Ecrit et le Document Nantes France (1996) 259-266.

[11]   Trupin, E. : A modified contour following algorithm applied to document segmentation. Intelligence Artificial and Pattern Recognition The Hague Netherlands (1992) 525-528.

[12]   Wahl, F., Wong, F., Casey, R. : Block segmentation and text extraction in mixed text/image documents. Computer Graphics and Image Processing 20 (1982) 375-390.

[13]   Trier, O.D., Taxt, T. : Evaluation of binarisation methods for document images. Pattern Analysis and Machine Intelligence 17-3 (1995) 312-315.

[14]   Watanabe, T., Luo, Q., Sugie, N. : Layout recognition of multi-kinds of table form documents. Pattern Analysis and Machine Intelligence 17- 4 (1995) 432-445.

[15]   Casey, R., Ferguson, D., Mohiuddin, K., Walach, E. : Intelligent forms processing system. Machine Vision and Applications 5 (1992) 143-155.

[16]   Dengel, A., Dubiel, F. : Clustering and classification of document structure - A machine learning approach. International Conference on Document Analysis and Recognition 2 (1995) 587-591.

Appendices

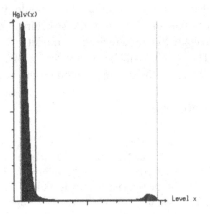

Fig.8. Form example

Fig.9. Histogram of grey level values

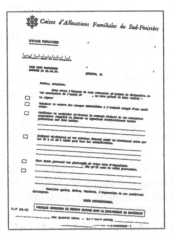

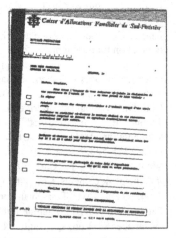

Fig.10. Binarized form with skew

Fig.11. Skew detection

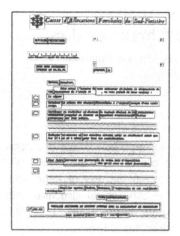

Fig.12. Segmentation result

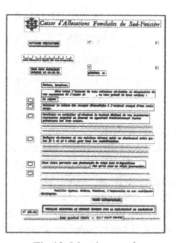

Fig.13. Merging result

Fig.14. Association result

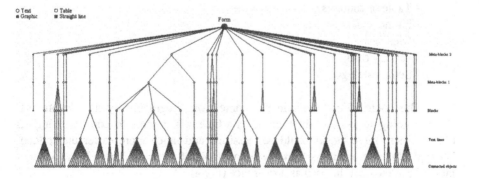

Fig.15. Representation of a form structure

Composite Document Analysis
by Means of Typographic Characteristics

Laurence Duffy - Frank Lebourgeois - Hubert Emptoz

E-mail : duffy@rfv.insa-lyon.fr

1 Introduction

Logical structure analysis of document is an important theme of research, indeed this logical structure stands between the content and the layout of the document. It is an inevitable way if we want to attain the way of « the printed document comprehension ». Generally, we link up this logical structure with a hierarchy of visual components which are observable by the reader. In a textual document, these components are titles, sections, sub-sections, lists, ...

Automatic recognition of logical structure, will only occur for document in which it is contained in graphical elements (In other terms, this logical structure will have to be extractable from the numerical image of the document). This extraction will be more easy if the publisher has created the document paste-up in order to accentuate this structure, which can help the reader to understand the text.

The document paste-up composition can be guided by typesetting rules which can consist in the description of the document organisation, in terms of sections and paragraphs. Niogy and Srihari [NIYOGI95] have written about this theme «Standard printed documents generally conform to a certain geometric structure that dictates that the document be composed of a set of interconnecting rectangular printed regions, or blocks. The layout of the blocks varies among different types of documents and depends on the layout conventions for each document.».

These typesetting rules can also act upon the use of different fonts. We could contrast the global organisation side (titles, paragraphs,...) with the most local and particular side carried by the use of different fonts. A list of some typographically rich documents follows :

- Table of contents
- Phone book
- Dictionary
- Library card
- Some catalogue

Our contribution stands in this context of typographically rich documents. We are going to suggest different tools which will permit to constitute logical blocks. Then, these blocks could be classified and much more easier understood. We will illustrate our remark by means of table of contents. The « News Week » table of contents will be used as a reference (Figure 1).

Our processes is different from the Satoh and al. method [SATOH95] which consists in blocks classical search in the IEEE table of contents. Satoh and al. have written « Table of contents images are firstly converted into binary images. Then the text blocks are extracted using RLSA method ».

Our processes will also be different from the Fisher and al. method [FISCHER95] which is applied to phone books. Our processes will be almost exclusively based on words, or groups of words, typographic location.

2 Typography : Use ... and Recognition

2.1 Use of fonts and of typography

In this contribution we will refer to table of contents. We will study the variety of graphic patterns of characters, which are used in these table of contents.

The News Week table of contents (Figure 1) will clarify our remark.

We have chosen table of contents because on one hand they are copious and well known. Their presentations are similar from a country to an other. On the other hand, in terms of electronic library, it is very important to be able to read a table of contents of which the logical structure has been located. It means that we can traduce the table of contents in a HTML or markers language.

We have studied more than hundred set of magazines table of contents. They are principally French or west European, and sometimes American.

These magazines are very different : scientific review (like IEEE), political journal or children magazines.

Each of them has its own layout and its particular typesetting rules but they show a typographical stability. For example, we have just find one magazine written by means of an unique police. It was the magazine of an association which has not money enough to buy a text editor. In a great majority, polices or fonts used to write the title, author name, page number, ... are different.

Cyclical use of different fonts seems convey much more logical information than the layout, in terms of sections or paragraphs.

We have moreover note that the use of typographical information, in order to mark the logical structure, is becoming the norm since text editor has replaced the typewriting.

When we use a text editor, we use all the range of functions which is proposed. So we traduce the logical structure by the layout and above all by the use of typography.

For a named magazine, the use of a font is equivalent to a hierarchical mark. Figure 1, we can see that « World Affairs, Special Reports, Business, ... » are written by means of the same font.

This is similar to the use of marks, in HTML for example, which indicates that some expressions are hierarchically identical.

Recognise the font name, wouldn't provide more information about the document logical structure. We are only interested in the fact that some words are written by means of the same font.

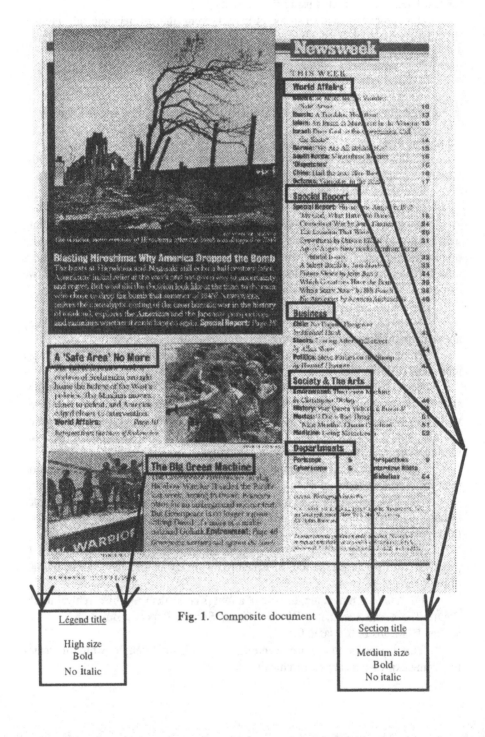

Fig. 1. Composite document

Légend title	Section title
High size	Medium size
Bold	Bold
No italic	No italic

2.2 Fonts Recognition, and Documents Typographic Analysis

There is few works concerning fonts recognition or identification by means of their characteristics. We haven't found works corresponding to the limits of our problem, i.e., to be able to identify that two or several words are written by means of the same font. Works concerning typographic analysis used data basis which contain all the existing polices as well as their declinations.

Systems are generally progressive [ZRAMDINI94, ANIGBOGU92], users can enrich data basis with new fonts. These systems are very profitable because they permit to adapt to typographical evolution. We could envisage to use this kind of system to settle our problem, even if they provide some information which don't interest us.

In fact, we can't use them. Firstly, this kind of systems needs an important sample of characters in order to learn a new font. Table of contents contain many polices or fonts which are few numerous. Their learning will be few reliable and so few profitable.

Secondly, an attribute label is very relative from a police to an other. For example, the Bookman police, size 12, not bold, appears bolder than the **courier police with the bold declination**. So the bold attribute can be not significant.

We haven't find a solution to our problem from literature in existence, so we have developed a new method which goal is to group together, words written by means of a same font.

3 Font Analysis Using Patterns Redundancy

In a first time, we want to group together words which have the same font, but we don't want to identify their font.

3.1 Utilisation of Patterns Redundancy in OCR

Patterns redundancy is a technique used to regroup printed letters which have a similar pattern, in order to compress documents, or to decrease the number of patterns to recognise [LEBOU92] [CASEY72].

It consists in comparison of each letter with other letters, by partial template matching. All letters having a similar pattern, are joined to the same label (a family) which corresponds to the same prototype. On the other hand, a letter could produce several prototypes. Figures 2.a 2.b 2.c illustrate this method. We could note that the letter « a » produces several prototypes. Each of them differs from the other, by few pixels.

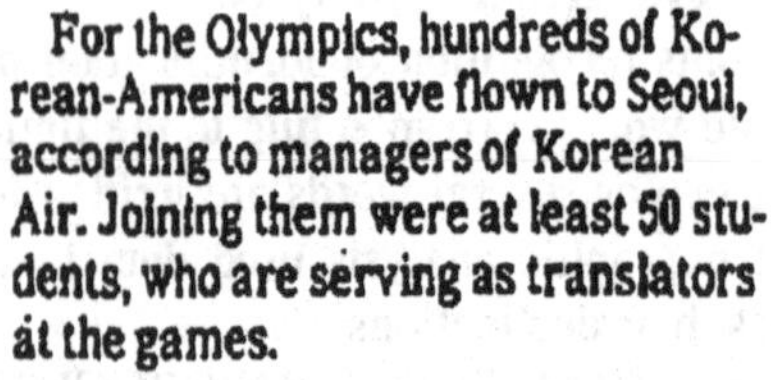

For the Olympics, hundreds of Korean-Americans have flown to Seoul, according to managers of Korean Air. Joining them were at least 50 students, who are serving as translators at the games.

Fig. 2.a Original text

Fig. 2.b. Prototype images

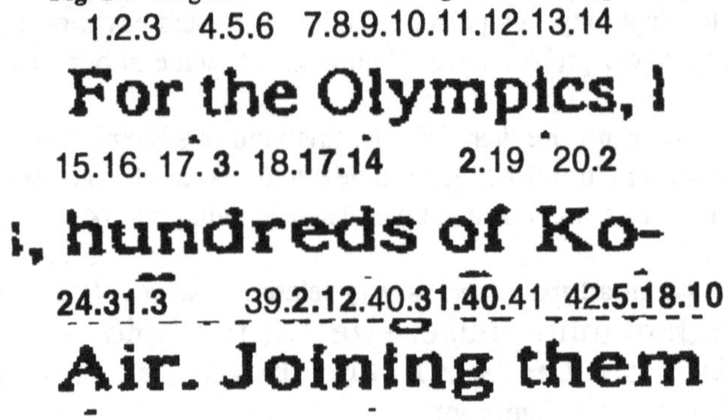

1.2.3   4.5.6   7.8.9.10.11.12.13.14

For the Olympics, 1

15.16. 17. 3. 18.17.14   2.19 20.2

ı, hundreds of Ko-

24.31.3   39.2.12.40.31.40.41 42.5.18.10

Air. Joining them

Fig. 2.c. Labelled text

This method which allows to regroup all letters having a similar pattern, simplifies OCR software job ; recognition of prototypes permits to recognise entire text.

3.2 Patterns Redundancy for Location of Different Fonts

We want to group into families, letters which are written with the help of a same font.

We propose to classify letters by means of the patterns redundancy method previously evoked, in order to be independent from OCR program results.

This method consists in comparison of letters pattern, and in the grouping of letters which have a similar pattern, under the same label. *In other words, by means of this method, we regroup letters which are written with the help of the same font.*

The figure 3 illustrates the different letters « e », which are extracted from the « News Week » summary (CF figure 1), and which are regrouped into families.

A letter, printed by means of different fonts, generates at least one family per font.

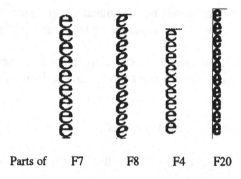

Parts of   F7   F8   F4   F20

Fig. 3. Different families of letter « e »

Noise problems and printing errors which distort letters pattern, falsify the partial template matching, so some families only contain one faulty letter (Cf. figure 4).

Fig. 4. Faulty letter

When we use a touched up image , we obtain a family per letter and per font, otherwise it is possible to obtain several families per letter and per font. This case is illustrated figure 5. Font Beta has generated two different families : F1 and F4 (ditto font Gamma).

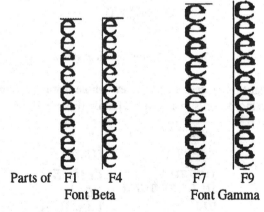

Parts of   F1   F4   F7   F9
        Font Beta   Font Gamma

Fig. 5. Construction of several families per font.

We have grouped together letters. Now we have to regroup words, in order to do that we state the following hypothesis : *A word only contains letters which are written with the help of the same font.*

We have therefore completed our treatment by grouping together words which have at least one common prototype, and so the same font, in homogeneous families.

In order to do that, we assign a number family to the first word, and then *we propagate this number family to each other word, which at least have one common prototype.*

The figure 6,7 and 8 illustrate this process.

- In a first time, the family is initialised with the first word of the text (Figure 6).

The

Fig. 6. Family content at step 1

- In a second time, the family contains all words which have at least one common prototype with the first word of text (Figure 7).

The
skeletal
eerie
remains
of Hiroshima
the
Refugees from
dropped
the

Fig. 7. Family content at step 2

- We repeat this process with the new whole words contained in the family until there was no more words to add to the family (Figure 8).

The   town
skeletal
eerie   rriors
remains   sail
of Hiroshima   against
the
Refugees from   bomb
                 was
dropped
the   bomb

Fig. 8. Family content at the end of the process

We repeat the same process, the following family is initialised with the first word which doesn't belong to a family.

4 Difficulties Link Up to the Context

4.1 Confusion Risks

The technique we have developed raised some difficulties.

Although we have fixed a high degree of similarity to avoid substitution and to not confuse two fonts, we have noticed that *some letters printed with a different style, look very similar and can be represented by the same prototype.*

Most important confusions take place when document contains some intermediate fonts, i.e., the same police but some different declinations.

With figure 9, we could see that *prototype «i» with an italic font looks similar in an italic-bold font, and that prototype «o» with an italic-bold font can be confused with bold font.* So, some fonts which never would have to be mixed up (italic and bold), are regrouped in the same family because it exits a word written by means of a bold-italic font. Nevertheless this case is rare, because few publishers choose near fonts.

Kong and Azriel   italic

Guest Editors: italic - bold

Guest Editors'   bold

Fig. 9. Examples of intermediate fonts

We also have noticed that *dots, commas and accents are very similar from one police to an other* (CF figure 10. So we haven't judged them significant for our analysis, and we have removed them.

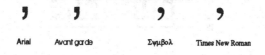

Arial   Avant garde   Σψμβολ   Times New Roman

Fig. 10 Identical commas of different polices

4.2 Importance of the First Letter

In a first time we use all word letters in order to classify words in families. But we have quickly noticed that when we use the first letter, which is often a capital letter, we group together capital and minuscule words of a same font.

It is very important to differentiate capital and minuscule fonts because their family could carry a different logical significance.

A solution had been to ignore the first letter in order to process the grouping in families.

But this raised a new problem : rejection of all the words which are composed of one letter, and of the big majority of words composed of two or three letters.

We just have one or two letters to classify words when our prototypment never have placed on the first letter.

This is insufficient because the more a word is long, the more the probability of one common prototype with of an other word is elevated.

This problem takes more importance according that our segmentation method is turned toward the over segmentation, so we often obtain some parts of words, that we can't systematically reject.

It was therefore essential to take this first letter in account when it doesn't entail risks of confusion between minuscule and capital letters of a same font.

In order to verify that the first letter of each word has the same type (capital, minuscule, number) *than the remainder of the word, we propose to calculate the letter and word body height.* If the height of the first letter is equal to the height of the following letters, it means that the first letter of the word has the same type than the word, so it can be used for the prototypment. Otherwise, we are in the case of one minuscule word beginning by a capital letter, and we reject this first letter.

In order to calculate words body height, and then differentiate fonts, it is necessary to *find very rigorously the word basis lines* (CF figure 11).

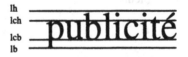

Fig. 11. Annotations regarding the word basis lines.

4.3 In Search of Basis lines

Most of basis lines detection method [LE96] are projection-based techniques and aren't sufficient to answer ours requirements. We must indeed be able to find words basis lines of words which aren't greater than three letters, and projection-based methods don't always produce good results with short words.

So we have formalised a technique of vertical probes throw which marks, the high (respectively the low) of the dark pixel met. This method compared to projection analysis, provides excellent results with words which don't exceed three characters, but could generate some mistakes with some words, specially if polices have serifs. Serif of some upstroke or downstroke characters, misleads us, and the retained basis lines are erroneous.

We have therefore interface our basis lines research with a validation technique, which consists to verify the consistency of results, and if necessary, to relaunch research with others parameters. This method is explained in [DUFFY97].

5. Results and Perspectives

The result of grouping together same font words in a family is illustrated figure 12. The grey level blocks mark one of the families which has been constructed during the process. We can notice that *this family corresponds to bold little titles* (right window), *but all bold little titles* (left window) *don't belong to this family*. This is attributable to the degree of similarity chosen for letters matching, which is fixed high to avoid substitutions.

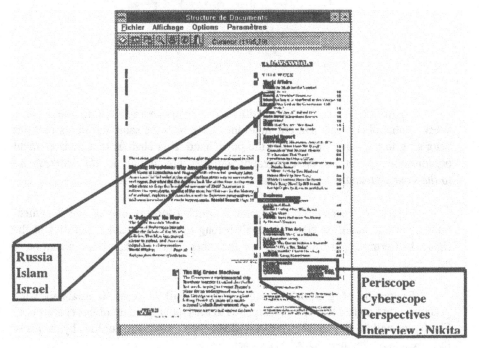

Fig. 12. One family

Adaptation of pattern redundancy method to our context has partially allowed us to group together words, which have the same font, into typographic families.

We have note that these families are connected with the document specific grammar.

Each typographic family carries a *logical information which corresponds to one terminal,* in terms of grammar, and *entities linking of different typographic families is a string of the document grammar.* Figures 13 and 14 illustrate this comment.

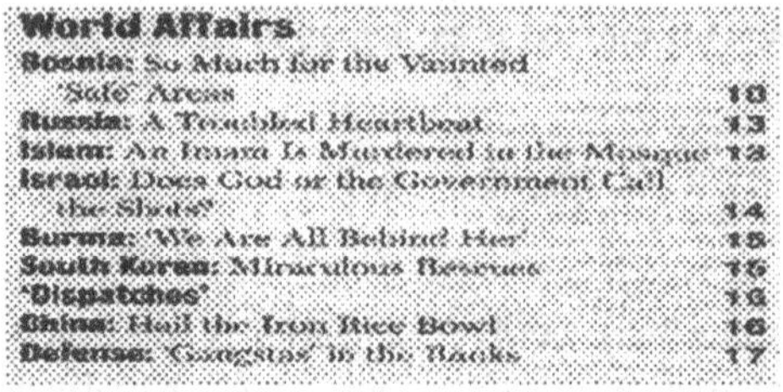

World Affairs
Bosnia: So Much for the Vaunted 'Safe' Areas 10
Russia: A Troubled Heartbeat 13
Islam: An Imam Is Murdered in the Mosque 13
Israel: Does God or the Government Call the Shots? 14
Burma: 'We Are All Behind Her' 15
South Korea: Miraculous Rescues 15
'Dispatches' 16
China: Hail the Iron Rice Bowl 16
Defense: 'Gangstas' in the Ranks 17

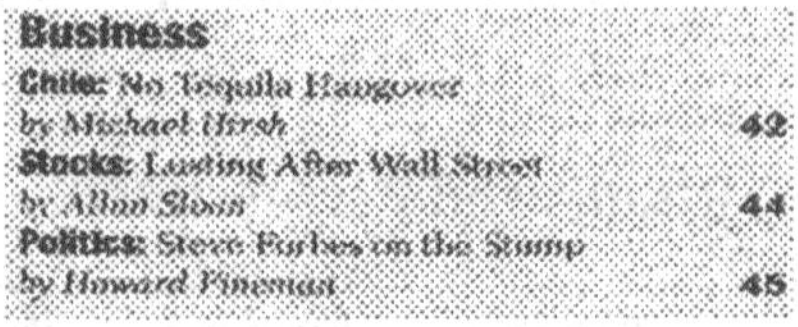

Business
Chile: No Tequila Hangover
by Michael Hirsh 42
Stocks: Lusting After Wall Street
by Allan Sloan 44
Politics: Steve Forbes on the Stump
by Howard Fineman 45

Fig. 13. Two examples of chapters

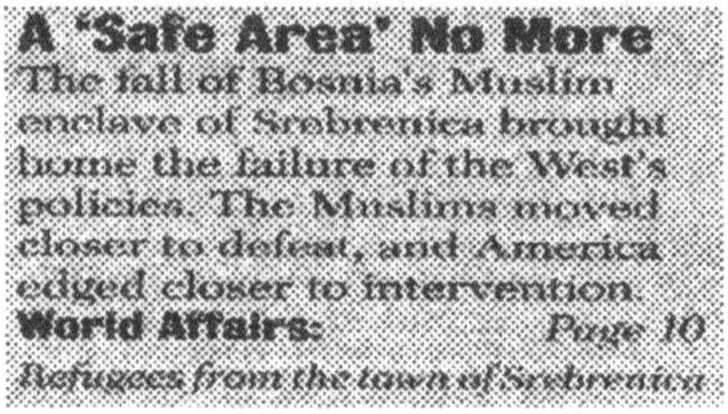

A 'Safe Area' No More
The fall of Bosnia's Muslim enclave of Srebrenica brought home the failure of the West's policies. The Muslims moved closer to defeat, and America edged closer to intervention.
World Affairs: *Page 10*
Refugees from the town of Srebrenica

Fig. 14. Example of legend

Figure 13, we can see two different chapters extracted from the « News Week » table of contents. On one hand they have both the same terminals (entities belonging to the same family), on the other hand, it is obvious that their element organisation type, and so the typographic families, are similar. *So, they correspond to the same grammar.*

Figure 14 we can see a « legend » which concerns one of the document image. On one hand, its elements don't belong to the typographic families of the figure 14 (terminals are different), one the other hand, the family organisation is different.

By means of grammar construction we will be able to establish what grammar, previously learned, corresponds to the typographic families organisation. Finally we will be able to label and/or validate the typographic families by means of the labels given by the selected grammar.

One of the essential requirements, if we want that our technique was outstanding, is that there was only one family per font.

If one family is excessively split, the string which describes the elements, will be erroneous and doesn't correspond to a grammar model. So, we must affect one unique typographic family per font.

So we are going to study the near setting of family words and the own characteristics of families in order to regroup identical families.

6 Conclusion

We have just presented a new method, of regrouping letters and words in homogeneous font families which doesn't necessitate to explicitly recognise the font.

This analysis, achieved with the application of one pattern redundancy technique, allows us to extract a part of the logical information which is carried by words typographic features.

After having differentiated, grouped together and compared the typographic families, we'll know :   - the cardinality of each family,
                            - its grease, slope and size compared to the others families.
The study of the typographic families organisation, and of their relative characteristics, will allows us to classify families according to their logical significance, and so to voice, when it will be possible, hypothesis concerning the logical signification of the families. A comparison between the constructed families and the learned grammar, will come to validate or correct the hypothesis, and to label families for which no hypothesis has been voiced.

The significance of the method, we have developed, is that each process only depend on the image ; it isn't depend on the document type or on fonts data basis. So this method can be applied to every document type, specially complex and typographically rich documents.

An other significance is that our text markers will be use for describing our document in HTML language

7 Bibliography

[ANIGBOGU92]   **ANIGBOGU J.C.** *Reconnaissance de Textes Imprimés Multifontes à l'aide de Modèles Stochastiques et Métriques.* Thèse Doct. Sci. : Université de Nancy 1, 1992

[CASEY82]   R.G. CASEY, G. NAGY Recursive Segmentation and Classification of Composite Character. *6th ICPR, Intenational Conference on Pattern Recognition*, Paris, France, 1982, vol.2, p.1023-1025

[DUFFY97]   **DUFFY L., LEBOURGEOIS F. et EMPTOZ H**. The Improve of Logical   Structure   Analysis by Typographic Characteristics Extraction. ICIAP97, *International Conference on Image Analysis and Processing*, Florence, Italie, 1997

[FISCHER95]   **FISCHER S., AMIN A. and DRIVAS D**. Segmentation of the Yellow Page. *Third ICDAR, International Conference on Document Analysis and Recognition*, Montréal, Canada, 1995, p. 605-609

[LE96]   **LE D.X., THOMA G.R. et WECHSLER**. Automated Borders Detection and Adaptative Segmentation for Binary Document Images. *13th ICPR, Intenational Conference on Pattern Recognition*, Vienne, Austria, 1996, p.737-741

[LEBOU92]   **LEBOURGEOIS F., HENRY H. et EMPTOZ H**. An OCR System for Printed Document. *MVA'92, IAPR Workshop on Machine Vision Applications*, Tokyo, Japon, 1992, p.83-86

[NIYOGI95]   **NIYOGI D. and SRIHARI S.N.** Knowledge-Based Derivation of Document Logical Structure. *Third ICDAR, International Conference on Document Analysis and Recognition*, Montréal, Canada, 1995, p. 472-475

[SATOH95]   **SATOH S., TAKASU A. and KATSURA E.** An Automated Generation of Electronic Library based on Document Image Understanding. *Third ICDAR, International Conference on Document Analysis and Recognition*, Montréal, Canada, 1995, p. 163-166

[ZRAMDINI94] **ZRAMDINI A. et INGOLD R.** Optical Font Recognition from Projection Profiles. *Third RIDT, International Conference on Raster Imaging and Digital Typography*, Darmstadt, Allemagne, 1994

Techniques for the Automated Testing of Document Analysis Algorithms

J. Sauvola, D. Doermann*, H. Kauniskangas and M. Pietikäinen

Media Processing Team
Machine Vision and Media Processing Group
Infotech Oulu, University of Oulu
FIN-90570 OULU, FINLAND
e-mail: jjs@ee.oulu.fi

*Language and Media Processing Lab
Center for Automation Research
University of Maryland
College Park, MD 20742-3275
e-mail: doermann@cfar.umd.edu

Abstract

This paper proposes a new approach to automate and manage the testing process for developing document analysis and understanding algorithms. A distributed test environment is proposed to assure visibility, repeatability, scalability and consistency during and between testing sessions. A variety of views are used to deal with multi-level operations in algorithm development. Tests are realized with dedicated test scenarios and events at different stages of the development cycle. Our main objective is to provide collaborating researchers with a flexible means of generating consistent ways to validate algorithm behaviour in a target environment. This is accomplished with a simulated environment and underlying resources for each test scenarios. A set of techniques to design test events for test scenarios (e.g. module, integration, regression) is proposed aimed at promoting a black-board style research approach. To demonstrate the functionality of this approach, we have implemented a prototype and trace an example algorithm through the development process.

Keywords: Document analysis, algorithm testing, distributed test management.

1 Introduction

Many applications that rely on image processing technology (IPT) are being realized as part of industrial and corporate systems. The IPT solutions are typically embedded as 'expert' modules in the larger automated systems, and these systems rely on the modules analysis and data validation for higher level processing. In document analysis systems, components rely on the image processing for low-level processing and feature extraction. Moreover, many document analysis systems operate in real-time environments, thus constraining system and module behaviour. In such environments, software testing is an important verification tool and the ability to match the design and implementation specifications is essential. For many systems, conventional test models and techniques are suitable primarily at the concept level.

The content, scope and focus of a document analysis (DA) system testing is usually limited to entities that demand a special attention. Currently, the DA technologies are used in postal automation [1], copying machines [2], office automation [3] and automated forms processing applications [4, 5], for example, all of which have strict performance requirements.

To reduce the time and cost of the development of a new system, a rapid prototyping is desirable, or even necessity [6]. During the development cycle existing algorithms are tested and new ones are implemented when necessary. The functionality of the developed prototype system must be validated systematically, whenever changes to its components are made. Similarly, the limitations of the prototype should be studied before it is transferred into a target environment. Document analysis applications are no exception. They are usually embedded in the systems that are comprised of many technologically different subsystems. It is therefore important to be able to validate the algorithm and subsystem functionality and performance to meet the requirement specifications set for the entire system.

In document analysis and understanding, the development of testing methods and approaches to test/project management have been given a little attention in the literature. Current testing methods are typically developed independently by research groups and lack consistency not only in the methods for testing, but also the way that results are managed and stored for later use. Generic software testing processes, on the other hand, have received a lot of attention over the last ten years in the field of software engineering.

We have previously described a test management system [7] that enables distributed, scalable and platform independent testing of document analysis and understanding applications. We have presented a collection of document images, a set of techniques to prepare the test cases and the means to control the entire testing process. In this paper, the objective is to consider the entire testing process in all stages of the development. We distinguish between three different factors that affect the document analysis system and overall test management construction: the application environment, the algorithm(s) and the overall system flow. Each one is described and its influence on the testing process is discussed. We propose a framework to enable design, creation and management of tests at different project and application levels. A distributed test environment is designed to assure visibility, repeatability, scalability and consistency in subsequent testing techniques.

The architecture and methods of the distributed test management system (DTM) are presented in Section 2. Document analysis algorithm and system testing techniques are outlined in the Section 3.

2 System overview

Document analysis and understanding techniques require a unique set of algorithms that are used in many cross-domain applications while embedded as expert modules. It is therefore important to be able to validate the algorithm performance and subsystem functionality to meet requirement specifications at an early stage. Furthermore, the target environment may be unavailable when research and design investigations/implementation takes place (due to, for example, costs, project timing, or nature of the project itself). One then requires a 'simulated environment' to verify the behaviour. This is a commonly used concept in other software engineering fields, such as real-time systems design.

For document image processing, the required environment and resources can easily be defined and realized using a distributed test management system (DTM) such as the one described in [7]. The DTM is built using current network computing technology to support remote operability, visibility and configurability needed in project organizations whose research is done in different locations. (see Fig. 1.)

To support different testing mechanisms and techniques with the requirements described above, the system must offer resources to provide test events with sufficient material to cover the test scenarios. These resources should be available to support remote computing and execution, thus special attention is focused on the network optimization issues in terms of the test visibility criteria. Since the central component of the document analysis algorithm test event is the selection and manipulation of the test document images, a set of techniques has been developed to offer graphical operability over networks using multithreading. We have integrated an object-oriented database, compression algorithms, transmission protocols and image presentation techniques to provide the remote user (client) with an optimized access to image and their manipulation result presentation. The images are managed and manipulated 'on-the-fly' to offer the DTM client smooth visibility over the system resources. The network traffic is optimized to contain only parameter values and control data, whenever possible.

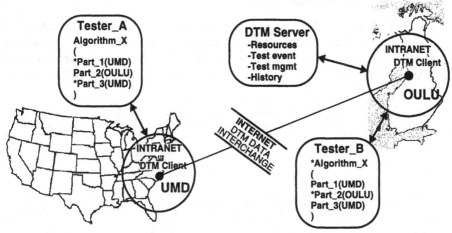

Fig. 1. Example of research-time test distribution in client/server environment.

The test management scenario provided by the DTM system covers phases from test project initiation and history management to test resourcing and execution. This is achieved utilizing distributed computing, i.e. different simulation environments that can easily be built and managed to the lowest level of the algorithm modularization. Resources for building such environments are located in a centralized graphical client application. The history and test functionalities are embedded features that take advantage of the systems database properties. An important part of the testing is the ability to re-engineer, to track and to (re)evaluate (e.g. benchmark) the test meta- and final results. For this reason, a set of dedicated evaluation and report generation tools are embedded components in DTM. Generic image and data conversion tools are also available to glue different functional domains together in the testing process.

Fig. 2. presents the design-implementation model used in the DAS testing management project that hierarchically supports the development of the different project phases and research on algorithms and their submodule(s). The model offers a scalable approach, using an object-oriented test bed and a simulated environment.

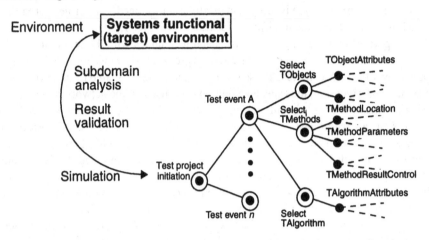

Fig. 2. Scenario design-implementation.

Fig. 3. depicts the DTM client user interface implemented in Java. An integrated presentation style combines all essential resources from the project initiation to the test execution with database access in a user interface, where the resources are requested and obtained from the DTM remote server in the Network Computing (NC) type transactions.

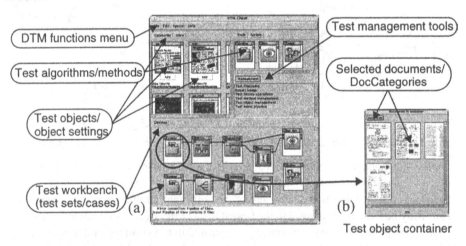

Fig. 3. The DTM graphical user interface.

Section 3 discusses the testing techniques that are implemented in the DTM client/server NC environment. The scope of the discussion is on the methodology and technical detail rather than on system level implementation.

3 Testing techniques

We differentiate between three different aspects that affect the document analysis system and test management construction: (1) the target environment, (2) the techniques and (3) the system structure/data flow (Fig. 4). The environment, E_t, defines a parameterization that controls the algorithm input, timing issues and the models used. These entities define the boundaries limiting the usage of document analysis algorithms. The techniques domain, T_d, describes the specific techniques exploited in DAS modules that affect the algorithm behaviour, e.g. the image filtering and the layout segmentation. These are scaled entities of the specific techniques that construct the DAS system functionality, S_d.

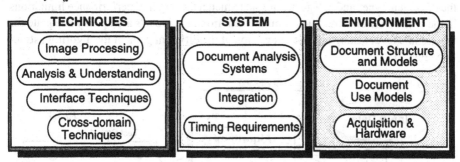

Fig. 4. A document analysis (development) system environment domain division.

3.1 Formal model

In the DTM system, support for DAS techniques is hierarchically organized starting from the systems level and having sublevels, events and scenarios. Each sublevel and event can be distributed under a different test scope. We define the document analysis testing process formally using Eq. (1-5):

$$J = \{E_t, S_d, T_d, C_d\}, \tag{1}$$

where

$$E_t = \{\Theta_1, \Theta_2, ..., \Theta_n\}, \tag{2}$$

$$S_d = \{P, L\}, \tag{3}$$

$$T_d = \{M, S\}, \tag{4}$$

$$C_d = \{T, A\}. \tag{5}$$

The J defines the testing process, E_t defines the environment and behaviour factors, S_d defines the data functionality, physical (P) and logical/semantic (L), T_d comprises techniques used for testing M in algorithm structure S and the document type and profile information is defined in C_d, where T defines the document type/categorization in a given scope A.

In the next subsection, this formal model is demonstrated with a test scenario, where the 'iterative testloop technique' (ITL) is utilized to tune the algorithm/module behaviour and to find limit criteria for the algorithm performance (Fig. 5). The proposed ITL is scalable from the module to system level, where the meta-result analysis can be given a significant control in loop decision making.

3.2 The ITL testing technique in DTM

In the given scenario, the environment effect E_t is used as an input parametrization in two entities: the input data preparation, and the algorithm/module behaviour setting. In practical terms, the input data, e.g. document image(s), the degradation levels/types and the document types and categories are set to match the required performance limitations in the target environment. Such simulated environment models are usually adequate for the module level testing and the integration of the different parts of the algorithm structure (Fig. 7).

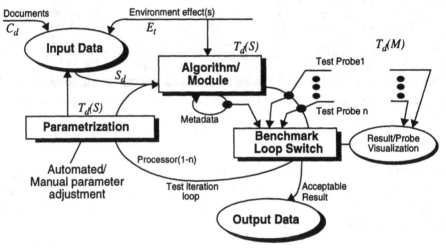

Fig. 5. The principle of the ITL technique.

In the first phase, the initial parameters are set either randomly, or defined a priori according to the demands of the test scenario. The parametrization module automatically changes the parameters according to testers settings, if more than one iteration is to be executed.

The iteration decision is made after each execution of the test module using a set of 'test probes' and a 'benchmark module'. Each probe measures one property that is utilized in the algorithm result visualization in a given scope, and making a looping decision. The loop decision can then be made (Fig. 6) - either the result is acceptable or the result is not acceptable. If the result is acceptable, the final test report is created and visualized, or higher (e.g. system) level tests are enabled with the (optimal) module results. If the result is not acceptable, the user can make use of the system in two different ways. First, the tester can use a set of ground truths that are used for the loop decision. The ground truth based loop decision is based on values with the limit(s) for acceptability of the overall (algorithm result) performance goal. Such ground truth can be an OCR performance for a document image, with a success ratio of the correctly recognized characters, for example.

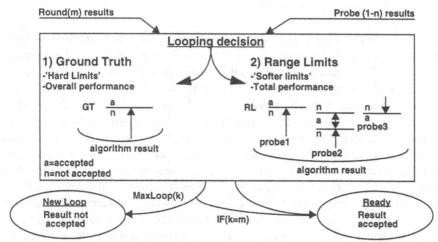

Fig. 6. Loop switching in ITL.

Second, the range limits based loop decision can be determined using performance indicators gained from probe measurements. The individual probe properties can be set in various ways as depicted in Fig. 6. In range limits decision, the 'new loop' decision is chosen, if the probe limits are not met. The decision is softer, since the tester can shrink or widen the acceptable limit(s), compared to ground truth decision. An example of the range limits probe set is the black-to-white ratio, the number of connected components and the average noise detected from the image.

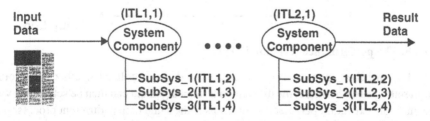

Fig. 7. System level view of ITL integration processes.

The loop values are computed and the tester can set a maximum number of iterations used. This also provides a timing constraint, although in DTM the different loops can be set to run concurrently in different processors.

When the looping decision is made and a 'new loop' is selected, reparametrization is performed. The current parametrization is adjusted automatically or manually depending on the testers settings. In parametrization, new processor(s) can be selected to perform preparation for new input data and algorithm execution, if needed. Manual adjustment is made using a set of hand tuned parameters, while the automated parameter settings follow a step-by-step series, or a defined function. Special ending criteria can be set, if the tester wants to manually modify the parameters when certain point in algorithm is reached.

3.3 The performance gain using ITL

Fig. 8. depicts a typical performance increase with the looping and parameter readjustment. The algorithm start performance is 50% with the ground truth set of input document images. The tester has limited the number of loops to k=7. In the second iteration (m=2) the performance decreases, and the system parametrization automatically initiates a new loop with 'different' parametrization. In the seventh iteration the performance exceeds the set acceptability threshold level, and the iteration ends with success. The tester executes new loops setting the parametrization to same direction (m>7), and tracks the result round after round. The performance decreases in the 9th round, indicating the 'peak' in performance in the 7-8th round parametrization. Since the result is gained using the test documents ground truth values, it only shows the overall performance that the algorithm is able to show using the given input documents. Still, this approach is enough to compute the overall system performance (algorithm by algorithm and integration latencies), or the algorithm's expected influence using the certain parametrization/need to change parameters with different inputs.

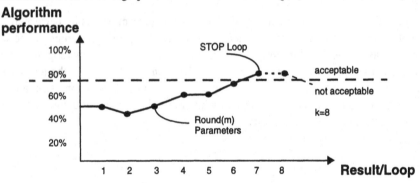

Fig. 8. The performance tracking in ITL.

The range limit approach is more complex, and can be utilized together with ground truth settings. Furthermore, the distributed test management can then be set to cope with the multiple execution/parametrization loops at the same time in different processors in a Unix network environment. The DTM server provides the necessary synchronization in loop decision phase for producing the simultaneous execution results.

3.4 Input data organization in DTM

The input data is typically comprised of document images from the acquisition or preprocessing unit, and attached attribute or environment information. The DTM supports natural document objects within and between properties. We have developed a document categorization with a set of ground truths for each category/subcategory presented in [6]. These are directly available as input (object) resources for test events. Furthermore, selected documents or entire categories can automatically be used in testing using a container arrangement (Fig. 3b). The documents in event containers are then fed into the testing process one by one, or concurrently when different processors are available.

---

1) Pure ground truth document images
2) 1)+different categories
3) 1)+one degradation in one degree
4) 1)+one degradation in several degrees
5) 2)+one degradation in one degree
6) 2)+one degradation in several degrees
7) 1)+several degradation types in one degree
8) 1)+several degradation types in several degrees
9) Variations 1-8)

Fig. 9. Examples of the different input object scenarios supported by the DTM.

Using object containers, several different input data scenarios can be utilized in different phases of algorithm development (Fig. 9). These enable use of a single (pure) document image up to utilization of several thousand documents with multiple cumulated degradations and phases. Our test image categorization includes synthetic document images and their ground truths designed for document image binarization algorithm benchmarking [8], and special cases for testing the multiple/special skewed documents.

3.5 Environment preparation

The degradation methods are important in determining the algorithm performance limits and to mimic the actual target environments effect on input data objects [9]. The DTM offers a special collection of different degradation/data enhancement algorithms, ranging from simple noise filters/addition to the effect produced by compression methods (Fig. 9). This section also includes such 'standard' document analysis tools such as binarization [10], skew detection [11, 12] and correction, and page segmentation and classification [13], for example. The utilization of these 'functional input objects' follows an object paradigm similar to that of the document objects. It is convenient to collect the similar functionalities into their own containers, e.g. image filters and the attribute extraction units, or probes for that matter. This helps to analyse and use the analysis information in context. The dedicated containers can be set to prepare the input documents for different processors at different times. This is advantageous especially in large tests, when hundreds or even thousands of documents with variations in degradation settings are used. The preprocessing can be very time demanding, and therefore input sets can be prepared and temporarily be stored to be used later, for example in ITL scenario, Fig. 10.

3.6 Results and experiments

The DTM system provides a means to design, control, and execute complex test scenarios and events. This is accomplished using the NC computing platform based on client/

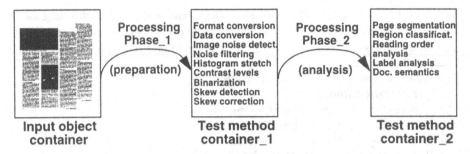

| Input object
container | Test method
container_1 | Test method
container_2 |

Fig. 10. An Example of input object container contents.

server Java. The data formats and methods used in tests are not dependent on the implementation platform, and different types of executable algorithms can be used. The DTM allows a seamless distributed computing over the internet and intranet networks. The client objects, and the test projects are controlled using pure object paradigms, and benefit from, for example, recursion as a test event.

When developing a new algorithm familiar concepts like object recycling and inheritance-persistence are important building blocks. These concepts help to reduce the research time substantially, after the initial arrangements have been done. The approach of experienced researchers is to simulate or realize the algorithms environment, e.g. tailor the environment from input objects to test result evaluation visualization into the DTM workbench. Then, the researcher is able to develop the 'missing pieces', and each time he/she gets one ready can domains automatically be tested for the performance and the robustness, having full control over the current/history result in the test database.

We have experimented with several research projects with the DTM platform and testing concepts. The main benefits include the reduction in piloting time, the visibility over the whole system in development, the automatic execution and testing routines, and especially the document analysis/image processing focused development platform, having all necessary resources available.

In algorithm development, the DTM was used, for example, in developing a document image binarization technique, described in [10]. Using the DTM test processes, we were able to reduce the number of parameters needed, by automatically detecting behaviour and the effects of each parameter change in the result image. Furthermore, we were able to track the effective range with certain parameters and limit their scale to those limits. The most important feature in this kind of controlled testing process approach is that the researcher is able to track the weaknesses immediately in complex algorithm procedures, using, for example the probe technique, and the mass testing of different documents from different categories with different degrees of degradations. Such complex algorithm constructions, having multiple features, scopes and computing conventions usually hide the errors, or the reason for poor algorithm behaviour. The errors are difficult to find from the code, while the DTM offers different metrics, and test result visualization in graphical format.

In the binarization algorithm development time, such visualization made a significant reduction, when analysing the shortcomings, and yet to be developed module properties.

4 Conclusions

In this paper we have proposed a new approach to manage document analysis algorithm and cross-domain testing. We have presented a distributed test management system (DTM) in earlier work that facilitates network computing, object-oriented technology and distributed computing concepts. Document processing systems demand special techniques for testing at different levels and connections to other domains. We have therefore developed a framework to enable design, creation and management of tests in different project and application levels. In this paper we have concentrated on presenting special testing techniques and their management on document analysis research projects. The testing issues are presented in different entities and at different levels; test project construction, test scope, visibility, result analysis and management and techniques to aid research and development of separate researchers working on a same project framework. Different techniques are presented to deal with the casing and scope of test and benchmarking that is aimed to refine research work and promote black-board type algorithm development from test management point of view.

Acknowledgements

The financial support of Academy of Finland and Technology Development Centre and the United States Department of Defense under contract MDA 9049-6C-1250 is gratefully acknowledged.

References

[1] Srihari S., Shin Y., Ramanaprasad V. and Lee D. (1995) Name and address block reader system for tax form processing. In: Proc. of the 3rd International Conference on Document Analysis and Recognition, Montreal, Canada, 1:5-10.

[2] Dengel A., Bleisinger R., Fein F., Hoch R., Hones F. and Malburg M. (1995) OfficeMAID - A system for office mail analysis, interpretation and delivery. In: Spitz L & Dengel A (ed) Document Analysis Systems, 1:52-75. World Scientific Press, Co.

[3] Sharpe M., Sutcliffe G. and Ahemd N. (1994) Implementation of an intelligent document understanding and reproduction system. In: Proc. of the IAPR Workshop on Machine Vision Applications, Kawasaki, Japan.

[4] Casey R. and Ferguson D. (1990) Intelligent forms processing. IBM Systems Journal, 29(3):435-450.

[5] Taylor S., Fritzon R. and Pastor J. (1992) Extraction of data from preprinted forms. Machine Vision and Applications, 5:211-222.

[6] von Mayrhauser A. (1990) Software Engineering: Methods and Management. Academic Press Inc., pp. 435-497.

[7] Sauvola J., Doermann D., Haapakoski S., Kauniskangas H., Seppänen T. and Pietikäinen M. (1997) A Distributed Management System for Testing Document Image Analysis Algorithms. To appear in ICDAR'97, Ulm, Germany.

[8] Nieminen S., Sauvola J. and Pietikäinen M. (1997) Benchmarking System for Document Image Binarization. Submitted.

[9] Kanungo T., Haralick R.M. and Phillips I. (1993) Global and local document degradation models, 1993 IEEE, pp. 730-734.

[10] Sauvola J., Seppänen T., Haapakoski S. and Pietikäinen M (1997) Adaptive Document Binarization. To appear in ICDAR'97, Ulm, Germany.

[11] Sauvola J. and Pietikäinen M. (1995) Skew Detection Using Texture Direction Analysis. In: The Proc. of the 9th SCIA, Uppsala, Sweden, 1:1099-1106.

[12] Sauvola J., Doermann D. and Pietikäinen M. (1997) Local Document Skew Detection. In: The proceedings of West-Coast SPIE, Document Recognition Systems IV, California, USA, 3027:13 p.

[13] Sauvola J. and Pietikäinen M. (1995) Page Segmentation and Classification Using Fast Feature Extraction and Connectivity Analysis. In: The Proc. of the 3rd ICDAR, Montreal, Canada, 2:1127-1131.

Graphical Tools and Techniques
for Querying Document Image Databases

J.Sauvola, D.Doermann[*], H.Kauniskangas,
C.Shin[*], M.Koivusaari and M.Pietikäinen

Media Processing Team
Machine Vision and Media Processing Group
Infotech Oulu, University of Oulu
FIN-90570 OULU, FINLAND
e-mail: jjs@ee.oulu.fi

[*] Language and Media Processing Lab.
Center for Automation Research
University of Maryland
College Park, MD 20742-3275
e-mail: doermann@cfar.umd.edu

Abstract

This paper describes document models and relations for the retrieval of document images. The underlying methodology was developed for the Intelligent Document Image Retrieval System (IDIR), which aims to extend document image database query capabilities. Traditional component type and keyword features are insufficient in describing logical and structural aspects of documents nature.

We have developed the necessary object-oriented document models to carry out complex multi-domain retrieval scenarios. In this paper we focus on retrieval capabilities and underlying methodology that supports different schemes. For these models and query schemes (QS), new graphical techniques are introduced. The IDIR allows complex combinations of different QS's, using the extended concept of 'frame logic', developed in our earlier work for the attribute management. Furthermore, a concept of document similarity is introduced with relations to QS's, document models, structure and role of use. Examples are shown for the retrieval of document images from University of Washington Database.

Keywords: Document image retrieval, document image query, query scheme, document similarity.

1 Introduction

Information retrieval systems are emerging as important tools in many different fields and for many different media. Much of effort in this area has been devoted to the development of new algorithms to provide the underlying functionality to deliver the information requested by users. Such capabilities are especially important with visual media such as document images, scene images and video, where traditional keyword type queries are not sufficiently expressive. Digital library initiatives are one focus area where researchers are trying to build more efficient and flexible mechanisms for information retrieval in a multimedia environment.

In the document community, there is a recognized need to advance beyond retrieval based solely on ASCII content. Although current document management and database products offer well-designed text search and retrieval functions, they typically ignore the document's physical and logical structure. Furthermore, even though documents have evolved into complex entities, few systems address the problem of how document analysis, image based querying and conventional database management and retrieval tools can be combined to form more robust searches.

In this paper, we lay the groundwork for a system which allows users to interactively define and combine queries using the structure and content of a document. In particular, how the underlying document models and relations between subdomains are presented, which aid in building different query schemes (QS). A new way to combine QS's is presented using frames. The use of frames and frame logic (FL) in the attribute level is described in our earlier work. We now extend the FL to QS level by introducing underlying techniques that enable different query schemes to be freely mixed.

The following sections describe the models, component relations, query schemes and examples with complex retrieval cases with different scopes and document image databases.

2 The system components

One of the primary focuses in the IDIR system has been on the development of a document image retrieval architecture for research and application development. The goal is to provide flexibility in the design to support a wide variety of document models and processing capabilities.

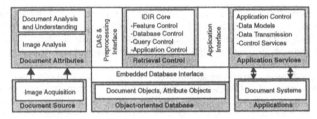

Fig. 1. Overview of IDIR architectural domains and components.

By doing so, the system will allow the custom design of modules and rapid application development in the later phases. The intelligent document image retrieval architecture, IDIR, consists of four functionally separate modules to allow easy modification. Fig. 1 depicts the basics of the architecture presented in [1].

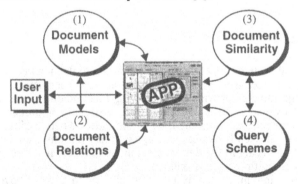

Fig. 2. Domain resource sharing and employment.

The underlying methodology is strongly object-oriented due to suitability to complex data presentation (data structure, relations and physical size). The document models and related topics are described in Section 3. We define and re-emphasize four underlying areas on top of the ones presented in our earlier work [1] that affect the query tailoring and retrieval components. These areas are (1) OO-document models, (2) OO-document relations, (3) document similarity, and (4) query schemes. Areas (1-2) center on the database and modeling aspects that dictate the degrees of freedom of the document (sub)entities used in a query formulation. (3) is related to the document models, to how to use (1-2) and the document analysis results in building an optimal description of the document entity. (4) involves multiple query schemes that use (1-3) in different models that are structured as frames. We then extend the 'Frame Logic' on two levels by considering (i) attribute frames, and (ii) query frames. The former was introduced in [1]. The latter forms a new concept, where multiple QS's can be used at the same time with a controlled inference into other defined features in multiple levels. The domains (1-4) are depicted in Fig. 2.

3 Databases and models

Applications in which document images are captured, organized, stored and retrieved, require relevant information to be maintained in databases. Object-oriented technology can enhance application development by introducing new data modeling capabilities and programming techniques [3]. They organize code into objects which incorporate both data and procedures, and provide natural query and retrieval mechanisms.

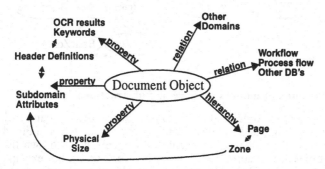

Fig. 3. Document entities vs. complexity.

In the IDIR system, the quality of the representation of a document is critical, since the query tools are dependent on knowledge obtained from the document image. Fig. 3 depicts document object entities and their complexity.

It is possible to provide reasonable responses to specific queries with an expressive query language and with an efficient document representation. By combining queries we can obtain generic and productive retrieval solutions to more complex retrieval problems. In IDIR, we use a native object-oriented database, Objectivity/DB, to store document and attribute objects. The following subsections describe the document database and the document models used.

3.1 Document database

Object-oriented databases are designed to support object models generated with object-oriented languages such as C++. There are two types of data stored - the document image data and the features computed by the processing modules. The document image and its features are stored in the OODB as objects, which has general management characteristics, such as persistence, concurrency and transactions, distribution, queries, a data dictionary and a namespace.

The image data is used not only by IDIR, but by external applications as well. The computed features include document, page and zone attributes, physical and logical structure and a content representation, for example. The document object relations are formed within the database. An example of such a relation is an attachment of a newspaper document to its category definition.

One of the most important properties of an object-orientation in the database organization is the support for user-defined abstract data types, where complex (aggregated) objects are formed from simpler ones by using inheritance mechanism. This approach is suitable for document images, since documents comprise several subcomponents in different levels of abstraction. Inter-document relations are supported in the document model. Their design comprises document objects with their own relations and characteristics.

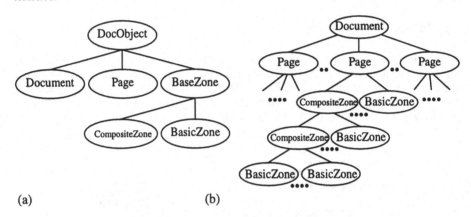

(a)   (b)

Fig. 4. IDIR document model structure and abstraction.

Document objects are organized using an inheritance hierarchy and include objects, such as document, page, and composite and basic zones as shown in Fig. 4. The corresponding document or component specific data is embedded to own hierarchy with defined properties and relations to other objects (see Fig. 3). For example zone data, such as ZONE_ID, ZONE_TYPE and FONT_TYPE are embedded to their own specific zone class, whose attributes and relations are defined for the page object and other zones in the object aggregation hierarchy. The data can then be accessed using the hierarchy relation by using the document structure model.

Provisional and new relations can be established using management tools. Since we use federated databases with pure OO-models, the new properties (attributes-relations), behavioural models and support for new query schemes are supported (Fig. 5).

3.2 Document models

The object-oriented document model encapsulates a set of data and associated methods, where the specifics of data structures are hidden. This simplifies access to the documents inner structure and contents, and provides independence for the object, since the internals can be modified without affecting external contents.

The objects are created and initialized according to predefined document models. The preprocessing step then defines the data and stores it with the corresponding object. The component addressability and identity are defined when creating objects and provide a means for access. In the IDIR document model these properties are specified for each object class and level, enhancing the way that documents and their subcomponents (page, zone, pixel, etc.) can be accessed and used. The identity property is unique to every document image and attribute object and is used to distinguish the object from the others in database. The feature information of each database image object is available for queries. An example document object instance is shown in Fig. 4b.

The document is divided into pages according to page content relations. Each page contains zones, whose properties and inter-relations are defined accordingly. The zone can include objects down to a pixel and supports the document analysis presentation paradigm (Section 2). An example of the page level parametrization (content attributes + structure parameters) is depicted in Fig. 7b. The abstract document model is designed to support not just conventional document structure and content, but also the document query formation approach, where the query specifics are inserted to utilize different document abstraction and content levels. The model, therefore, is built to support flexible integration of representation data, such as different hierarchical instances, Fig. 5.

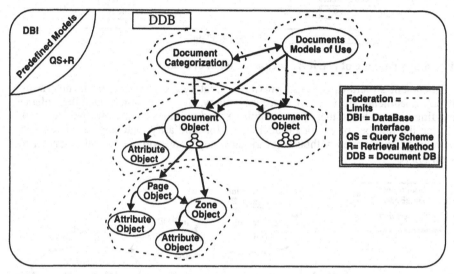

Fig. 5. Document model in Object-oriented database.

4 Query and retrieval

Given the framework for a document database system, the similarity between documents can be defined at many levels with respect to both the physical and logical structures, relations between objects, and the content.

More formally, similarity is a measure of relatedness between two objects, and is a function of their structural similarities and the content similarities. Content similarity is a measure of relatedness between the properties or attributes of the content of two basic objects, and structural similarity is a measure of relatedness between the organizations of two composite objects. The computation of content and structural similarity measures are typically dependent on the features describing them and the representation scheme used for their structures respectively [6]. Fig 6. depicts a retrieval scenario construction. The scenario includes constituent frames having their own query scheme, or simply query parameters. Each scheme is built using similarity groups, or classes for defined properties.

To support content queries, images are analysed to extract quantitative and qualitative descriptions of their content to be stored as index information in the database, together with the image. These descriptions are application domain-dependent and are searched to quickly narrow down a number of candidate images which images may satisfy the given query, Fig. 7. Further processing may be performed to find images that satisfy specifics of the query. Ultimately, the effectiveness of an image database system depends on a number of factors including image content representations, the similarity or dissimilarity measures, the range of image queries allowed, the effectiveness of the search interface, and efficiency of the implementation [4].

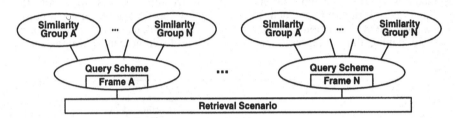

Fig. 6. A principle of retrieval scenario construction.

Document structures can also provide a great deal of information in determining which documents are relevant to a given query. Structure-based matching enhances existing content-based matching capabilities, and provides an effective way to quickly reduce the number of candidate documents for similarity matching. Structure-based matching is performed without a priori knowledge, and without optical character recognition (OCR).

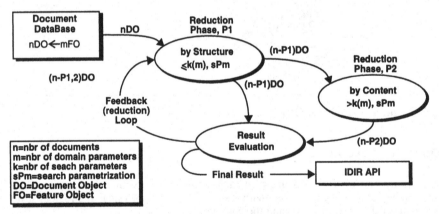

Fig. 7. Principle of retrieval image search base reduction.

In [1], we defined an approach to measuring similarity. The structural similarity of two documents can first be approximated by their constituent regions and their types (text, graphics or image). For each region in the query image, we match each query region to each region of the database image of the same type and overlapping it. Once this first correspondence has been established, an evaluation mechanism is used to refine and measure the quality of the match. It is clearly possible for a single region in the query image to be mapped to multiple regions in the database image and vice versa. There are several situations where such a mapping is not desired, and must be refined.

The first restriction is that no region should be mapped to two or more regions in the horizontal direction. This would occur, for example, if a page with a single block of text were mapped to a page with two columns of text. Splitting a block horizontally may occur between paragraphs, for example, but vertical splitting is typically an intentional structural occurrence. For query regions which map to more than one corresponding database region, a subset of regions which have maximal intersection but do not neighbour horizontally, is chosen and the remaining regions are removed from the mapping. For query image regions which overlap a single database region, the correspondence is trivial (but we must later consider a symmetric case where multiple query regions correspond to a single database region).

Once this condition is satisfied for all query regions, the symmetric case, where a single database region corresponds to multiple query regions, is still possible. Using the restricted mapping, a reverse mapping is constructed from the database image to the query image and the condition is checked again. Regions which violate the vertical split are again evaluated and the subset of maximal overlap is kept. This is the technique used for query by sketch in Section 5. A more complete discussion of structural similarity of document images is presented in [6].

5 The graphical user interface and tools

We have developed a graphical user interface and a set of related tools to take full advantage of the processing capabilities, the database and the architecture of IDIR system, Fig. 4.

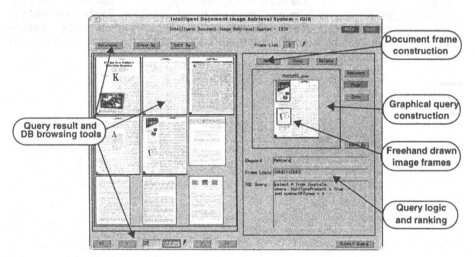

Fig. 8. The IDIR graphical query user interface.

The main design concept in graphical user interface development is centered on functionality, where different interface objects can be made explicit and combined interactively to form visual query specifications for the retrieval of document images.

It is not sufficient to provide access to document attributes alone. A system should also provide the ability to combine selected properties graphically, especially when these properties have spatial meaning. In our interface the user can visualize and construct complex queries which may extend over multiple levels of the document hierarchy. The interface consists of resources which are used to construct and execute queries, view the results and browse the resulting images. Each component works interactively, and iterative refinement of the query can be realized. The main components (or interface objects) are the image browsing tools, the query results tools, query construction and the execution tools.

5.1 Query formulation

The user can move information (including images and attributes) between the browsing and query construction components by selecting the target object. The query construction is based on the formation of document image frames, that can be adapted from an imported document (for query by example), a database object (for component query), or an empty template (for query by sketch). From these frames, attributes can be set by selecting regions in free hand draw mode, by selecting document, page or zone level attribute definition objects or by selecting readily stored database attribute object to specify the properties of the documents to be retrieved. When the frame is set and its properties are defined, it is used to form the database query. Different query types, such as query by example, query by user example and query by selected attribute(s) can therefore be managed efficiently and combined to build complex queries.

We have developed a way to further refine the query information specified in image frames. Frame logic offers simple logical operations used to define relationships between image frames. The user can combine defined multiple properties or query schemes by using, for example, the logical "and" operation ('&') between frames A and B (the properties defined in both frames must be valid to satisfy the match) or with the logical not operation ('!') as in A&B&!C (the properties in A and B must be valid and the properties defined in frame C must not be satisfied). Fig. 7a shows a possible query construction scenario. In the example, the user begins with an empty instantiation of a document to be queried (Document), then defines the zones and attributed within the first frame (Frame {A}).

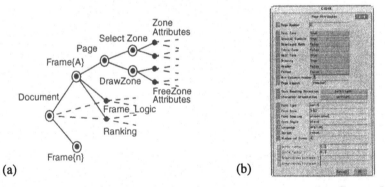

(a)   (b)

Fig. 9. Examples of options for integrated query construction flow and page level attribute definition window.

Other frames are set, using zone, page or document level properties. The page level attribute window is shown in the Fig. 9b. A single document frame may contain a lot of information set by user and exploit a unique query scheme. With the help of multiple frames the query definition can be simplified since the user can divide the complex query settings to several frames with simpler content. The frame logic can then be used to equalize these constituents of a query by applying the logical '&' between the frames. When the desired attributes are set, the frame logic operations, i.e. relations between image/query frames, can be set as described earlier. If desired, a set of keywords can also be incorporated into the search. The keyword queries ideally operate using a fuzzy matching on the often borderline acceptable OCR results, making it possible to perform content-based searches and retrieve the image. We have also included an SQL-syntax query interface to our user interface, allowing the user to type queries directly.

5.2 Ranked retrieval

A ranking procedure further emphasizes the definitions set to form the query definition object (QDF), Fig. 8.

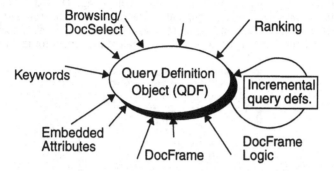

Fig. 10. Schematics of 'query definition object' with GUI.

Ranking can be accomplished in a number of ways [5]. For structural queries, the rank of a document is simply based on the numerical similarity between constituent components. For keyword queries, a standard way to rank documents is to provide a boolean retrieval system using a weighted vector model. Each document is represented by a term vector which is weighted by the inverse of the number of times the particular term occurs in the entire collection. The rank of the document is simply the sum of the normalized frequencies of the terms which appear in the keyword list. We are also considering an extended boolean model which may be better suited for large collections which have poor quality [2].

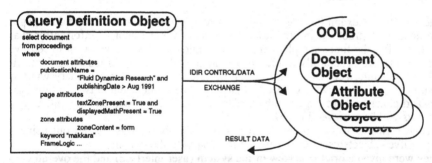

Fig. 11. IDIR query definition object, QDF contents.

The query definitions are formatted in SQL and represented as a query as depicted in Fig. 7. The properties and their relations are presented in SQL-format, using our extension (see [1] for details). An example of the query object is shown in Fig. 11. After the query definitions are set, the user submits the query for processing. The FCM (feature control module) receives the query and transmits the query to selected database using the database interface connection, (see Fig. 5)

The query result is shown in the browsing section of the graphical user interface. The user can limit the number of images to be fetched from database. In the current implementation, all results are first fetched as thumbnail document images.

6 Example retrieval scenarios

To test the robustness of the system, we have formulated several example retrieval scenarios. The queries are run against a database populated using the technical article images from the University of Washington Document Image Databases and a collection of multi-lingual memos. The technical article images are not full documents, so they are treated as individual, independent images. Each image has associated zone information, attributes such as content type and content properties for text regions.

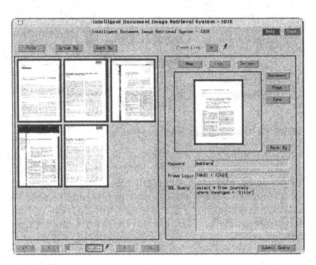

Fig. 12. Example results.

The first set of examples was designed to reduce the number of candidate pages using document structural information, before any further semantic information was considered. For the first example, consider a simple query where we want to retrieve title pages that are structurally similar to an example title page specified using the IDIR graphical user interface. One simply pulls the example image into the query window and submits it. Suppose we then decide that we are only interested in two column title pages and pages without page numbers. The query can be refined and submitted, and the results will be retrieved Fig. 10.

Second series of tests, similar to those described above, was carried out using a group of five subjects/users who were not familiar with the system.

They were given a brief overview of the system (user interface) and the overall contents of the database, i.e. what kind of documents were stored and could be retrieved. They then used the IDIR GUI to visually form the queries on the document database consisting of 1,079 document images generated from University of Washington Databases. Examples of the retrieved results are shown in Fig. 11a and 11b. As seen from the verbal description of the original queries, the document descriptions include both semantic and structural components. The latter definitions are easy to present with numeric or alphabetic information. On the other hand the semantic content is difficult to set with numeric information. The graphical user interface objects and functions were utilized to perform semantic mapping. In 'Case 2' one user prepared two document frames. The frame A was an example document from database document image, where the page number and graphics region was selected. Their zone tags were inverted with frame logic, e.g. '!A'. Then, the frame B was constructed from empty frame template. The user selected and defined title and abstract zones, defined column number attribute in user interface, and typed in the keyword 'keyword'. The frame logic for frames A and B was then !A&B.

Case 1: "I want to find documents with more than one graphical areas, where first one is located on top, and second one is located in the middle of page. The figure in the middle has a text region that describes the figure contents."

Case 2: "Find me documents that has two columns, a title, a text keyword with keyword fields and no page number, the title is in the upper part of the page and abstract (preferably) in the middle."

(a)

Case 3: "Find documents, where first a table is present, under which a graph is drawn from the numbers in the table. The describing text is one column. The language is English."

(b)

Fig. 13. Example results of queries performed using IDIR.

Conclusions

We have proposed an approach and underlying mechanisms for querying document databases. An architecture and set of graphical tools was presented for dealing with query formulation and complex document image retrieval issues. We use object-oriented methodologies and techniques for architectural design, implementation and document modeling. An object-based document model is presented which specifies document attributes at the document, page and zone levels, offering efficient retrieval definitions for document's structure and content.

The document similarity issues and methods are discussed for querying document databases and document abstractions are presented to form efficient object-oriented databases. We have demonstrated the system functionality, capabilities and performance with a series of retrieval examples using a database of more than a thousand page images created from the University of Washington test collection.

Acknowledgements

The financial support of Academy of Finland, Technology Development Centre, GETA and the United States Department of Defense under contract MDA 9049-6C-1250 is gratefully acknowledged.

References

[1] Doermann D., Sauvola J., Kauniskangas H., Shin C., Pietikäinen M. and Rosenfeld A. (1997) The development of a general framework for intelligent document image retrieval. A book chapter in Document Analysis Systems II, Series in Machine Perception and Artificial Intelligence, 28 pages.

[2] Information Retrieval: Data Structures and Algorithms, William B. Frakes, and Ricardo Baeza-Yates (Eds.), Prentice Hall, Englewood Cliffs, NJ, 1992.

[3] Rao B.R. (1994) Object-oriented databases: technology, applications, and products. Database Experts' Series, McGraw-Hill, 253 pages.

[4] E.G.M. Petrakis and C. Faloutsos. Similarity searching in large image databases. Technical Report CS-TR-3388, University of Maryland Institute for Advanced Computer Studies and Dept. of Computer Science, Univ. of Maryland, December 1994.

[5] Gerald Salton, and Michael J. McGill, Introduction to Modern Information Retrieval, McGraw-Hill, New York, 1983.

[6] Christian Shin, David Doermann, and Azriel Rosenfeld, Querying Document Image Databases using Structural Similarity, Technical Report, Center for Automation Research, University of Maryland at College Park (in preparation).

Handwritten Numeral Recognition via Fuzzy Logic and Local Discriminating Features

Natanael Rodrigues Gomes[*]
Lee Luan Ling
Decom-Feec-Unicamp
13083-970 Campinas, São Paulo, Brasil

Abstract: This paper describes a system to recognize disconnected handwritten numerals based on the concept of fuzzy logic and discriminating local features extracted from numeral images. Initially, the skeleton of an unknown numeral is obtained and decomposed into several segments called branches. The branches, due to their nature, present fuzzy characteristics in terms of their straightness and orientation. Precisely the three fuzzy sets were defined and used to classify branch segments into straight line segments, parts of circles and circles. The membership grade functions are built for character branches and their values are computed for the sequences of pattern branch features which represent numerals. A numeral image is classified to sequence of branch pattern features with the largest overall membership value. In the case of tie, some local topological features such as the number and the position of end points, intersection points and bend points, are used for the classification.

Keywords: Fuzzy logic, discriminating local features, disconnected handwritten numerals and image processing.

1 Introduction

Automatic handwritten character recognition is by no mean a trivial task due to a great amount of variations in the form presented even by the same characters. In literature, a considerable number of character recognition algorithms have been proposed. Many of these algorithms distinguish one to each other mainly on extracted features and/or the classification procedures. Among proposed recognition methods, there are statistical analysis [AU92], contour analysis [PA75], transform techniques [SB84], neural approaches [FM82] and fuzzy logic [Ped90], [SC74], [WW94].

In this work, we describe the implementation of a recognition system for disconnected handwritten numerals based on the concept of fuzzy logic and local discriminating features extracted from numeral character images. Each numeral to be identified is initially reduced to its skeleton format and then segmented into primitive elements, so-called branches [SC74]. Here we introduce fuzzy characteristics to the degree of straightness and orientation. Fuzzy sets, therefore, were used for classifying branches of the character into a straight line segment, an open curve or a circle according to membership grades. The character recognition then is performed by finding one of many sequences of branch patterns previously saved in the system memory, which best matches to the sequence of branch patterns of the being tested character image. In other words, the numeral image is assigned to the numeral class which provides the largest membership grade. If two or more sequences of branch

---

[*] Work supported by São Paulo State Foudantion - FAPESP

pattern features simultaneously hold the same largest membership grade, the final classification will be then based on character's local discriminating features, such as number of end points, of intersection points and many others.

The character recognition method proposed in this work can be divided into the following steps:

1. Numeral character image acquisition;
2. Image pre-processing and thinning;
3. Detecting and labeling the nodes and branches in the skeleton of the numeral image;
4. Calculation of the membership grade of each branch segment;
5. Extraction of local discriminating features;
6. Classification.

Fig. 1 shows the procedure of numeral character recognition in form of a block diagram. Next, we describe briefly each one of these steps involved in the recognition process.

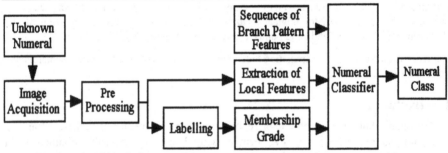

Fig. 1. The functional block diagram of the proposed handwritten character recognition.

2 Handwritten Character Image Acquisition

The digital images of numeral characters considered in this work are binary under 180 dpi resolution. Fig. 2 shows some numeral sample images. After being digitized, each numeral image is undergone to an enclosing process which determines the smallest rectangle capable of covering all black pixels in the input image. Fig. 3 shows two images of numeral "7" before and after enclosing.

3 Pre-Processing

The image pre-processing procedure in our numeral recognition system consists of the elimination of noises present in the digitized image. These image processing operations have the goal of improving the quality of the numeral images, therefore, obtaining high quality in the extracted features and segmented branches.

Fig. 2. Representative samples of numeral images.

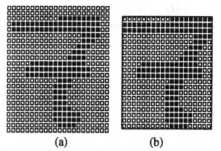

(a)   (b)

Fig. 3. Enclosing of numeral "7". a) Original character. b) Character after enclosing.

3.1 Elimination of Noise

The routine for noise elimination consists of removing a black pixel (bit 1) or a white pixel (bit 0) in numeral images under certain pre-defined conditions. In other words, a black pixel in an image is removed if an extracted 3-by-3 window from the image with a black pixel in the center of this window matches perfectly anyone of three masks shown in Fig. 4.a, 4.b and 4.c. Conversely, a white pixel is removed if a 3-by-3 window from image matches anyone of the three masks presented in Fig. 4.d, 4.e and 4.f. Note that the mark "X" in the masks symbolizes that the pixels may be either black or white. In addition, the patterns obtained from rotating the masks in Fig. 4.b, 4.c, 4.e and 4.f in 90°, 180° and 270° degrees, are also used for the elimination of noise.

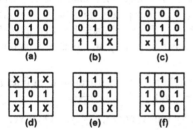

Fig. 4. The 3x3 masks for noise removing.

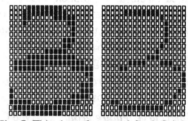

Fig. 5. Thinning of numeral 3. a) Original character. b) Character after thinning.

3.2 Thinning

The thinning algorithm is used to obtain the skeleton of the handwritten numeral strokes. The operational principle of the algorithm is to remove sequentially suitable contour elements of the given numeral strokes without losing the connectivity [AB81]. Fig. 5.a and 5.b show an image of numeral "3" before and after thinning.

4 Detection of Nodes and Branches

Once obtaining the skeleton of the numeral image, the system proceeds the detection of nodes and segments in the skeleton image. The detection procedure has its fundamental role in the recognition procedure since it defines basic elements used to identify primitive feature components in the recognition system. Fig. 6 shows types of nodes in a numeral "3", denoted by marks "E", "B" and "T", and branches.

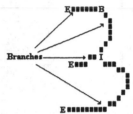

Fig. 6. Branches and nodes in a numeral 3.

4.1 Node Detection

Three types of points are defined and can be found in a character image [BK88]:

- *End points ("E"):* points which have only one black pixel neighbor.
- *Bend Points ("B"):* points which have two black pixel neighbors and where an abrupt change of line direction occurs.
- *Intersection Points ("I"):* points which have at least three black pixels in their contacting neighborhood.

4.2 Branch Detection

In general a branch in a character is regarded as a line segment connecting a pair of points as illustrated by the example of Fig. 6. A branch that does not present nodes is called *circle* and a branch with only one node is called *one-node circle*. In this work, we establish that a branch must possess at least four pixels. The classification of branches considered in this work, as shown in Fig. 7, has already been adopted by other researchers [BK88],[JR72]. Each branch was denoted by an alphabet which has the following meaning:

1) *Straight line feature* group: H= horizontal straight line; V= vertical straight line; P= straight line with positive slope; N= straight line with negative slope.

2) *Open curve feature group:* C= type C curve; S= type S curve; D= type D curve; A= type A curve; U= type U curve; Z= type Z curve.

3) *Circle feature group:* OL= circle on the left of the node; OR= circle on the right of the node; AO =circle above the node; OB= circle below the node; OO= circle without node.

5 Character Branch Classification

The character branches are classified according to the line segments defined in Fig. 7. For this end, the membership grade with respect to each of branch groups is evaluated. The membership grade is calculated considering the fuzziness of two conditions:

1. Evaluation of a branch with respect to its straightness: the membership grades are calculated with respect to a straight line segment and a curved segment.

2. Evaluation of a branch with respect to its orientation : membership grades are calculated with respect to the slope inclination. (0°, 45°, 90°, 135°, 180°, 225°, 270°, 315° and 360°).

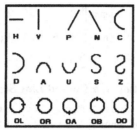

Fig. 7: Classification of branch for handwritten numeral characters.

5.1 Evaluation of the Straightness of a Branch

The evaluation of the straightness of a branch is based on the mean square error between the branch and a straight line candidate. The evaluation involves all black pixels in the branch and the minimum mean square error is used to select the best fitted straight line candidate [BK88].

Once obtaining the best fitted straight line segment, we calculate the distance between the branch segment and the found line via a factor of closeness (d_{SL}) defined as follows:

$$ dsl = \frac{1}{n} \cdot \text{Min} \left\{ \sum_{i=1}^{n} \left[1 - \frac{\left(\frac{x_i}{0.5}\right)^2}{1 + \left(\frac{x_i}{0.5}\right)^2} \right], \sum_{i=1}^{n} \left[1 - \frac{\left(\frac{y_i}{0.5}\right)^2}{1 + \left(\frac{y_i}{0.5}\right)^2} \right] \right\} $$

where x_i and y_i denote the differences between the coordinate position of ith sample of the branch and the straight line segment in the system's horizontal and vertical coordinate orientation, respectively, and n is the number of black pixels in the branch segment.

Based on the measure of closeness d_{SL} above, we define the measure of closeness d_c between the branch segment and a curved line as $d_c = 1 - d_{SL}$.

5.2 Evaluation of Orientation of a Branch

The evaluation of the orientation of a branch consists of finding the inclination of a branch in terms of the following angles: 0°, 45°, 90°, 135°, 180°, 225°, 270°, 315° and 360°. Next, we introduce three procedures which explicitly quantify the orientation of the branches previously classified as straight lines, open curves and circles, respectively [SC74].

a) Evaluation of the Orientation of a Straight Line Segment

In this case only the straight line feature group {H, P, V, N} is considered. The membership grade of a candidate branch with respect to each element of the straight line feature group is calculated as $\theta = tan^{-1}(a)$, where a is the angular coefficient of the straight line segment. For each element of the straight line feature group we find the membership grade as follows:

$$H(\theta) = 1 - \min\left\{\min\left[|\theta|, |180 - \theta|, |360 - \theta|\right] / 45, 1\right\};$$

$$V(\theta) = 1 - \min\left\{\min\left[|90 - \theta|, |270 - \theta|\right] / 45, 1\right\};$$

$$P(\theta) = 1 - \min\left\{\min\left[|45 - \theta|, |225 - \theta|\right] / 45, 1\right\};$$

$$N(\theta) = 1 - \min\left\{\min\left[|135 - \theta|, |315 - \theta|\right] / 45, 1\right\}.$$

b) Evaluation of the Orientation of a Curved Line Segment

For curve line segments, we can further divide the curve feature group, $\{C, D, A, U, S, Z\}$ into three sub-groups, namely the vertical curved feature group $\{C, D\}$, the horizontal curved feature group $\{A, U\}$ and the inclined curved feature group $\{S, Z\}$. To obtain the membership grade of a branch segment with respect to a curved feature, we first analyze the angular value $\theta = tan^{-1}[(y_n-y_1)/(x_n-x_1)]$. Note that (x_1, y_1) and (x_n, y_n) are, respectively, the coordinates of two end points of the being observed branch segment. Therefore, the angle θ is exactly the slope angle of the straight line segment connecting the two end points.

The membership function which characterizes horizontally and vertically curved branch segments is given respectively by:

$$HC(\theta) = 1 - \min\left\{\min\left[|\theta|, |180 - \theta|, |360 - \theta|\right] / 90, 1\right\};$$

$$VC(\theta) = 1 - \min\left\{\min\left[|90 - \theta|, |270 - \theta|\right] / 90, 1\right\}.$$

Note that an inclined curved branch, S or Z, may be either classified into a horizontal or vertical curve depending how inclined is the branch.

After a branch being classified into a horizontal or vertical curve, it can be assigned to one of the elements of the groups: the inclined curved feature group $\{S, Z\}$, the vertical curved feature group $\{C, D\}$ or the horizontal curved feature group $\{A, U\}$, by considering the following no-fuzzy information [SC74]:

- the number of intersection points between the branch segment and the straight line segment (LN) connecting two end points;
- the number of intersection points between the branch segment and some straight line segments parallel to the LN segments as shown in Fig. 8.

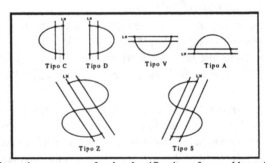

Fig. 8. Information necessary for the classification of curved branch segments.

c) Evaluation of the Orientation of a Circle Branch Segment

In this item, we describe the procedure for determining the membership of a branch segment with respect to the element *OL, OR, OA and OB* of the circle feature group. Once a branch segment is classified as a circle element with a node, the orientation (θ) of this circle branch is given by the slope angle of the straight line segment connecting the node and the center (x_c, y_c) of the gravity of the circle. Therefore, the membership grade with respect to four circle features are calculated as:

$$OL(\theta) = 1 - \min\left\{\min\left[\,|180 - \theta|\,\right]/\,90, 1\right\};$$

$$OR(\theta) = 1 - \min\left\{\min\left[\theta, |360 - \theta|\,\right]/\,90, 1\right\};$$

$$OA(\theta) = 1 - \min\left\{\min\left[\,|90 - \theta|\,\right]/\,90, 1\right\};$$

$$OB(\theta) = 1 - \min\left\{\min\left[\,|270 - \theta|\,\right]/\,90, 1\right\}.$$

Example: Through an example we elaborate three tables of membership grades for handwritten numeral "3" as shown in Fig. 9. Note that the handwritten numeral 3 has 4 nodes and 3 branch segments.

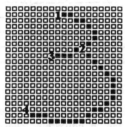

Fig. 9. The skeleton of a numeral "3".

Table 1. Membership values with respect to the degree of straightness.

Membership Values		
Nodes	Straight Line	Curve Line
1-2	0.3341	0.6659
2-4	0.1650	0.8350
2-3	0.8130	0.1870

Table 1 and Table 2 present the membership grade with respect to the degree of straightness and the orientation of the branch segments. Table 3, obtained from combining Tables 1 and 2, shows the relative membership grades of the orientation of character branches. In other words, the membership grades with respect to the branch orientation for the curved feature elements, {C, D, A, U, S, Z}, are multiplied by the membership grades of curved branches, while those of straight feature elements, {H, P, N, V}, are multiplied by the membership grades of straight branch segments. The handwritten character recognition therefore will be based on the information of Table 3 and some morphological discriminating features.

Table 2. Membership grades with respect to the orientation of branch segments.

Sets	Nodes			Sets	Nodes		
	1-2	2-4	2-3		1-2	2-4	2-3
H	.2	.455	.713	S	0	0	0
V	0	0	0	Z	0	0	0
P	0	.55	.288	OR	0	0	0
N	.822	0	0	OL	0	0	0
C	0	0	0	OA	0	0	0
D	.625	.533	0	OB	0	0	0
A	.374	0	0	OO	0	0	0
U	0	0.47	0	-	-	-	-

Table 3. Relative membership grades

Sets	Nodes			Sets	Nodes		
	1-2	2-4	2-3		1-2	2-4	2-3
H	.067	.075	.58	S	0	0	0
V	0	0	0	Z	0	0	0
P	0	.091	.234	OR	0	0	0
N	.275	0	0	OL	0	0	0
C	0	0	0	OA	0	0	0
D	.416	.445	0	OB	0	0	0
A	.249	0	0	OO	0	0	0
U	0	.389	0	-	-	-	-

6 Local Discriminating Features

The set of features described in this section will be used in parallel with the branch membership information for the handwritten character recognition. The local information in numeral character images has the goal of complementing the membership information and eliminating ambiguity among numeral classes in the classification process. As mentioned before, three types of local discriminating features are used:

• *Number of end points, intersection points and bending points.*

• *Positions of endpoints, intersection points and bending points with respect to the principal vertical and horizontal axes:* The vertical and horizontal principal axes are straight vertical and horizontal line segments passing at the center of gravity. The relative position of the endpoints with respect to the principal vertical axis can be either *left* or *right* while with respect to the principal horizontal axis can either *up* or *down*. Fig. 10 shows how the relative position of the endpoints in a numeral "5" are classified. It may happen that the vertical and/or principal axes pass exactly through an endpoint, to eliminate this kind of ambiguous situation here we assume that the endpoint relative positions with respect to the vertical and horizontal principal axes are right and up, respectively.

• *Number of b lack pixels in a specific region:* The thinning of numeral images reduces the thickness of handwritten strokes to one pixel. Under such a circumstance, the amount of pixels in some specific regions of the skeleton image of handwritten numeral is able to distinguish numeral classes from each other. We used this characteristic to distinguish numeral "3" from numeral "7" and numeral "5" from numeral "6". In Fig. 11 the regional feature is used to distinguish numeral "3" from numeral "7". In other words, there is considerable concentration of black pixels in the bottom line of numeral "3" while few black pixels are found for numeral "7". Fig. 12 shows how the regional feature is used to distinguish numeral "5" from numeral "6". There is no concentration of black pixels on left side of the down left end point in numeral "5", while some black pixels are found for numeral "6".

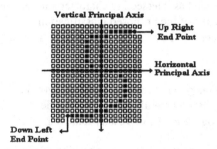

Fig. 10. Principal axes of a numeral 5.

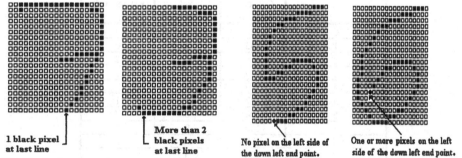

Fig.11. A situation where a regional feature is used to distinguish numeral "3" from numeral "7".

Fig. 12. The Regional features used to distinguish numeral "5" from numeral "6".

7 Classification

Once a numeral character image is decomposed into nodes and branches, as described in Section 2, the classification of the numeral character is based on the comparison between the character's branch pattern sequence and the reference characters' branch pattern sequences previously stored in the recognition system. All the reference branch sequences used to classify unknown numerals are presented in Appendix 1.

7.1 Classification Procedures

Next we summarize the procedures involved in numeral character classification:

Step I: Build a sequence of patterns which contain all branches extracted from an unknown character image;

Step II: For each extracted branch obtain its individual membership grade relative to each fuzzy set;

Step III: Classify the numeral character to the class of feature pattern sequences offering the largest class membership grade and also the best match in terms of the number of endpoints and intersection points;

Step IV: If two or more reference pattern sequences from different numeral classes present the same largest membership value, the local regional features of the character, such as the positions of end points, intersection point and bend points of unknown character, as well as intersecting characteristics are analyzed.

With respect to the Step I of the proposed classification algorithm, note that the order of appearance of branches in the constructed sequence does not matter.

7.2 Experimental Results

The handwritten numeral character recognition algorithm proposed in this paper is tested by 1258 handwritten numeral characters provided by 100 subjects. The testing result shows that a recognition rate of 92% is achieved.

Table 4. Summary of the testing results.

Numeral Classes	Number of Characters		
	Correctly Recognized	Not Recognized	Erroneously Recognized
0	120	3	9
1	133	1	0
2	115	4	7
3	139	1	9
4	116	4	2
5	127	2	1
6	104	0	7
7	102	9	5
8	77	20	8
9	120	5	8
Total	1153 (92%)	49 (4%)	56 (4%)

8 Conclusion

Due to fuzzy nature of features presented by the branches extracted from handwritten numerals, the fuzzy sets were used during the feature extraction process. The proposed feature extraction method establishes with respect to each fuzzy set a certain membership grade. This grade allows the classification of unknown handwritten numeral characters to a sequence of branch patterns. Since each branch extracted from a numeral image present a certain membership grade for each one of the considered fuzzy sets as illustrated in Fig. 7, it is possible to characterize the numeral character via a sequence of branch patterns in spite of variations of form that the numeral presents. Via fuzzy set representation, it is possible to reduce the number of pattern sequences used as well as the number of numerals not classified. Another important contribution of this work is a complement classification procedure using local or regional discriminating features.

9 References

[AB81]   C. Arcelli and G. S. Di Baja. "A thinning Algorithm Based on Prominence Detection". Pattern Recognition, Vol. 113, N° 3, pp. 225-235, 1981.

[AU92]   Al-yousefi, H. S. Udpa. "Recognition of Arabic Characters". IEEE - Trans. Syst., Man and Cybernetics, Vol. 14, N° 8, 1992.

[BK88]   G. Baptista and K. M. Kulkarni. "A High Accuracy Algorithm for Recognition of Handwritten Numerals". Pattern Recognition, Vol. 21, n° 4, pp. 287-291, 1988.

[FM82]   K. Fukushima and S. Miyake. "Neocognitron: A New Algorithm for Pattern Recognition Tolerant of Deformations and Shifts in Position". Pattern Recognition, Vol.15, pp. 455-469,1982.

[JR72]J. T. Tou and R. C. Gonzalez. "Recognition of Handwritten Characters by Topological Feature Extraction and Multilevel Categorization". IEEE Trans Comput., vol. C-21, pp. 776-785, July 72.

[PA75]   T. Pavlidis and F. Ali. "Computer Recognition of Handwritten Numerals by Polygonal Approximations." IEEE - Trans. Syst., Man, Cybernetics, Vol. SMC-5, N° 6, pp.610-614, November, 1975.

[Ped90]   W. Pedrycz. "Fuzzy Sets in Pattern Recognition: Methodology and Methods". Pattern Recognition, Vol.23, N° ½, pp. 121-146,1990.

[SB84]   M. Shridar and A. Badreldin. "High Accuracy Character Recognition Algorithm Using Fourier and Topological Descriptors". Pattern Recognition, Vol.17, N°5, pp.515-524,1984.

[SC74]   P Siy and C. S. Chen. "Fuzzy Logic for Handwritten Numeral Character Recognition". IEEE - Trans. Systems, Man, and Cybernetics, November, 1974.

[WW94]   G. Wang and J. Wang. "A New Hierarchical Approach for Recognition of Unconstrained Handwritten Numerals". IEEE Trans. Consumer Electronics, Vol. 40, N° 3, August, 1994.

Appendix 1

Table of Pattern Branch Feature Sequences

Sequence	N° of branches	End Points	Intersection points	Characters
(oo)	0	0	-	0
(od+oe+oa+ob)	1	0	-	0
(c+a).(u+d)	2	0	-	0
(c+u+a+d).(oa+ob)	2	0	-	8
(d+u).(c+a).(h+p+n)	2	0	-	8
(d+u).(c+a).(od+oe+oa+ob)	2	0	-	8
(d+u).(h+n).(c+a).(d+a+u)	4	0	-	8
(d+u).(c+a).(oa+ob+od+oe).(h+p+n)	4	0	-	8
(v+p+c+a).(od+oe+oa+ob)	2	1	-	0/6/9
(v+p+c+a).(d+u)	2	1	-	0/6/9
(c+a+p).(s+d+a).(c+u+p)	3	1	-	6
(v+p+d).(od+oa).(h+p+n)	3	1	-	9
(d+u).(c+a).(d+u+p+v)	3	1	-	9
(p+c+a).(s+d+a).(c+a).(h+p+n)	4	1	-	6
(v+p+n+c)	1	2	0	1

Sequence	Nº of branches	End Points	Intersection points	Characters
(z)	1	2	0	2
(s)	1	2	0	5
(c)	1	2	0	6/0
(v+p+n).(v+p+n)	2	2	0	1
(z+c+d).(v+p+n)	2	2	0	2
(c+a).(d+u)	2	2	0	2/5
(c+a).(v+p+n)	2	2	0	4/9
(s).(h+n)	2	2	0	5
(p).(c+u)	2	2	0	5
(c).(h+p+u)	2	2	0	6
(v+p).(h)	2	2	0	7
(v+p+d).(h+a).(c+a+u)	3	2	0	2
(z).(h+p).(h+u)	3	2	0	2
(c+a).(d+u+p).(h+n)	3	2	0	5
(h+p+n).(v+p+c).(d+u)	3	2	0	5
(h+n).(s).(h+n)	3	2	0	5
(c).(u).(a)	3	2	0	6
(c+a).(d+a).(d+a)	3	2	0	9
(v+p+d).(h+a).(c+a).(h+c+d)	4	2	0	2
(c+a).(h+n).(v+p).(h+p+n)	4	2	0	5
(h+p).(v+p+c+u).(h+p+a).(d+a+u)	4	2	0	5
(v+p+n).(h+n)	2	2	1, 2 or 3	1
(d+a).(h+p+u)	2	2	1, 2 0r 3	2/5
(d+a+u).(d+a+u)	2	2	1, 2 or 3	3
(h).(v+p+n)	2	2	1, 2 or 3	7
(u).(c+a)	2	2	1, 2 or 3	9
(z).(h+p+n+u).(a+u+h+p)	3	2	1, 2 or 3	2
(d+u). (h+p+n+u).(oe+ao+ob)	3	2	1, 2 or 3	2
(d+u).(d+u).(h+p+n)	3	2	1, 2 or 3	3
(c+a).(d+u).(h+n)	3	2	1, 2 or 3	5
(d+u).(c+a).(u)	3	2	1, 2 or 3	8
(v+p).(h+a).(c+a).(h+p+u)	4	2	1, 2 or 3	2
(d+u).(h+p+u).(c+a).(p+d+u)	4	2	1, 2 or 3	2
(d+u).(h+p+n+u).(p+d+u).(oe+ao+ob)	4	2	1, 2 or 3	2
(h+p).(p+c+u).(d+a+u).(v+p+n)	4	2	1, 2 or 3	5
(p+c+a).(p+c+a).(d+u).(h+p+n)	4	2	1, 2 or 3	6
(c+a).(n+c).(d+u).(c+a)	4	2	1, 2 or 3	8
(d+u).(p+c+a).(h+p).(h).(h+n)	5	2	1, 2 or 3	2
(v+p+n).(v+p+n+h).(h+p)	3	3	1, 2 or 3	1
(d).(h+p+n).(h+n+u)	3	3	1, 2 or 3	2
(a+d+u).(h+p+n).(a+d+u)	3	3	1, 2 or 3	3
(v+p+n).(c+u).(v+p+n)	3	3	1, 2 or 3	4
(c+a+p).(d+u+a).(h+p+n)	3	3	1, 2 or 3	5
(d).(h+n).(v+p)	3	3	1, 2 or 3	7
(c).(v+p).(d+u)	3	3	1, 2 or 3	9
(d).(h+p+a).(h+u).(h+n)	4	3	1, 2 or 3	2
(d+a+u).(h+p+n).(v+p+d+a).(h+p+n)	4	3	1, 2 or 3	3
(v+p+n).(v+p+n).(h+n).(v+p+n)	4	3	1, 2 or 3	4
(h+n+a).(v+p+d).(h+n).(v+p+n)	4	3	1, 2 or 3	7/1
(h+n).(h).(h).(p)	4	3	1, 2 or 3	7
(v+p+n).(v+p+n).(v+p+n).(h).(h)	5	3	1, 2 or 3	4
(h+c+a).(h).(h).(v+p).(v+p)	5	3	1, 2 or 3	7
(v+p+n+c).(v+p).(v+p).(h+p+n)	4	4	1, 2 or 3	4
(v+n+c+a).(h).(h).(v+p+n)	4	4	1, 2 or 3	7
(v+p+n).(v+p+n).(v+p+n).(h).(h)	5	4	1, 2 or 3	1/4
(h+c+a).(h).(h+p+n).(v+p+n).(v+p+n)	5	4	1, 2 or 3	1/7

Unsupervised Learning of Character Prototypes

Annick Leroy

IRISA, Campus de Beaulieu, 35042 Rennes cedex - aleroy@irisa.fr

Abstract. In the framework of handwritten word recognition, the use of characters extracted from words instead of written in isolation, is essential to train recognizers. We propose a segmentation method which relies on anchor points such as the ascenders or the descenders, but also on certain kinds of loops. We do not use a manually segmented prototype set to initialize our incremental learning process, but instead we use an a priori knowledge about the alphabet characters. This knowledge is introduced as the encoding of the descending movements of the pen and of loops. From a set of words written by the same writer, we evaluate the different possible segmentations for each word and use the ones superior to a certain threshold. In the beginning this threshold is rather high. At each step of the segmentation of the words, its value decreases in order to register new prototypes. The characters already accepted are used in the next steps. The confidence rate is maximum for 3 steps.

1 Problematic

Training recognizers on isolated characters does not allow us to take into account the simplifications resulting from the ligatures, nor to differentiate some particular shapes occuring at the beginning, the end or in the middle of the words.

Our application concerns on-line handwriting, which means that the words have been written on a digitizing tablet or on a pen computer. Our data are the coordinates of the pen together with its pressure. We dispose of sets of labeled words written by the same writer. For each handwritten word we establish segmentation points hypothesis. The problem is to spread the different segments among the characters and to evaluate the possible distributions.

One solution consists of proposing different segmentations to a human operator and to ask him to choose the best one. This is called supervised segmentation. This implies an enormous amount of work however and it doesn't offer guarantees of stability in the choice of the segmentation points, nor does it guarantee the choiceof the recognizer will be the same. Another solution consists in introducing a knowledge about characters and eventually about ligatures. This knowledge might come from a recognizer already trained on manually segmented characters.

This is the case for the SPS Algorithm used by Burges et al. which uses an already trained neural network, [Ba92]. They establish the optimal segmentation by constraining the length of the path in the evaluation graph to the number of characters to recognize. With a 40% rejection rate the system reaches 3.6% error per character. The neural network is trained in a continuous way.

Hans Leo Teulings and Lambert Schomaker in [TS92] propose an evaluation method based on an energy function. They initially have a set of allographs for each character and a set of examples for each allograph. The experiment consists of moving the beginning and the end of the characters and of measuring the new energy for the allographs (which correspond to the characters of the word). When they add a new sample, if the global energy of the system decreases, the new shape is accepted. At first the temperature is high and new shapes are easily accepted. It decreases little by little.

Laurent Duneau carries out an incremental clustering, [DD94]. At first he has, for each writer, a set of 50 words manually segmented (for about a thousand words to segment). He progressively removes well-segmented words from the set of words. The segmentation evaluation consists of maximizing a function such that: the highest number of characters is recognized, there are a minimum of hypothetical characters and a minimum of dissimilarities with the recorded prototypes. The similarity is measured through a K nearest neighbors algorithm. Results show 20 mistakes for 1595 words, that is 1.2%, with 20% rejected.

All these methods require a set of prototypes to initialize the process. They are relatively costly, computer wise. They need either an important volume of data to stabilize, or, after each insertion, to recompute the center of the new allograph classes, or to reevaluate a global energy function for the samples of an allograph. Except for a high rejection rate, the acceptation of new prototypes can include mistakes and cause a poor evaluation of the word segmentation in the future.

We shall present a method which will let us segment the words using elementary a priori knowledge about the alphabet characters. We rely on anchor points such as the ascenders or the descenders, but also on particular loop shapes. We shall first present a signal synthesis which uses ellipse quarters. Next we shall extract features from this representation and encode it with strings. We shall also present an encoding of the alphabet with the same features. It will be used to split the features between the characters and evaluate the best segmentation of a word. We shall conclude by a 2 phase algorithm for progressively recording prototypes. It starts from a high requirement about the quality of the prototypes and slowly decreases it.

2 Encoding the written words

The encoding of the written words relies on a signal synthesis which uses ellipse quarters and lines. We are not looking for a synthesis very close to the original but instead to keep the information readable and easy to manipulate. We mainly need to know the direction of the pen and the amplitude of the shapes.

2.1 Signal synthesis

Once the words have been made horizontal and the slant corrected, we proceed to a detection of the x and y minima and maxima. Between two adjacent x extrema or two adjacent y extrema we observe an inflection point, which is a curvature change. We code such a portion of the writing with X or Y according to the nature of the extrema. We bind two extrema of a different nature with an ellipse quarter. The code depends on the tangent at the extrema. The ellipse quarters going clockwise (or positive direction) are described by a code between 1 and 4, and the negative ones are described by a code between 5 and 8 (see figure 1). The "D" code corresponds to the beginning of a stroke or a pen down while the "F" code corresponds to the end of a stroke or a pen up.

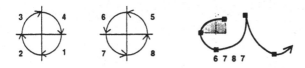

Fig. 1. Signal synthesis through ellipse quarters

The beginning and the end of the strokes, although they correspond to an ellipse arc whose size is inferior to an ellipse quarter, are synthesized with an ellipse quarter because it is homogeneous and the difference is hardly visible.

A verification phase lets us correct possible mistakes in the arc number successions. This is in general due to a bad estimation of the tangent at an extremum.

2.2 Choice and extraction of the features

The notion of silhouette has been mentioned in [Ler96] for a lexicon reduction. We use the same kind of representation here. Handwritten words are encoded by strings which describe the succession of features extracted from left to right. We chose to encode each descending segment (between a y maximum and a y minimum) and to eventually generate several features for each, together with a confidence value in the detection.

We use 2 kinds of features :

- The *default features* which correspond to the descending movements of the pen. They are coded "i" if they belong to the body of the word, "|" to the ascenders, "y" to the descenders and "f" when dealing with an "f". Very short segments are coded "-". Whether a segment belongs to the body, the ascenders, the descenders or the "f", is the fusion of the answers of three experts, each of which provides a fuzzy membership to these subsets (or features). The limits of these fuzzy subsets are adapted to the writer. The first expert bases its answer on the height of the descending segments, the second one on the distance of the extrema to the baseline while the third one takes into account the difference of height between two adjacent extrema of the same kind (maxima or minima).
- The *looped features* cover a descending segment and the preceding or the following ascending segment. They are extracted thanks to a syntactic method. For each feature the rules check the arc number succession and the relative position of the extrema. Here too the encoding is continuous and is based on fuzzy rules. The limits of the fuzzy subsets are adapted to the writer. The loops which go clockwise are described : by the code "o" if they are closed, "e" if they are shaped like "e", "c" if they are open and "l" if looped ascenders. The loops

written counter clockwise are coded "s" if they belong to the body and "j" if looped descenders.

For each looped feature a default feature is also generated. The detection confidence values sum to 1 for each descending segment.

2.3 Example

Fig. 2. Encoded parts of the writing which describe the word "and". The numbers correspond to the descending movements.

Here is an example of generated code for the word "and" written in figure 2:

```
segment 1 : (c   0.8)(i   0.2)
segment 2 : (i   1.0)
segment 3 : (s   0.6)(i   0.4)
segment 4 : (i   1.0)
segment 5 : (e   0.9)(i   0.1)
segment 6 : (c   0.2)(i   0.1)(|   0.7)
```

A silhouette is built by concatenating one code for each segment. Its confidence value is equal to the product of the confidence values of the corresponding features.

All possible silhouettes can be divided in 2 sets which correspond to distinct default silhouettes (which contain only default features). In fact they correspond to different numbers of ascenders and descenders.

$$E_1 = \{ \; (cisiec \;\; 0.0864) \;\; (cisiei \;\; 0.0432) \;\; (ciisic \;\; 0.0096) \;\; (ciisii \;\; 0.0048)$$
$$(iisiec \;\; 0.216) \;\; (iisiei \;\; 0.156) \;\; (iisiic \;\; 0.0012) \;\; (iisiii \;\; 0.0006)$$
$$(ciiiec \;\; 0.0576) \;\; (ciiiei \;\; 0.0288) \;\; (ciiiic \;\; 0.0064) \;\; (ciiiii \;\; 0.0032)$$
$$(iiiiec \;\; 0.0144) \;\; (iiiiei \;\; 0.0072) \;\; (iiiiic \;\; 0.0016) \;\; (iiiiii \;\; 0.0008)\}$$

$$E_2 = \{ \; (cisie \mid \;\; 0.3024) \;\; (ciisi \mid \;\; 0.0336) \;\; (iisie \mid \;\; 0.756) \;\; (iisii \mid \;\; 0.0042)$$
$$(ciiie \mid \;\; 0.2016) \;\; (ciiii \mid \;\; 0.0224) \;\; (iiiie \mid \;\; 0.0504) \;\; (iiiii \mid \;\; 0.0056)\}$$

3   A priori encoding of the alphabet

Instead of starting the segmentation process with the help of characters manually segmented, we chose to describe all different possible codes with the previously described features. We take into account even the most uncommon shapes. If some are forgotten, it will be compensated by the default silhouettes which describe the minimum encoding, that is to say the descending movements of the pen. We classify the codes into 4 categories:

- VFREQ : the most frequent number of descending segments allied to the presence of looped features (if possible for the considered characters).
- FREQ : the most frequent number of descending segments and default features.
- AVER : a less common number of descending segments allied to the presence of looped features (if possible).
- RARE : a less common number of descending segments and default features.

Table 1 gives the different codes for each letter.

4   Evaluation of the different possible segmentation

For a given word we have a list of silhouettes from the richest (with looped features) to the poorest (with only default features). They are each encoded by a string.

We call decomposition the operation which, with the help of the alphabet codes, consists of getting a possible distribution of the features among the characters. For each decomposition we evaluate a quality factor equal to the product of the quality factors of each of the characters silhouettes.

1. Some of the generated silhouettes correspond to the right number of ascenders and descenders. This is due to the uncertainty introduced in the coding of the signal with features, in order not to miss the right one. We filter this set to obtain a set of possible silhouettes for the current word.
   In the previous encoding example, only the silhouettes from set E_2 may correspond to "and".

2. Among these silhouettes we keep the richest possible silhouette, that is to say the one which contains the highest number of looped features possible for the current word. This verification eliminates the silhouettes which contain features erroneously detected.
   In the example, inside set E_2 the silhouettes which contain an "s" cannot be matched. This reduces to 4 the number of remaining silhouettes:

   $$E_3 = \{ ciiie \mid \quad 0.2016) \, (ciiii \mid \quad 0.0224) \, (iiiie \mid \quad 0.0504) \, (iiiii \mid \quad 0.0056)\}$$

   The richest possible silhouette is "$ciiie$ |".

3. For the richest possible silhouette we evaluate all possible distribution of features among the characters. We compute the quality factor of this decomposition (equal to the product of the quality factors of each character).
   In the example the decomposition of the silhouette "$ciiie$ |" for the word "and" will give "ci ii e|". The quality factor for this decomposition is equal to VFREQ since all characters quality factors are equal to VFREQ. In contrast the decomposition of the silhouette "$ciiiie$ |" would provide several decompositions: 'ci iii e|" and "ci ii ie|" with a quality factor equal respectively to
   $\sqrt[3]{VFREQ \times RARE \times VFREQ}$ and $\sqrt[3]{VFREQ \times VFREQ \times RARE}$.

If the highest score decomposition is unique, it can be used. In the opposite case, only a comparison with prototypes can help desambiguate. To keep the best decomposition we shall use a criterion which varies according to the prototype recording phase.

5 Splitting the written word into characters

When associating a feature succession with a character, it doesn't always tell the limits with the adjacent characters since only parts of the signal are encoded. We need to establish cutting rules. At the same time the recording of a new prototype involves comparison with the prototypes already accepted. Segmentation points have to be stable. Ligatures with the previous or the next character modify the shape of the character. To get a description as independent as possible of the context, we chose to divide the characters in 3 parts:

- *the central part* which is the most stable
- *the previous part* which ensures the connection with the previous character
- *the following part* which ensures the connection with the next character

The previous and following parts might be empty in some cases. If not, they always correspond to an ascending zone since every descending movement belongs to a feature.

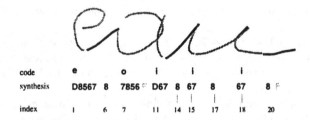

Fig. 3. Matching the code and the writing: in grey are the central parts, in black the uncertainties which belong to ligatures. The code "F" is a marker and does not correspond to an index.

Let us consider the word "eau" described in figure 3. Let us suppose we extracted the silhouette "e o i i i" in the synthesis of the signal. To each code we associate the first and last index of its units (ellipse quarters or lines).

$$(e, 2, 5) \rightarrow D8567$$
$$\text{liaison} \rightarrow 8$$
$$(o, 7, 10) \rightarrow 7856$$
$$\text{liaison} \rightarrow FD$$
$$(i, 12, 13) \rightarrow 67$$
$$\text{liaison} \rightarrow 8$$
$$(i, 15, 16) \rightarrow 67$$
$$\text{liaison} \rightarrow 8$$
$$(i, 18, 19) \rightarrow 67$$
$$\text{liaison} \rightarrow 8F$$

From this information, since we know the decomposition of the word "eau" for this code (e oi ii), we can assign the central part of the characters and leave an uncertainty zone in between.

$$\text{uncertainty} = D$$
$$"e" \quad \rightarrow \quad e \rightarrow (e, 1, 5)$$
$$\text{uncertainty} = 8$$
$$"a" \quad \rightarrow \quad oi \rightarrow (a, 7, 13)$$
$$\text{uncertainty} = 8$$
$$"u" \quad \rightarrow \quad ii \rightarrow (u, 15, 19)$$
$$\text{uncertainty} = 8F$$

Because of the segmentation points chosen for the features, the uncertainty zones always correspond to an ascending movement. Except if empty, these zones end with a y maximum or a pen up. If there is a backward movement or an undulation in this zone, we observe one or several x extrema. The distribution of the uncertainty zone between 2 characters is done thanks to cutting rules according to the previous character. In our example, cuts are simple and occur at y minima before the characters. We assign what follows a cutting point to the previous part of each character. The remaining of the uncertainty zone is assigned to the following part of the previous character. In this example it is always empty. Here is the result:

character	previous	central part	following
"e"	D	8567	
"a"	8	7856FD67	
"u"	8	67867	8F

We split the characters in two groups depending on the complexity of the boundaries that might follow:

- group 1 : a, c, d, e, h, i, k, l, m, n, r
- group 2 : f, g, j, p, q, s, y, z, t, x, b, o, v, w

The cutting rules of the uncertainty zones are the following:

rule 1: if there is a pen up, assign the end of the stroke to the previous character and the beginning of the next stroke to the next character.

rule 2: if the letter belongs to group 1, cut at the y minimum before.

rule 3: if the letter belongs to group 2, look for a x minimum before and if there is none, look for a y minimum.

Example: figure 4 illustrates different cutting rules depending on the previous character. In the example (*a*) the cuts occurs at y minima, when in examples (*b*) and (*c*) cuts occur at a x minimum after the "b" and at the pen up after "i". According to the shape of "g", a looped feature has been extracted (*a* and *b*) or not (*c*). It implies a different distribution between the central and the following parts.

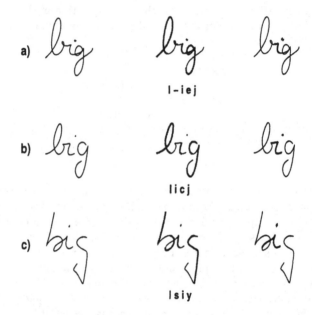

Fig. 4. Application of the cutting rules between characters : the encoded parts of the signal are shown in the center and the result of the segmentation on the right.

6   Algorithm for accepting characters

The recording of the characters occurs in 2 phases:

phase 1 : is based only on the quality of the decomposition. In this phase the chosen decomposition is the one whose quality factor is maximum. We need this decomposition to be unique and its quality to be superior to FREQ for each character. This means that we consider only words whose characters correspond to the most common number of descending movements. To be unique, it must contain looped features where there could be ambiguities.

phase 2 : relies on the prototypes recorded during the first phase. The chosen decomposition is selected according to the following criteria:

- there is no uncertainty zone left after splitting the handwriting among the characters.
- the number of recognized or partially recognized characters as well as their score is maximum. Partially recognized means that the central part has been found but not necessarily the previous part or the following part. The score is the sum of each character prototype quality (see further).
- if there is more than one possible decomposition, we choose the one with the highest quality of decomposition.

To each character prototype we associate the quality of the decomposition it was extracted from (VFREQ, FREQ, AVER or RARE). This is how we evaluate the recognition score of the word.

To accept a new prototype, the rules are the following:

- it is surrounded by 2 already known characters.
- it is at the beginning of the word and followed by a known character or it is at the end of the word and preceded by a known character. As a consequence it is better to start with short words.
- it is preceded by a known character and followed by a pen up.
- it is preceded by a new prototype and followed by a known character or a pen up.
- it can be derived from a known prototype by rewriting rules (not yet implemented).

When the central part has been already encountered, we record the following or the preceding parts if they are new. This second phase is repeated by lowering the minimum quality required until there are no more characters to record.

The recording of new characters requires a minimum global quality for the word and a minimum quality for each character. This requirement is progressively lowered after each step. Experiments showed that 2 steps offer the maximum guarantee:

1. Minimum quality superior or equal to FREQ for the word and to FREQ for each character. This will solve ambiguities met during the first phase thanks to some recognized characters.
2. Minimum quality equal to FREQ for the word and AVER for each character. This will accept uncommon silhouettes for the characters, but only if there are looped features detected.

The acceptance of RARE characters would necessitate a detailed comparison between characters. We do not accept prototypes from words with a quality lower than FREQ to keep a certain security. For writers who have many uncommon shapes, we will not be able to record as many characters. The results speak for this choice since there are very few errors. When observing the segmentation results and in particular zones which have not been used inside certain words, we can see that the segmentation did not offer enough guarantee.

Actually the comparison is limited to a verification of the arcs and inflection successions. This is rather rough, but extremely fast since there are only a few string comparisons to compute (compared to a K nearest neighbor method it is very quick). We are developping a more precise matching of the characters based on our representation. We could imagine a verification phase to eliminate outsiders among the prototypes. This can be done by looking at the number of occurences for each allograph or with the help of rewriting rules (transformation of peaks into humps...). The limited number of steps in the second phase avoids error propagation.

The recording of the prototypes can take indifferently the format of the rough signal, of the synthesis or of the string codes.

7 Experimentation conditions and results

We segmented about 50 words written by 10 different writers. The rejection rate raises 25% and the number of mistakes is of 4 for 343 words segmented, which means 1.5%. The total number of prototypes recorded is 556 with a maximum of 60 for certain letters.

8 Conclusion

The good points of our method are the following:

- great simplicity of the encoding which relies on stable elements : the extrema.
- no normalization
- no need for a manual segmentation phase or for a trained recognizer.
- very fast since based on string comparisons.
- no threshold in the segmentation decisions.
- quick adaptation to a writer.
- cuts are not always done at y minima or at pen up, but also at x extrema.
- in contrast to a left to right segmentation, only parts of a word may be recognized and produce prototypes. Some uncertainty zones may remain. In the same way some ambiguous words, or words whose quality is too low, will be rejected.
- boundaries are learned at the same time as the central part of the characters.
- we can distinguish characters which start or end with a pen up.
- From a rough encoding we produce a prototype database with different possible representations: rough signal, synthesis or string codes

In comparison to other methods, it is very fast. Only the second phase relies on prototypes already recorded, which limits error propagation. This is generally controlled by costly procedures which involve distance computation. We use only prototypes from the same user, which also limits the confusions. We can compare this method to a puzzle in which we would first put the pieces we are certain of and then lean on them to fill up the holes. The uncertainty zones would remain vacant.

References

[Ba92] C.J.C. Burges and al. Shortest path segmentation: A method for training a neural network to recognize character strings. In *IJCNN*, pages 165–172, vol 3, Cambridge, Ma, 1992.

[DD94] L. Duneau and B. Dorizzi. Incremental building of an allograph lexicon. In *Advances in handwriting & drawing : a multidisciplinary approach*, pages 39–63. C. Faure. Keuss P., Lorette G. and Vinter A., 1994.

[Ler96] Annick Leroy. Progressive lexicon reduction for on-line handwriting. In *IWFHR 5*, Colchester, UK, 1996.

[TS92] H.L. Teulings and L. Schomaker. Unsupervised learning of prototype allographs in cursive script recognition. In *From Pixels to Features III*, pages 61–73. S. Impedovo and J.C. Simon, 1992.

a	"oi"	VFREQ	"ci"	VFREQ	"ie"	VFREQ	"ei"	VFREQ	"ce"	FREQ	"ii"	FREQ
	"o-i"	AVER	"-ii"	AVER	"-i"	AVER	"oii"	AVER	"iii"	RARE		
	"o"	RARE	"i"	RARE								
b	"|"	VFREQ	"|-"	VFREQ	"|i"	VFREQ	"s"	VFREQ	"|i"	FREQ	"|"	FREQ
	"-li"	AVER	"-|"	AVER	"-|"	RARE	"ili"	RARE				
c	"c"	VFREQ	"i"	FREQ	"e"	FREQ						
	"-c"	AVER	"-i"	RARE	"ii"	RARE						
d	"o|"	VFREQ	"e|"	VFREQ	"c|"	VFREQ	"i|"	FREQ				
	"|c"	AVER	"|o"	AVER	"|i"	RARE						
	"-e|"	AVER	"ii|"	RARE	"oi|"	AVER	"-i|"	RARE	"-o|"	AVER		
e	"e"	VFREQ	"c"	FREQ	"i"	FREQ	"-e"	VFREQ	"-i"	FREQ		
	"-i"	AVER	"-i"	RARE								
f	"f"	VFREQ	"f-"	VFREQ								
	"|"	AVER	"|"	RARE	"|-"	AVER	"|-"	RARE	"-fi"	RARE	"-|i"	RARE
g	"cj"	VFREQ	"oj"	VFREQ	"ej"	VFREQ	"ij"	FREQ				
	"cy"	FREQ	"oy"	FREQ	"ey"	FREQ	"iy"	FREQ				
	"j"	AVER	"y"	RARE	"-ij"	AVER	"o-y"	AVER	"-iy"	RARE		
	"-y"	RARE	"iij"	RARE	"iiy"	RARE						
h	"li"	VFREQ	"|i"	FREQ	"-li"	AVER	"|"	RARE	"|"	RARE	"-|i"	RARE
i	"i"	VFREQ	"-i"	AVER	"ii"	RARE	""	RARE				
j	"j"	VFREQ	"y"	FREQ	"-j"	AVER	"-y"	RARE	"iy"	RARE	"ij"	RARE
k	"li"	VFREQ	"|i"	FREQ	"-li"	AVER	"-|i"	RARE	"|-"	AVER	"|-"	RARE
	"-|-"	RARE	"i|ii"	RARE								
l	"|"	VFREQ	"|"	FREQ	"-|"	AVER	"i|"	RARE				
m	"iii"	VFREQ	"ii"	AVER	"-iii"	AVER	"ii-"	AVER	"i-"	AVER	"iiii"	RARE
n	"ii"	VFREQ	"-ii"	AVER	"i-"	AVER	"i"	AVER	"iii"	RARE		
o	"o"	VFREQ	"i"	FREQ	"o-"	VFREQ	"i-"	FREQ	"-"	AVER	"-i"	AVER
	"e"	AVER	"o-"	RARE	"iii"	RARE	"-i-"	RARE				
p	"yi"	VFREQ	"ys"	VFREQ	"-yi"	AVER	"yi-"	RARE	"iyi"	RARE	"y"	RARE
q	"oy"	VFREQ	"cy"	VFREQ	"ey"	VFREQ	"iy"	FREQ	"-ey"	AVER	"-iy"	RARE
	"o-y"	AVER	"oy-"	AVER	"iiy"	RARE	"-yi"	RARE	"iyi"	RARE		
r	"i"	VFREQ	"-i-"	VFREQ	"ii"	FREQ	"-i-"	AVER	"iii"	RARE		
s	"s"	VFREQ	"i"	FREQ	"-s"	AVER	"-i"	RARE	"ii"	RARE	"-"	AVER
	""	RARE										
t	"|"	VFREQ	"|i"	RARE	"-|i"	RARE						
u	"ii"	VFREQ	"i-"	AVER	"-ii"	AVER	"iii"	RARE	"i"	RARE		
v	"i"	VFREQ	"i-"	AVER	"-i-"	AVER	"ii"	RARE				
w	"ii"	VFREQ	"ii-"	VFREQ	"iii"	RARE						
x	"sc"	VFREQ	"si"	VFREQ	"ic"	FREQ	"ie"	FREQ	"ii"	FREQ	"-ii"	RARE
y	"ij"	VFREQ	"iy"	FREQ	"-iy"	AVER	"j"	AVER	"y"	RARE		
z	"j"	VFREQ	"y"	FREQ	"i"	FREQ	"ii"	RARE	"ij"	AVER	"iy"	RARE

Table 1. A priori encoding of the characters of the alphabet using our features.

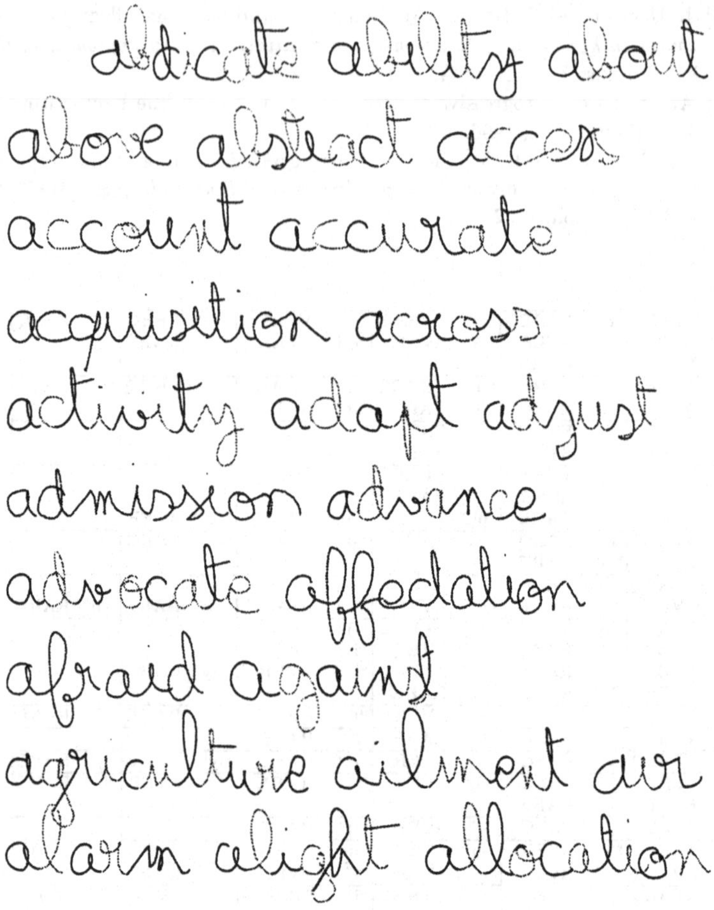

Fig. 5. Example of segmented words. Colors alternate to emphasize the segmentation result. Four words have not been segmented because of ambiguities (acquisition,admission, afraid and affectation). The words presented here have been synthesized with the help of ellipse quarters and lines.

Off-Line Cursive Word Recognition with a Hybrid Neural-HMM System

Zsolt Wimmer*, S. Garcia-Salicetti°, B. Dorizzi*, P. Gallinari**

*Institut National des Télécommunications, Dépt. EPH; 9 rue Charles Fourier, 91011 Evry, France
°Laboratoire LIS, Tour 66-56, Université Paris VI; 4 Place Jussieu, 75252 Paris Cedex 05, France
**LIP6, Tour 46-00 Boite 169, Université Paris VI; 4 Place Jussieu, 75252 Paris Cedex 05, France
e-mail addresses: Zsolt.Wimmer@int-evry.fr, ssalicet@worldnet.fr, Bernadette.Dorizzi@int-evry.fr, gallinar@laforia.ibp.fr

Abstract

In a recent publication [1], we have introduced a neural predictive system for on-line word recognition. Our approach implements a Hidden Markov Model (HMM)-based cooperation of several predictive neural networks. The task of the HMM is to guide the training procedure of neural networks on successive parts of a word. Each word is modeled by the concatenation of letter-models corresponding to the letters composing it. Successive parts of a word are this way modeled by different neural networks. A dynamical segmentation allows to adjust letter-models to the great variability of handwriting encountered in the words. Our system combines Multilayer Neural Networks and Dynamic Programming with an underlying Left-Right Hidden Markov Model (HMM). In this paper, we present an extension of this model to off-line word recognition. We use on-line data in these off-line experiments, generating a binary image from trajectory data. The feature extraction module then turns each binary image into a sequence of feature vectors, called 'frames', combining low-level and high-level features in a new feature extraction paradigm. Some results for word recognition are presented.

1 Introduction

Off-line handwriting recognition has a wide variety of applications ranging from adress recognition for mail sorting to checks reading and forms interpretation. Even if each application has its own specifications and thus requires a specific processing, a common need appears : being to able to recognize words, in an omniscriptor framework, even when some letters are badly written. Much of the models proposed for such recognition tasks involve the use of Hidden Markov Models (HMM), which are able to absorb some of the variability that appears in this context. The main differences between all these works relie on the coding of the drawing and the type of HMM used : discrete vs continous one [9][10].

In a previous work, we have proposed a new system for word-recognition in the "on-line" framework [1]. This system is an hybrid system in the sense that it implements a cooperation of predictive neural networks in the HMM formalism. This system has been shown to provide good results on a word database composed of 8781 words from 9 writers, written on a digitizing tablet.

Following the same idea as [4], we wanted to know if the same system could be suitable for off-line recognition. This is the object of the present paper. For that, as an intermediate step, we tested it on off-line words coming from our previous on-line data. This has the effect of suppressing the need of preprocessing which is so difficult and costly on the "real" off-line environnements.

The other interest is that it will be possible to compare the performance of the system on both type of data : on-line and off-line and probably to make them cooperate.

The main part of the work presented below concerns the definition of an appropriate set of features, for off-line words which allows our model to be efficient.

This paper will be organized as follow.
Section II briefly describes our databases and the way they have been built from the on-line drawings. Then Section III is dedicated to a recall of the functioning of our hybrid neural-HMM system. In Section IV are detailed the different feature extractions that we have tried. They are thus compared in Section V through the word recognition rates they produce on our database.

2 Databases

The databases used in this work are word databases from 9 writers, composed of 8781 words chosen from a 1000 words' vocabulary. Originally, words were written with a fixed scale on a digitizing tablet and the pen trajectory was sampled at the frequency of 200 points per second. Such data appeared as a sequence of (x,y) coordinates, that we first convert into a binary pixel map. Afterwards, successive points belonging to each word's trajectory are connected to each other by a straight line. The latter results in a kind of smoothed skeleton and no heavy preprocessing is needed.

To each word is associated a sequence of "frames". A frame is a vector of parameters which describe a section of the word bitmap. Several types of frames can be considered they will be described in Section IV.

3 System Overview

3.1 System Specification

The system has been already detailed in several publications [1][2][3]. We will just recall its general structure. A handwritten word is modeled by the concatenation of neural predictive letter-models, corresponding to the sequence of letters composing the word. Each letter-model is composed of m states, each one consisting of a multilayer neural network of fixed dimensions. A Left-Right topology [5] rules transitions between the set of neural predictors modeling a word. Only transitions from each one to itself or to its right neighbor are permitted (see *Figure 1*).

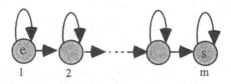

Figure 1. The simple left-right topology considered for an m-states letter model
e: entry state; s: exit state

During learning, *only* the neural predictors in letter-models corresponding to the letters composing a given word, give as output a non-linear prediction on each frame of the word. This prediction is computed frame per frame, each time from a left context of 3 frames [1]. The structure of each neural network is the same: if each frame is described by n parameters, $3n$ input units in the first layer encode the 3 frames-left context of the current frame, n units follow in the hidden layer, and n units in the output layer encode the prediction given by the network on the current frame.

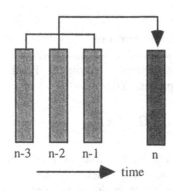

Figure 2: Predictive context of 3 frames

When a word is presented to the system, a prediction error is computed on each of its frames (the euclidean error between the output of each neural network and the current frame). The resulting prediction errors' matrix is processed through Dynamic Programming: every authorized path in the matrix has a global cost, the summation of prediction errors corresponding to that path. Segmentation is performed by the search of the *optimal* path, representing the *least global cost*. This is done thanks to a Virebi algorithm. Afterwards, frame per frame, the weights of the networks belonging to the optimal path, are modified by back-propagating its prediction error on the current frame.

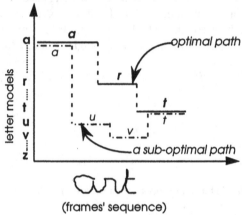

(frames' sequence)

**Figure 3. The optimal path computation and its letter-labelling process (labels "art")
during recognition**

During recognition, a prediction error matrix is computed on the presented word, but considering this time, *all* letter-models in such matrix. The optimal path represents the least global cost path and allows to attribute a letter-model to each frame composing the word, that is a letter-label (see *Figure 3*). Any other path has a higher global cost and is therefore sub-optimal (see *Figure 3*).

3.2 Statistical background

We consider an approximate Maximum Likelihood criterion for the model parameters' learning process [1]:

$$\max{}_\lambda \left[\max{}_q \left[P(O, q \mid \lambda) \right] \right],$$

where λ is the N-states Hidden Markov Model of a given word presented during learning to the system, and q denotes the hidden state sequence. As detailed in a previous work, this criterion leads (under precise assumptions) to consider as cost function, in the neural predictive framework, the summation of prediction errors per frame along the optimal path [1]. Optimization is thus performed first on segmentation through Dynamic Programming, followed by the reestimation of the word-model's parameters via the Back-Propagation algorithm.

During recognition, our criterion becomes:

$$\max {}_q [P(O,q \mid \Lambda)]$$

where Λ denotes now the *global* parameter set, including *all* letter-HMMs. A sequence of labels is this way obtained. The latter represents only a first step in word recognition, since a word from a lexicon must be proposed by a recognition system. In this framework, our statistical criterion is:

$$\max {}_W [\max {}_q [P(W,q \mid O)]],$$

where W denotes a word in the considered lexicon. This can also be written as:

$$\max {}_W [\max {}_q [P(W \mid q, O) . P(q \mid O)]],$$

and now the two steps of our recognition process clearly appear: first, the second term's maximization leads to the optimal state sequence q*; then, the first term's maximization permits, through an editing distance, to find the optimal word W* in the considered lexicon.

3.3 The Neural Lexical Post-Processor

The system that has been described so far allows to make a correspondance between a word and a sequence of letters labels. Of course, it happens very frequently that this labels sequence does not correspond to any word in the dictionary because of letter insertion or deletion by example. Even for a human being, the task of recognition at the letter level without any knowledge of a dictionary is very difficult and results in ambiguities. We have thus introduced a post-processing of the letter label sequence when a dictionary is avalable. In fact, we began by a simple edit distance which revealed to be unsufficent and arrived to the system which is described below :

It works on the sequence of labels given by the recognizer by means of 2 modules. The first module is an edit-distance between that sequence and each word in the dictionary. Its output is a list of 50 words, ordered according to their scores. The second module presents the frames of the word to recognize, to the trained neural word-model corresponding to each of those 50 words (see *Figure 4*). For each of them, a prediction error matrix is computed as well as an optimal path. The global cost along each optimal path obtained, is this way associated to the considered neural word-model. In this framework, as *Figure 3* shows, the neural word-model giving the least global cost permits to determine the winning word in the lexicon.

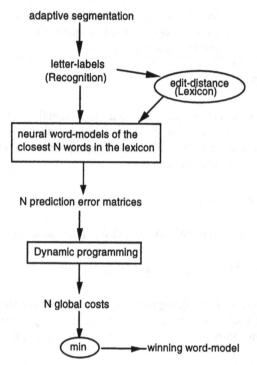

Figure 4. The recognizer and the neural lexical post-processor (N=50)

4 Feature Extraction

4.1 Grey-level pixel map

First, we were inspired by the features extraction method described in [10]. The grey-level pixel map generated for each word in the database is normalized in height to 40 pixels, and in length of a constant value independently of the word's own length. Parameters are extracted for each column of the normalized image, resulting in one frame or feature vector. Therefore, long words are encoded in more frames than short ones, but each character is approximately encoded by the same number of frames.

We also used the lower and upper base lines, recovered from the original on-line data. In our normalization procedure a 10 pixels height is kept between those two lines, that is in the word's core region. The ascenders and descenders regions are thus both fixed to a 15 pixels height. We first considered as features the grey-level value of each pixel within the column, leading to a feature vector of dimension 40.

4.2 Skeleton Coding

We also considered another set of features, partly inspired by those used in [6]. The area covered by the unnormalized word corresponding to the initial binary map is first divided into a grid of rectangles, as shown in *Figure 5*.

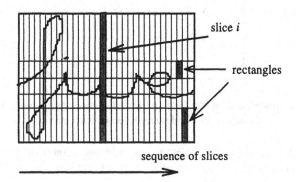

Figure 5. Skeleton with grid for parametrisation

Frames are then extracted from the slices obtained. Each slice has a width that is fixed in order to encode one character approximately by 11 frames (see *Figure 5*). The vertical resolution of the grid is chosen in order to divide the word's pixel map into 5 regions. For each rectangle in the grid, 5 parameters are computed :

❑ 3 invariant pattern moments (m_{20}, m_{02}, m_{11}) of order 2 are computed to represent different directions. They are defined as the mean-square x and y deviations about orthogonal x and y axes through the centroid of the invariant pattern [7][8], and given by the following expressions:

$$m_{pq} = \frac{1}{n_{20} + n_{02}} n_{pq}$$

where

- $n_{pq} = \sum_{x=1}^{N} \sum_{y=1}^{M} (x - x_g)^p (y - y_g)^q I(x, y),$

$$p, q = 0, 1, 2$$

- $I(x, y)$ is the state of the pixel (black or white),

- (x_g, y_g) is the centroid of the image.

❑ Junctions indicate points where two strokes meet or cross in the skeleton. They are easily found in the word with a set of masks of 3x3 pixels. *Figure 6* shows some possible configurations :

```
X - X      - X -      - - X
- X -      - X -      - X X
- X -      X - X      X - -


X - X      - X -      X - X
- X -      X X X      - X X
X - X      - X -      X - -
```

Figure 6. Six possible 3x3 masks for junction's detection. "x" denotes the pixels of skeleton and "-" denotes those of background.

❏ <u>Endpoints</u> are points in the skeleton with only one neighbour and mark the ends of strokes. They can be some potential points of segmentation.

This feature extraction procedure leads to a parameter vectors of dimension 25 (frame) to encode each slice.

It is worth noticing that the latter features are local since they code only on the information from a small area of the image. We then considered that global features were necessary to encode information about the structure of the whole word or at least of the letter to which the corresponding slice belongs (see *Figure 5*). Therefore, we also introduced two types of global features :

♦ <u>A low-resolution bitmap image</u> for the description of each slice's spatial context. The context bitmap is computed on a window that slides along the whole word's image (see *Figure 7*). The window's width is approximately the one of a character. The number of points in each of the 15 rectangles composing the window is counted, leading to a grey-level representation that is also normalized.

slice *i*

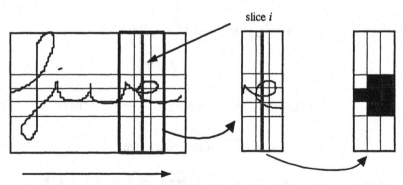

Figure 7. The context-bitmap

♦ <u>The sine of the angle,</u> between the horizontal and the line defined by the two mean points of the profile's points belonging to two successive slices (see *Figure 8*). We consider the top and bottom word's profiles, so we have two values of sine in each slice. The top profile (respectively bottom) is the first transition from the top (respectively from the bottom) in successive columns of the binary image (see *Figure 8*).

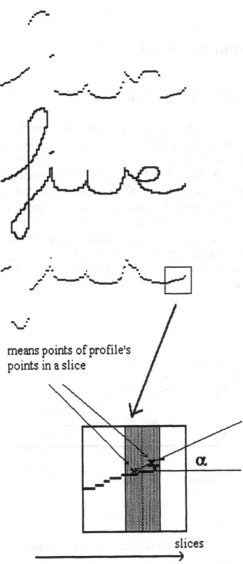

means points of profile's
points in a slice

α

slices

Figure 8. Top and bottom word's profiles

The context bitmap description and the profiles were then added to the local features. This new feature set leads to a significant improvement of the results, as shown in the next section. Moreover, the calculations involved remains relatively simple and tractable while encoding high-level contextual information. This approach has therefore a main advantage relatively to structural methods [6], which are highly costly in calculations.

5 Experimental Setup and Results

5.1 Initializing the Learning Procedure

In the databases, no information is available about the "correct" segmentation of words into letters. One of the strengths of this system is that it finds segmentation by itself without any learning even on a small training set of an appropriate segmentation. To that end, the optimal path computation on each word had to be constrainted directly on the cost function. The optimal path is indeed forced to remain inside a "diagonal" band around the prediction errors's matrix (only letter-models corresponding to the letters composing the current word are involved during learning) (see *Figure 9*). This constraint makes almost equal parts of the word's trajectory correspond to every network in the word-model. The weight of this forced segmentation is progressively lowered during training. After 3 iterations, learning is totally unrestricted, and segmentation evolves freely.

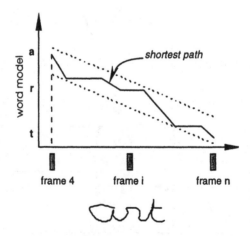

Figure 9: The initial constraint on the optimal path

5.2 Results

Our system was trained on the multi-scriptor framework on 7000 words from 9 writers and tested on 1500 words from the same writers. Three dictionaries are used: the 1000 words dictionary (D1) corresponding to the words in the database, and 2 other dictionaries consisting of the same 1000 words plus a random selection of additional words from the UNIX Spell dictionary. Their sizes are 10600 (D2) and 20200 words (D3).

Because we have approximately 11 frames per character, we chose a HMM of 7 states per character (cf. III.1.), after some experimentations.

Results obtained with the different features described above are in *Table 1* using the three dictionaries. The first 3 rows show recognition rates for the first 3 proposals given by the system, when using only the pixel map's grey-level representation. The following 3 rows show the same recognition rates considering local features only (invariant pattern moments, endpoints, junctions), and the last 3 rows correspond to the latter local features combined with global ones (context bitmap and top and bottom word's profiles : cf. IV.2.).

		D1 (1000 words)	D2 (10600 words)	D3 (20200 words)
Grey-level representation	1st proposal	55.87 %	35.59 %	29.18 %
	2 first proposal	63.70 %	38.43 %	33.45 %
	3 first proposal	65.48 %	41.64 %	35.59 %
Local features	1st proposal	66.00 %	55.07 %	52.40 %
	2 first proposal	74.27 %	64.53 %	61.67 %
	3 first proposal	77.73 %	68.73 %	66.40 %
Local and global features	1st proposal	86.87 %	77.00 %	73.13 %
	2 first proposal	91.07 %	81.67 %	78.80 %
	3 first proposal	92.40 %	84.13 %	81.33 %

Table 1. Comparison of the results obtained with three different feature sets on a 1500 words database with three dictionaries.

6 Conclusions

The introduction of local features (invariant pattern moments, endpoints, junctions) clearly improves the first results obtained from a simple grey-level representation of the word's image. Moreover, we noticed a significant improvement in recognition with our new feature extraction paradigm that permits to combine low-level features to high-level ones in an original and computationally unexpensive way. These results are encouraging, since they show that the hybrid model combining predictive neural networks and Hidden Markov Models which was shown performant on on-line word recognition, is certainly suitable for off-line word recognition.

Moreover, on this database, the results obtained by our system on off-line representation are approaching those obtained previously on the on-line representation. It will thus be interesting, following [4], to make the two recognition systems cooperate in order to improve the global recognition rate.

Another direction will also concern the optimization of the parameters of the system (size of the frame vector, number of states per models) and of the system itself (introduction of discrimination between the states by example).

Finaly, the results obtained with our coding, tested in a quite "easy" off-line task, renders us optimistic about the ability of our global system to deal with "real" off-line data as the one of NIST form database by instance which are more representative of the application problems for which such a system should be usefull.

7 References

[1] S.Garcia-Salicetti, B.Dorizzi, P.Gallinari, Z.Wimmer, "Adaptive Discrimination in an HMM-Based Neural Predictive System for On-line Word Recognition", *Proceedings of ICPR'96*, pp.515-519, Wien, 1996.

[2] S. Garcia-Salicetti, B. Dorizzi, P. Gallinari, A. Mellouk, D. Fanchon, "A Hidden Markov Model extension of a neural predictive system for on-line character recognition", *Proceedings of ICDAR 95*, 1995, pp. 50-53.

[3] S. Garcia-Salicetti , P. Gallinari, B. Dorizzi, Z. Wimmer, S. Gentric, "From Characters To Words: Dynamical Segmentation And Predictive Neural Networks", *Proceedings Of Icassp 96*, 1996, Atlanta.

[4] R.Seiler, M.Schenkel, F.Eggimann, "Off-line Cursive Handwriting Recognition Compared with On-line Recognition", *Proceedings of ICPR'96*, pp.505-509, Wien, 1996.

[5] M.Schenkel, I.Guyon, D.Henderson, "On-line cursive script recognition using Time Delay Neural Networks and Hidden Markov Models", *Proceedings of ICASSP'94*, pp. II637-640, Adelaide, 1994.

[6] Andrew William Senior, "Off-line cursive handwritting recognition using recurrent neural networks», *Thesis of University of Cambridge*, 1994.

[7] R.G. Casey, "Moment Normalization of Handprinted Caracters», IBM J. Res. Develop., 1970, p. : 548-557.

[8] M.K.Hu, "Visual invariant pattern recognition by moment invariants", *IRE Transactions on Information Theory,* vol.8, pp. 179-187, 1962.

[9] M.Gilloux, M.Leroux, J-M. Bertille, "Strategies for Handwritten Words Recognition Using Hidden Markov Models», *Proceedings of ICDAR 93*, 1993, pp. 299-304.

[10] Kjersti Aas, Line Eikvil, "Text Page Recognition using Grey-level features and Hidden Markov Models», *Pattern Recognition*, vol.29. No.6, pp.977-985, 1996

A Fuzzy Perception for Off-line Handwritten Signature Verification

C. SIMON[1], E. LEVRAT[1], R. SABOURIN[2], and J. BREMONT[1]

[1] CRAN, URA821, Equipe PRAISSIH, Fac. des Sciences, Bd des Aiguillettes, BP.239, 54506 Vandoeuvre FRANCE

[2] LIVIA, Departement de génie de la Production, Ecole de Technologie Supérieure, 1100 Rue Notre-Dame Ouest, Montréal (Québec), H3C 1K3, Canada

Abstract. The purpose of this paper is the assessment of a family of shape factors for off-line signature verification. The initial method, suggested by Sabourin [1], which extracts geometric features, is modified to assess the performance of verification in the general case of forgery. We contribute to the information coded by adapting the coding method to the image and by integrating a spatial distance into the shape factor definition. Moreover, we include the coding of an information related to the dynamic of the signature. To use these two types of information, we propose a fuzzy technique to combine and then obtain one kind of information. We evaluate this new coding operator with two types of forgery : random forgery and photocopy simulation with some adapted protocols.

1 Introduction

Financial and legal transactions imply the necessity of verifying the identity of the persons. This verification meets with two kinds of problems : identity usurpation or simulation of an usurpation. We note that this is a specific problem inherent to the notion of identity.

All verifications are carried out utilizing different attributes. A hand-written signature is a personal and a universal attribute for which the achievement is independent of the level of knowledge. It is not anonymous, characterizes the individual, and presents a certain reliability. Moreover, it enables the verification of the identity of a person on paper documents.

However, several conditions can alter the morphology of the signature. These modifications can be caused (among other factors) by age, habits, psychological state of mind, and practical conditions. All these flaws in the act of signing produce uncertainties concerning the features of the signature.

There are several types of forgery including : random forgery (i.e., the signature of other writers enrolled in the verification system); simple forgery, with a totally different signature; forgery by transfer, obtained by transfer techniques; freehand forgery, made by a skilled forger using visual memory. Verification is, therefore, a two-class pattern recognition problem that comes down to one question : " *Is the signature true ?* "

The verification in the case of random forgery remains an unsolved and critical issue. The generalization of the verification for all types of forgery is the most as difficult. Within this framework, defining the shape factor (i.e., pointing out the features of the signatures) is always a crucial point in the design of the verification system. Our approach is one of global recognition processes. The shape of the signature, whose geometric and dynamic properties are intrinsic characteristics of the writer, is preferred to the signature structure. A multi-scale approach leads us to the use of local as well as global geometric and dynamic signature properties. The aim is, of course, to obtain the highest possible efficiency and to generalize the verification of forgery.

In this paper, an improved method of extracting geometric features as suggested by R. Sabourin [1], is modified to assess the performance of verification in the general case of forgery detection. We contribute to the information coded by adapting the coding method to the image and by integrating a spatial distance into the shape factor definition. Moreover, we include the coding of an information related to the dynamics of the signature. We propose to combine these two types of information in one kind of information, by a fuzzy technique. This resulting information can be use by only one classifier. We evaluate this new coding operator with two types of forgery : random forgery and photocopy simulation.

2 Shape Factor Definition

The definition of a verification system needs the definition of a shape factor which allows the transformation from the image space to the feature vectors space. Thus, we should define a *perception mechanism* to obtain an appropriate transfer of pertinent information related to the discriminant features.

2.1 The Principle of the Shape Factor

To obtain a set of characteristics on the signature, while respecting the constraint of insensitivity for the text and the draw, Sabourin [9] has proposed an original method based on a multi-scale approach that collects a set of geometric characteristics. This set is then exploited by a vectorial discrimination.

Coding Technique The shape factor suggested, the so-called Shadow Code [8], was initially put forward by Burr to extract global characteristics out of hand-written characters. By extending the Shadow Code (ESC) to the whole signature, Sabourin proposed the definition of the Extended Shadow Code as a global characteristic extractor for off-line hand-written signatures verification [10]. This technique consists of projecting the signature information onto a bar mask lying on the binary signature image. The shape factor is made by the repetition of one basic pattern in a multi-scale approach (Figure 1).

Each pattern is made of three types of bars (i.e., vertical, horizontal, and cross bars where the two directions are taken into account). Each bar is made of

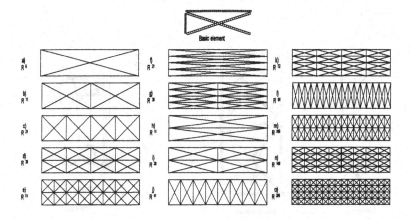

Fig. 1. Multi scale shape factor

a finite number of basic elements (pixels). A value expressing the characteristic of the coded signature is attributed to each basic element. The goal of this method is to make it possible to code several characteristics of the signature (e.g. position, orientation, gray levels).

The interest such an approach is to offer different points of view on the signatures. Large bar masks give a global view where only the general characteristics are usable. A fine view gives a perception close to the text without being attached to the spelling. This technique eliminates the first step of segmentation in letters that have not been solved for handwritten signatures. It allows us to particularize one or more scales for the specific draws of a writer. Indeed, the scale of perception (fine or large) should be relativized to the inner volume of the signature of every writer.

The inconvenience of this coding method for handwritten signature images is the poor definition of the geometric characteristics of the signatures (Figure 2a). This is why we have proposed, in [11][13], some modifications in the nature of the information taken into account in the coding process.

Distance Coding The distance in coding allows us to code the proximity between the pixels of the signature and the mask bars. Thus, the farther from a bar a pixel is, the weaker its influence on this bar (Figure 2b). Several points may be coded on the same pixel. To find the coding value, the maximum operator is chosen. This maximum value provides the shape of the pieces of the signature inside the pattern. For a large scale pattern, the coding signal provides the external shape of the signature that is close to the shape factor suggested by Nouboud[12]. The new pattern is more descriptive thanks to a multi-scale approach.

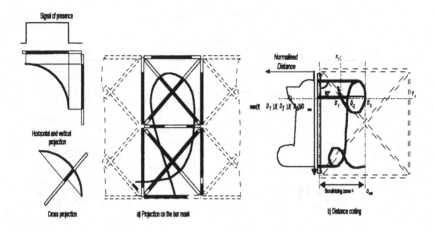

Fig. 2. Projection principle on the bar mask a) Coding of presence b) Coding of distance

Pseudo-dynamic Signature Information To generalize an automatic off-line signature verification system, we should use geometric and pseudo-dynamic information on the signatures. The introduction of the distance in the coding tends to improve the shape factor because it takes into account the pertinent information necessary for hand-written signature verification in the geometric sense. Now, it becomes necessary to introduce the dynamics of the hand-written signature.

To use geometric and dynamic signature information, several methods can be used : a sequential, a parallel, or a simultaneous. The sequential method needs to make the verification procedure on one type of information in order to verify the second. Two kinds of drawbacks are noted. Every type of forgery implies a verification with specific information like pseudo-dynamic, in the case of optic transfer forgery, and geometric in the case of random forgery. Yet, the type of forgery is never known, so we cannot define the sequence of treatments to be used. Moreover, sequential treatments are longer than other analysis systems. It is for this reason, we generally prefer different kinds of architecture. The parallel treatment needs to construct two verification systems based on every type of information. The first one treats geometric signature information and the second treats pseudo-dynamic signature information. Then, it is necessary to combine the results from the two systems to obtain the final decision. This combination is hard to be define and the verification needs twice as many resources. A simultaneous verification needs the use of signature features which directly combine geometric and pseudo-dynamic signature information. For this, we should establish a combination operator that mixes the information from a heterogeneous nature. Yet, the pseudo-dynamic signature information is specific to each writer. So, it becomes very difficult to propose an analytical combination operator, but it is easier to give a logical qualification for which it should operate. This last comment induces the possibility of using the fuzzy technique.

To define a qualitative combination operator, we must make a qualitative analysis of pseudo-dynamic and geometric signature information.

First, to note the different qualification of the pseudo-dynamic signature information, we should analyze the different combinations of speed and pressure during the writing process. We can dissociate three global components of a signature : the "pen down", the "body", and the "pen up" [13].

The "pen down" is an initial event where the writer puts his pen in place to begin or to resume again his gesture. This action is characterized by a low speed and a high pressure in a short time. The gray levels corresponding to this event are dark. After the "pen down", the writing speed increases, whereas the pressure decreases, to reach a fluctuating medium level corresponding to the "body" of the signature that we can separate into plains and looses. Lastly, the "pen-up" is a part of the signature where the writer prepares himself to stop his gesture and finish his signature or to resume again in another position. "Pen-up" events are very fast with a decreasing pressure that gives clear gray levels in the signature. In addition, the passage between every event is gradual with the harmony of the gesture. This helps to qualify, in a continuous way, the pseudo-dynamic signature information.

These events have a relative importance in their repeatability in time. The "pen-down" is an important event because it is the starting point of the signature. The writer pays a particular attention at this moment because this starting point plays a role on the expression of his identity. The "body" is a succession of plains and looses, then it's a succession of acceleration and deceleration of the gesture where the attention of the writer is less important. During the "pen-up", the speed increases and the writer unfolds his attention. Then, it's a less important event due to its time repeatability. Sometimes, we note some degeneracy because the writer completely unfolds his attention in this part of the signature.

From this description and the order of importance of the different execution steps of the signature, we can define that the "pen-down" is a crucial event but it's less present. The "body" represents the major part of the geometry, but it seems to be certain for the dynamic. And last, since the "pen-up" is less certain in the geometry and the dynamic, then it's the less important event in the signature.

2.2 Combination Operator

From this last linguistic description relating to the importance of the event and the notion of proximity usable to describe the geometric characteristics of the signature, a simultaneous coding can be defined. The relation between the characteristics and the coding can, for example, translate the importance of the events of the signature, but other relations can be expressed.

To realize this relation between the different notions of pseudo-dynamic ("pen down", "body", "pen up"), the proximity and the importance of the events, a fuzzy inference formulation appears to be well adapted.

To support this relation, we retain the most simple and the most known principle, the Mamdani [14] model which allows the integration of a linguistic formulation from the fuzzy characterization of geometric and pseudo-dynamic signature information (Figure 4).

Such a model requires a preliminary step of fuzzification or symbolization of the inputs. The obtained characterization should be defined on semantic fundamentals. In our case, the universe of discourse is a combination of speed and pressure, associated for the pseudo-dynamic, with some particular events ("pen down", "body", "pen up") corresponding to gray level ranges.

These three terms concern the gray level range of the signature. A threshold determines the edge between the gray levels of the signature and of the background. This threshold allows the placement of the three terms concerning the signature (i.e., "pen up", "body", "pen down") and a fourth term placed to represent the background. The linguistic terms are taken by trapezium to express an imprecise, but more certain, characterization of the terms. Simultaneously, we have chosen to traduce the notion of proximity by three linguistic terms ("close", "medium" and "far") in a distributed way so as to have a good definition of the level of proximity. The establishment of inference rules is intuitively done to express the relation of importance of the pseudo-dynamic events in the global analysis we have made, while taking into account the proximity notion that should stay a characteristic of the signature geometry. Thus, the "pen down"/"close" is considered as very important information, as is the "body"/"close". These two combinations are associated to a "large" coding. A "pen up"/"close" is less important due to its low repeatability. Simultaneously, the "body" is essentially characterized by the proximity in order to keep the discrimination potential of random forgeries by traducing the geometry of the signature. Otherwise, the "pen down"/"far" and the "pen down"/"medium" are more important than a "body"/"far". Also, the "pen up" is considered less important than other notions in the same case of proximity. The linguistic definition of the inference rules which we gave can be expressed in an "IF-THEN" implication way.

IF the proximity is "far" and the pseudo-dynamic is "pen up" THEN the coding is "low".

IF the proximity is "close" and the pseudo-dynamic is "pen down" THEN the coding is "large".

Other combinations are derived from this extreme case and the notion of pseudo-dynamic is placed as an alteration of the notion of proximity. A point "close" is globally important. Yet, if the gray level of a point becomes light, it is less certain that it is so important for the signature then its interest decreases which implies the decrease of the coding. We have chosen the minimum and maximum for the operator of combination/projection, and the product of the inputs is accomplished by the minimum operator [14].

In this way, where many rules can be fired in a fuzzy model, the smoothing of the real behavior towards the behavior we want is allowed and many output terms can be activated. But, the coding on the bar mask needs a unique value.

The defuzzification operator is obliged to aggregate the different output propositions in only one numerical value. It is done by the centroïd that provides a consensus on the final output. Figure 4 shows the fuzzy formulation of the coding operator.

3 Validation

To evaluate our shape factor, we are using a standard signature database of 800 images completed by 20 writers (40 signatures per writer). Signatures were written on a white paper with a pilot fine liner pen in a 3 x 12 cm rectangle. Signatures were digitized to produce an image of constant size (512 x 128 pixels) and quantified in 256 gray levels. Every signature is moved so that its gravity center corresponds to the image center. This translation is allowed because it does not modify the characteristics we will use. Furthermore, the local variations of position, related to the gravity center, are reduced isotropically and allow only the use of the right characteristics.

3.1 Numerical Experiment : The Case of Random Forgeries

To evaluate the discriminant potential of each bar mask, we define a reference set and a test set for each writer and each bar mask. For a writer (class ω_1), his first 20 signatures have been attributed to the reference set and his last 20 ones to the test set. We introduce 5 signatures, chosen randomly from every other writer (class ω_2) within their first 20 signatures, for the reference set and a same operation is done for the test set but with their last 20 signatures.

For a correct evaluation of the discriminant power of each bar mask associated to our coding technique, we use a k nearest neighbors classifier with vote. This choice can be justified by the non-parametrical nature of the signature verification problem. Moreover, this classifier is often used in this kind of problem and provides the lower bound of the error with only one neighbor, if there is enough data [15]. An overall stage of evaluation consists of applying the protocol defined by Sabourin [10], on the same database.

Each discrimination produces two types of error reported in terms of type I [E_1, false rejection of genuine signatures] and type II [E_2, false acceptance of random forgeries] error rates evaluated for the 20 writers. The mean total error rate E_t of the test is expressed as the average of E_1 and E_2 which considers the same level of importance of each error type.

We can note that a statistical assessment is obtained by repeating this evaluation 25 times. By choosing different random forgeries, the final error rates are independent of the choice of forgeries.

To evaluate the performance of the coding process, we cannot directly use the entire quantity of information given by our coding technique in the discrimination system. The volume of information is reduced by describing the coding signals on the average. Some experiments, referenced in [13], show that the average of the coding signal gives the best performances. Then, the dimension of each characteristic vector space is given by the number of bar in each mask; as for example, R^6 for the mask a, and R^{276} for the mask o (Figure 1).

The results obtained are presented in table 1 and we can compare them to the results of Sabourin's method with the same database and protocol. In this case of random forgeries, our coding operator demonstrates better results than Sabourin's method for large bar masks such as mask a, h, b and f. For the largest bar mask a, the performance increases 40%. But, for fine bar masks, the performances are close to the Sabourin method results. Yet, we should relativized these last results because the dimension of the representation space is very large compared to the number of presented samples. Indeed, Bellman [15] defines this problem as a curse of dimensionality which is why it is very hard to prejudge the exact value of the performances if the number of samples increases. Nevertheless, we note that the global performance of our coding operator is better than Sabourin's for the discrimination of random forgeries by 23%, even if this method is made not only made for this kind of detection.

3.2 Numerical Experiments : The Case of Optic Transfer Forgeries

To evaluate the performance of our coding operator with this type of forgery, the problem is obtaining optic transfer forgeries because the database is not defined in this way. Thus, we apply an image treatment close to the photocopier optic operation. This operation can be viewed as a contrast modification that enforces the homogeneity of the signature gray levels. To achieve this, we propose the following look up table (Figure 3).

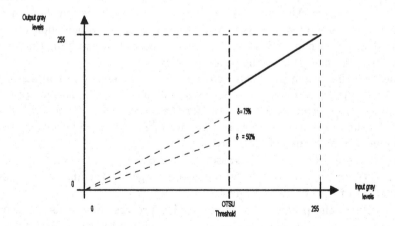

Fig. 3. Contrast modification look up table (LUT)

The factor δ allows us to define the rate of contrast modification. Its value can be directly associated to the photocopy quality or the difficulty level of detection. Indeed, the higher δ is, the closer a photocopy is to a genuine signature. Nevertheless, the geometry of the photocopy is the same as the genuine signature, whatever is the value of δ. Figure 5 shows a genuine signature and its photocopy with $\delta = 50\%$.

The last protocol cannot be used directly for the performance evaluation of optic transfer forgeries detection. The definition of the reference set from ω_1 and ω_2 is the same because we are not able to integrate all the types of optic transfer forgeries. On the other hand, the codification of the protocol focuses on the definition of the test set, which should now integrate many photocopies simulation of genuine signatures and of random forgeries. This new test set is constituted in the same manner as the precedent protocol. Then, we have not two kinds of affectation during the evaluation, but only one for the forgeries set.

That's why the error E_1 is replaced by an error E_{rec1} that characterizes the detection rate of genuine signatures photocopied. E_{rec1} should be maximum to define a good recognition of this type of forgeries. Simultaneously, E_2 should remain a minimum. Then, the evaluation of the performances, done by the nearest neighbor classifier, should be calculated by $E_t = \frac{(100 - E_{rec1}) + E_2}{2}$, where E_{rec1} is a success rate of detection of a genuine signature photocopied.

The evaluation of the discriminant power of the coding operator, to show the difference between genuine signatures and their photocopies, is done by a nearest neighbor classifier with the last defined protocol. Moreover, this protocol is repeated twice with the different values of δ (75% and 50%). Table 2 shows the results of experiments with E_t and the values of E_{rec1} according to the values of δ.

These results show that the detection rate E_{rec1} with $\delta = 50\%$ is 7 - 27% and it demonstrates a good detection of photocopied random forgeries. The decrease of δ is directly observed by the increase in the detection rate from 27% to 45%.

The results are low, but not totally inadequate because the protocol used in this evaluation is directly issued from one of the random forgeries. Since the shape factor based on fuzzy inference rule is only able to do a small transformation of the coding signal because the geometry of the signatures is the same, the photocopies are naturally closer to genuine signatures than random forgeries. Indeed, random forgeries have different geometric characteristics. So, in the characteristic vector space, photocopies will be close to genuine signatures. A k nearest neighbors without distance rejection can yield only low performances. With the demonstration that photocopies, viewed by the shape factors, are effectively different from a genuine signature, we have defined a new protocol that integrates only 5 photocopies of genuine signatures (randomly chosen from the set ω_1) into the class of forgeries of the reference set. These 5 photocopies will to characterize a domain in the characteristics space to which photocopies can be attached.

The performances obtained from this second protocol are shown in table 3 with error rates E_t, and E_{rec1}, according to the value of δ. We note that the detection of photocopies becomes very interesting according to the value of E_{recl}. With $\delta = 75\%$, the performances increase to the range of 48% to 68%. With $\delta = 50\%$, an increase to the range of 57% to 87% of photocopie detection is noted. This progression of the performances by the introduction of only five photocopie references shows that it is a local problem in the characteristic space. Consequently, the use of a classifier to make a verification decision should not be an extrapolator classifier such as classifier by compilation (e.g, Fuzzy C means, probability estimator,...).

Moreover, we can see that the performances are less dependent on the shape and the resolution of the bar masks than in the case of random forgeries. This is due to the nature of the gray level transformation. Because it is global and linear, the pertinence of the FESC is representative with large bar mask. We can found an equivalence with the performances of fine bar masks.

We note that the simulation of a photocopy is a particular case of optic transfer forgeries and another technique of copying will make more general modifications on signatures, both geometric and dynamic signature information. So, the shape factors can show better performances, with this case of modification, but they remain difficult.

4   Conclusion

We have shown that the use of geometric and pseudo-dynamic signature characteristics of signatures allows the increase of the performances of the verification in the case of random forgeries. These performances are better than those of the Extended Shadow Code. Moreover, it permits the treatment of other kinds of forgeries. By the acquisition of real forgeries of every type, we will be able to finetune the different parameters of our operators such as fuzzy membership functions and inference rules.

In future works, we will develop the step of multi-scale discrimination to operate on our shape factor in order to define a relevant decision from the different points of view of perception and the knowledge of verification problems given by samples of signatures in each scale.

References

1. Sabourin R. Genest G., " Definition and evaluation of a family of shape factors for off-line signature verification ", Traitement du signal, pp. 585-596, (12)6, 1995.
2. Evett I.W., Totty R.N., " A study of the variation in dimensions of guenine signatures ", Journal of the Forensic Science Society, 25, pp. 205-215, 1985.
3. Gupta S.K., " Protecting signatures against forgery ", Journal of the Forensic Science Society, 19, pp. 19-23, 1979.
4. Moon H.W., " A survey of handwritting styles by geographic location ", Journal of the Forensic Sciences, 4, 1977.
5. Totty R.N., " The dependence of slope of handwritting upon the sex and handedness of the writer ", Journal of the Forensic Science Society, 23, pp. 237-240, 1983.

6. Locard E., " Les faux en écriture et leur expertise ", Editions Payot, Paris, 1959.
7. Sabourin R., " Une approche de type compréhension de scène appliquée au problème de la vérification automatique de l'identité par l'image de la signature manuscrite ", Thèse de Philosophiae Doctor, Ecole polytechnique de Montréal, 1990.
8. Burr D.J., (1988), "Experiments on Neural Net Recognition of Spoken and Written Text.", IEEE Transaction on Acoustics, Speech and Signal Processing, 36(7), pp. 1162-1168.
9. Sabourin R., Cheriet M. and Genest G., " An extended-shadow-code based approach for off-line signature verification ", 2nd IAPR Conference on Document Analysis and Recognition, Ibraki, Japan, pp. 1-5, October 20-22,1993.
10. Sabourin R. and Genest G., " An Extended-Shadow-Code based approach for off line signature verification : Part-I- Evaluation of the bar mask definition ", 12th ICPR International Conference on Pattern Recognition, pp 450-453, Jerusalem, Israël, October 9-13, 1994.
11. Simon C., Querelle R., Levrat E., Bremont J., Sabourin R., (1996) " Codage d'images de signatures manuscrites pour la vérification hors ligne ", CNED'96, pp. 171-177, 3-5 Juillet, Nantes.
12. Nouboud F. et al., (1988)," Authentification de signature manuscrites par programmation dynamique ", PIXIM, Paris, pp. 345-360.
13. Simon C. " Vérification automatique de l'identité par perception multiéchelle et discrimination intégrée floues de l'image de la signature manuscrite ", Thèse en Automatique de l'Université Henri Poincaré Nancy I, 1996.
14. Foulloy L., Gallichet S., (1992), " Représentation des contrôleurs flous ", 2^{ieme} Journées Nationales sur les applications des ensembles flous, pp. 129-136, Nîmes, 2-3 Novembre.
15. Bellman R., (1958), " Dynamic Programming and stochastic control processes. ", Information and control, 1(3), pp 228-239.

Table 1 : Results of coding methods with the nearest neighbor algorithm

Bar mask	a	h	b	i	c	j	d	k	e	l	f	g	m	n	o
ESC	2.4	1.2	0.6	0.4	0.2	0.1	0.2	0.2	0.1	0.1	1.5	0.5	0.2	0.01	0.02
FESC	1.3	0.6	0.5	0.4	0.1	0.1	0.3	0.1	0.2	0.2	1.2	0.5	0.3	0.1	0.09

Table 2 : Results of FESC coding applied to photocopies

Bar mask	a	h	b	i	c	j	d	k	e	l	f	g	m	n	o
$E_t, \delta(75\%)$	40	40	42	37	43	46	39	43	39	47	39	39	40	43	44
E_{recl}	21	20	17	27	13	8.4	23	14	21	6.7	22	24	20	13	12
$E_t, \delta(50\%)$	32	31	33	27	38	42	27	34	30	45	31	30	20	13	12
E_{recl}	38	40	36	47	25	16	47	32	40	9.5	40	42	38	27	24

Table 3 : Results of FESC coding applied to photocopies with some photocopies in reference sets

Bar mask	a	h	b	i	c	j	d	k	e	l	f	g	m	n	o
$E_t, \delta(75\%)$	23	17	21	16	22	23	16	20	16	26	17	39	18	19	21
E_{recl}	55	66	59	68	57	53	68	60	67	48	67	64	65	63	58
$E_t, \delta(50\%)$	15	9.5	13	8.7	14	16	16	9.7	9.7	21	8.2	9.1	11	12	14
E_{recl}	72	82	75	84	73	68	87	81	81	58	85	83	79	76	73

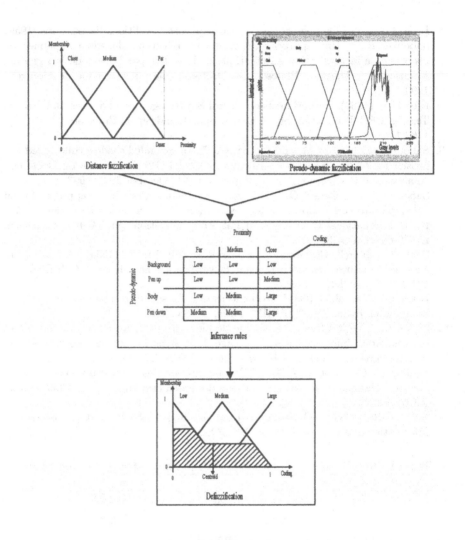

Fig. 4. Fuzzy coding operator

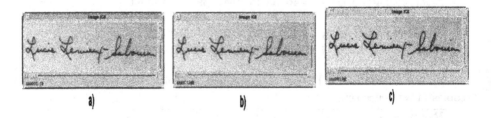

Fig. 5. An example of genuine signature (a) and its photocopie with a contrast factor of 75% (b) and 50% (c)

An Evaluation of Different Handwriting Observation Techniques from a Signature Verification Point of View

Ronny Martens, Luc Claesen
IMEC vzw
Kapeldreef 75
3001 Heverlee, Belgium
Tel.: +32(0)16-281211
Fax.: +32(0)16-281501
E-mail: martens@imec.be

Abstract.

In this paper, we discuss the relative importance of different signal types from the viewpoint of on-line signature verification. The signals that are evaluated are forces on the writing devices tip in 3 dimensions and angles of its shaft relative to the writing surface in 2 dimensions. These signals are captured by the SmartPen™, which is the heart of a completely operational biometric identity verification system. The results of the experiments that have been carried out are not just interesting for this peculiar identity verification system, since they allow to evaluate the implications of potential simplifications of the SmartPen™-concept, but they are undoubtedly useful to future developers of on-line signature verification systems.

1. Introduction.

One of the most typical characteristics of our current society is the steadily growing number of tasks that is performed by machines, without any kind of human intervention. This poses us to the relatively new problem of automatic identity verification.

Until now, almost all automatic identity verification approaches have been non-biometric. This means that we use some kind of secret, shared between the subject that has to be verified and the object that is performing the verification task. Typical examples of this approach are the use of simple physical keys, badges, pass-words, PIN-codes etc. These techniques are currently still extremely popular in the identity verification sector, because of the low complexity and cost of the systems that are based on their use. It is however clear that these non-biometric identity verification systems are not offering a very high degree of safety, since the key that is used for identification can be stolen or simply lost. However, the main reason why non-biometric techniques have become infeasible nowadays is that they require a non-trivial user-effort like remembering a pass-word or carrying a key.

To get rid of these disadvantages has been the main reason why during the last few decades more and more research effort has been put into the development of biometric identity verification systems. In this alternative approach some kind of physiological (fingerprint, iris pattern) or behavioural (typing rhythm, signature) characteristic is used to decide about the identity of the person who the system is interacting with.

An overview of the most popular biometric identity verification approaches can be found in [1].

In this paper, we will limit ourselves to biometric identity verification by analysing human signatures. A general overview of the activities in this domain during recent years can be found in [2] and [3].

Two different trends exist in signature verification. Off-line signature verification has a very general applicability, but results in a rather poor system performance. On-line signature verification, on the contrary, guarantees a much better system performance. However, since it requires the use of some kind of special input instrument to capture the dynamics of the signing process, it has a limited application area. Our work has to be situated in the on-line signature verification domain. Even here several trends exist. Some researchers ([4], [5], [6]) use a conventional tablet to measure pen-tip positions and pressure over time while a person is signing. Others ([7], [8]) try to build special purpose input-instruments, hoping to end up in a better cost-benefit situation by either improving the verification performance or by using a system that can be manufactured cheaper than a tablet. We follow this second approach.

Section 2 focuses on our input instrument - the SmartPen™ - and describes how its signals are used to do verification. Our experiences about the relative importance of the different types of signals are summarised in section 3. Conclusions are drawn in section 4.

2. Verification Process.

The complete signature verification process is usually performed in 3 sequential steps. These are the data-collection and pre-processing, the feature extraction and the actual classification. They are described briefly in sections 2.1-2.3.

2.1 Data-collection.

The instrument used in this work - the SmartPen™ - (see figure 1 and [9]) offers a very complete and interesting view to a human signature. The device itself looks like an ordinary pen, but is packed with electronics that collect, process and transmit the signals from several different sensors. These sensors serve 2 purposes:

- Measuring the forces on the pen-tip in 3 dimensions.
- Measuring the angles of the pen-shaft, relative to the writing surface.

These 5 signals are slightly oversampled [2] at 100 Hz. Since the measurements are done relative to the pen, we are faced to the problem of eliminating the effect of rotations of the pen around its own axis, as this is obviously not a characteristic of the signing process. This elimination is done by rotating the reference axes so that the new X and Y axes are respectively the ones with maximal and minimal energy contents. The actual orientation of the axes is done by considering the average values of the force or angle signals that are under study.

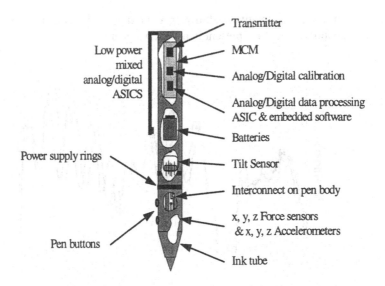

Transmitter

Low power
mixed
analog/digital
ASICS

MCM

Analog/Digital calibration

Analog/Digital data processing
ASIC & embedded software

Batteries

Power supply rings

Tilt Sensor

Interconnect on pen body

x, y, z Force sensors
& x, y, z Accelerometers

Pen buttons

Ink tube

Figure 1: Schematic representation of the SmartPen.

2.2 Feature-extraction.

As described in [2] and [3] feature extraction can be done in many different ways. According to us [10] one of the most essential characteristics of handwriting is the fact that it is a non-linear process. By non-linear we mean that there exist small, non-homogeneous timing differences between several trials to write the same word or signature. The existence of these non-linearities is illustrated in figure 2a.

In the signal processing and pattern recognition community, several different techniques to deal with these timing-differences are known. In this work we opt for an approach that has been proving its value during the last few decades, especially in the area of speaker specific isolated word recognition. The approach is called dynamic time warping (DTW, see [12]). The aim of the algorithm is to remove the timing differences that exist between the patterns that have to be compared. An example of the output of the DTW-algorithm is shown in figure 2. Part b shows how the patterns from figure 2a resemble to each other after the timing correction. This information is generally called form information. Part c reveals the relation between the time axes of both signals. In literature, this information is referred to as motion information.

One of the first successful attempts to use DTW in the signature verification area has been reported by Sato and Kogure [13]. Sato and Kogure used 2 very simple parameters x_{Form} and x_{Motion} to do the actual classification. These parameters are defined as the (Euclidean) distances between the aligned patterns (x_{Form}) and the warping path and the diagonal (x_{Motion}). In [8], a way to extract much better form and motion parameters is described. The most important difference in the feature extraction scheme is that in the alternative approach from [8] the DTW-step and the actual parameter determination are no longer allied unrelentingly. The main advantage of this slight complication is that it allows us to take into account information about the relative stability in both time and frequency domain of the different phenomena that are present in a certain signature.

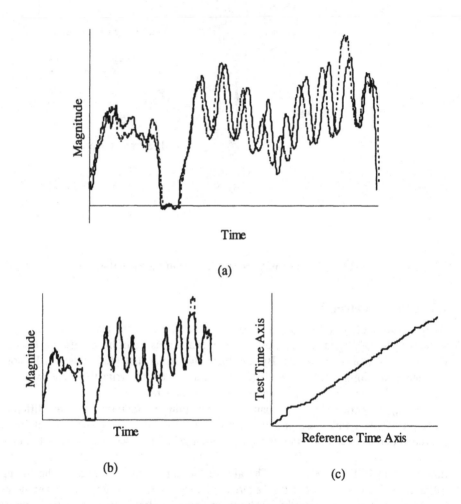

Figure 2: Non-Linear Timing Differences.
(a) Their presence in a pressure signal.
(b) Form information.
(c) Motion information.

In this work we use a similar feature extraction scheme. The only relevant difference with [8] is that instead of using a simple Gabor transform to extract our parameters, we use the discrete wavelet transform (DWT). As has been illustrated in many practical situations, this transform is very well suited to describe most natural signals, as it allows us to "see the forest and the trees". By this we mean that high-frequent, mostly unstable, phenomena are being observed using a small time-window and - remember the uncertainty priciple - a low frequency-resolution. To study stable, low frequent phenomena on the contrary, a wide time-window, guaranteeing a high frequency resolution, is used. The energy distribution of the wavelet coefficients and the filterbank represention of the DWT is depicted in figure 3.

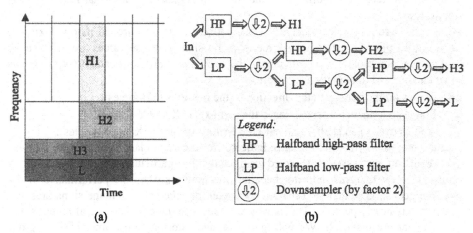

Figure 3: Wavelet Transform.
(a) Energy localisation of coefficients.
(b) Filterbank representation.

As illustrated in figure 3b only the low frequency outputs of a certain filtering stage are split further, while high frequency outputs remain unchanged. Of course, since the frequency range of each filter output is only half the range of the input signal, we can sample the output signals down by a factor of 2. A more complete discussion about the use of alternative transformation schemes in the parameter extraction stage of the on-line signature verification process can be found in [14].

2.3 Classification.

The last step in the verification process is the condensation of the extracted parameters into a single binary constant that indicates whether or not a signature is a forgery. The function that does this mapping from the complete feature space into Z_2 is called a discriminant function. There exist a lot of different ways to construct these discriminant functions, ranging from Mahalanobis classifiers, over kernel functions and artificial neural networks, to nearest neighbour classifiers. The latter techniques are obviously more general, in the sense that less assumptions about the statistical properties of the populations involved are required. The price to pay for this generality is the rather high amount of training that is needed in order to construct a good classifier. As illustrated in [11], the quantity of available training data is in practical situations insufficient to use these methods effectively. This results in a better system performance if we use simpler classification schemes like Mahalanobis decision making. In this paper we opt for simplified Mahalanobis decision making [11]. This simply means that we ignore the existance of cross-correlations between the different features.

3. Signal Comparison.

Now that we have built a complete signature verification system, we can use it to evaluate the relative importance of the different signals that the SmartPen™ puts at our disposal.

Since we have to eliminate the effect of rotations of the pen around its own axis, we can not do this evaluation with force or angle signals whose values are affected by these rotations separately. This means that we have 3 different categories of signals that can be used to classify a signature. These are:

- Forces on the pen-tip in the direction of the pen-axis (=Z-forces).
- Forces in a plane perpendicular to the pen-axis (=X and Y-forces).
- Angles of the pen shaft, relative to the writing surface (=X and Y-angles).

As the most important performance indicator for a certain class of signals we will use the EER[1] of our system, achieved using only this particular type of input. It is however clear that this indicator becomes meaningless when the performance of the system gets so high that the separation between the classes of genuine signatures and forgeries is nearly perfect. This is why we don't use the complete set of features that we have at our disposal. We will ignore motion information parameters, basing our classification completely on form information. The interested reader can get an idea about how the performance of the complete system will improve when we include motion information into the parameter extraction process from [8].

A very important factor when discussing signature verification results is the database that has been used. In the construction of our database 57 people have been involved. 18 of them provided us with 20 genuine signatures each. These signatures have been collected in 3 sessions with varying writing conditions, spread over 3 months. Our system is trained using 10 genuine signatures for each person. As forgeries for a certain signer we use the signatures produced by the other 56 signers. This means that we evaluate our system using what in literature is called random forgeries. In general, using random forgeries is certainly not the best thing one can do to evaluate the performance of an on-line signature verification system. In our situation, however, there is no real objection, since we are only comparing the value of different signal types.

3.1 Original Signals.

In a first experiment we use the original SmartPen signals for classification. The only pre-processing is the elimination of rotation effects, as has been described in 2.1. The performance of the system using all possible combinations of the 3 signal classes is summarised in Table 1.

---

[1] EER = Equal Error Rate. This is the percentage of misclassifications that occurs when the False Acceptance Rate (FAR) and the False Rejection Rate (FRR) are equal.

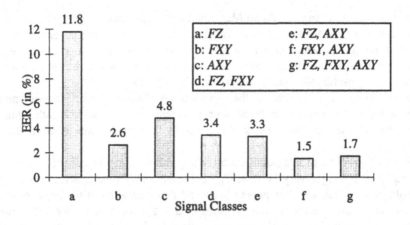

Table 1: Evaluation of signal classes

Table 1 reveals that all types of signals are useful on their own. Force and angle signals are obviously complementary, since classification by using a combination of them is always better than classification by using only one of these signal types. Looking inside the class of force signals, however, we observe a rather strange phenomenon. Adding Z-force based parameters to a system that is already using XY-forces reduces the system performance instead of improving it. A straightforward conclusion would be that Z forces carry no information that is additional to the one present in the XY-forces. As a result we should be able to simplify our sensor-concept, ignoring Z-forces from the scratch. In the next section we will clarify why this conclusion is not valid.

3.2 Decorrelated Signals.

A basic pattern recognition error in section 3.1 is that we ignored the existence of cross-correlations between different signals. Table 2 reveals their magnitude.

	FX	FY
FY	0.0	-
FZ	0.97	-0.33

Table 2: Cross-correlations between force signals.

The lack of cross-correlations between FX and FY should be clear from the way we redefined our reference axes (section 2.1). The huge cross-correlation between FX and FZ can be understood easily from this pre-processing stage as well. The most important component of the forces that are observed is the pressure between pen and paper. This pressure component is almost an order of magnitude bigger than the friction forces produced during signing. Since for normal handwriting the pen-shaft is far from perpendicular to the writing surface, it is clear that this pressure component represents a lot of the energy in both the FZ and the FXY class of signals.

The effect of these cross-correlations can be understood rather easily. Suppose we have 2 features x and y, with an almost equal importance. A 3rd parameter z is highly correlated to y. Using simplified Mahalanobis decision making [11], the introduction of z into the classification process will result in falsely stressing y, relative to both x and to the new information introduced by z. This means that in practical situations, the performance of the system can diminish, even when we are introducing intrinsically useful information into the verification process.

Table 2 suggests to decorrelate the signals to be classified before time-warping and parameter extraction. Since the number of samples in a signature is rather high, this can for instance be done using the generally known Karhunen-Loève transform (KLT).

As should be clear at this point, decorrelating the 3D force signals results in 1 force component having a very large energy contents, and 2 others having only a rather small energy. The first component is a good approximation of the pressure between paper and pen-tip, while the others correspond to friction forces. We will denote the friction forces by Fw, and the pressure pattern by P. Table 3 illustrates the EER's after decorrelation. For completeness, we have added once again the angle signals.

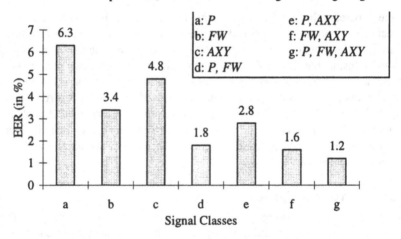

Table 3: Evaluation of forces after decorrelation.

By comparing Table 1 to Table 3 we can conclude that measuring forces in 3 directions instead of 2 is indeed useful. This conclusion is illustrated more clearly in Figure 4, where the real evolution of FAR and FRR as a function of the classification offset is shown. While the number of false acceptances remains almost the same for a certain classification offset, the number of false rejections declines obviously after decorrelating the force signals.

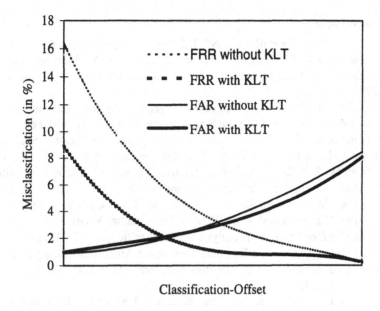

Figure 4: System performance using force signals with and without decorrelation.

4. Conclusion.

One of the most obvious conclusions of this study is that both information about forces between pen-tip and paper and angle signals of the pen-shaft are relevant when considering the problem of on-line signature verification.

Within the class of force signals, the situation is at first sight a little more complex. Using simple force signals, measured relative to the pen, it seems like forces in the direction of the pen-shaft do not provide information that is complementary to force-signals in a plane perpendicular to this axis. By decorrelating the 3D force signals before time-warping and parameter extraction, it is however demonstrated that each component of the force signal adds some information relevant to identify the signer. As could be expected however, friction forces have a higher value than pressure patterns.

When we consider the relative importance of force and angle-signals, we notice that a person is identified more explicitly by forces on the pen-tip than by angles of the pen-shaft. However, both types of information are clearly complementary.

5. Acknowledgement.

This research was supported by a scholarship from the Flemish Institute for the promotion of the scientific-technological research in industry (IWT).

6. Bibliography

[1] B. Miller, "Vital Signs of Identity", IEEE spectrum, pp. 22-30, feb. 1994.

[2] R. Plamondon, G. Lorette, "Automatic Signature Verification and Writer Identification - State of the Art", Pattern Recogn., vol. 22, no. 2, pp. 107-131, 1989.

[3] F. Leclerc, R. Plamondon, "Automatic Signature Verification: The State of the Art – 1989-1993", Int. J. Pattern Rec. Artif. Intell., vol. 8, pp. 643-660, 1994.

[4] R. Plamondon, "The Design of an On-Line Signature Verification System: From Theory to Practice", Int. J. Pattern Rec. Artif. Intell., vol. 8, pp. 795-811, 1994.

[5] J. Bromley, J. W. Bentz, L. Bottou, I. Guyon, Y. Lecun, C. Moore, E. Sackinger, R. Shah, "Signature Verification using a "Siamese" Time Delay Neural Network", Int. J. Pattern Rec. Artif. Intell., vol. 7, no. 4, pp. 669-688, 1993.

[6] B. Wirtz, "Stroke-Based Time Warping for Signature Verification", Proc. of the 3rd Int. Conf. on Document Analysis and Recognition, pp. 179-182, 1995.

[7] J. S. Lew, "An Improved Regional Correlation Algorithm which Permits Small Speed Changes Between Handwriting Segments", IBM J. R.&D., vol. 27, no. 2, pp. 181-185, 1983.

[8] R. Martens, L. Claesen, "On-Line Signature Verification by Dynamic Time-Warping", Proc. 13th. Int. Conf. on Pattern Rec., vol. 3, pp. 38-42, 1996.

[9] L. Claesen, D. Beullens, R. Martens, R. Mertens, S. De Schrijver, W. de Jong, "SmartPen: An Application of Integrated Microsystem and Embedded Hardware / Software CoDesign", ED&TC'96 User Forum, pp. 201-205, 1996.

[10] R. Martens, L. Claesen, "Dynamic Programming Optimisation for On-line Signature Verification", Proc. of ICDAR'97, to be published.

[11] R. Martens, L. Claesen, "On-Line Signature Verification: Discrimination Emphasised", Proc. of ICDAR'97, to be published.

[12] H. Sakoe, S. Chiba, "Dynamic Programming Algorithm for Spoken Word Recognition", IEEE Trans. on Acoustics, Speech and Signal Processing, vol. 26, no.1, pp. 43-49, 1978.

[13] Y. Sato, K. Kogure, "On-Line Signature Verification Based on Shape, Motion, and Writing Pressure", Proc. 6th Int. Conf. on Pattern Rec., pp. 823-826, 1982.

[14] R. Martens, L. Claesen, "Optimised Feature Extraction for On-Line Signature Verification", Proc. 9th Kolloquium "Signaltheorie" Bild- und Sprachsignale, 1997, pp. 305-308.

Generation of Signatures by Deformations

Claudio de Oliveira[1], Celso A Kaestner[1,2], Flávio Bortolozzi[1] and Robert Sabourin[3]
Pontifícia Universidade Católica do Paraná - PUC-Pr.[1]
R. Imaculada Conceição,1155 - 80215-901 – Curitiba -Pr.
Centro Federal de Educação Tecnológica do Paraná - CEFET-Pr[2]
Av. 7 de Setembro, 3165 - 80230-901 - Curitiba, Pr. Brasil
École de Technologie Supérieure[3] - 1100, rue Notre-Dame Ouest
Montréal (Québec) H3C 1K3, Canadá
claudio@rla01.pucpr.br[1], fborto@rla01.pucpr.br[1] sabourin@gpa.etsmtl.ca[3]

Abstract: The techniques of automatic classification of hand-written signatures have been studied and some of them are based on the application of neuronal nets or statistical methods. Nevertheless, the great number of samples required by these methods turns many of its practical applications unfeasible. This article describes a technique for automatic generation of signatures originated from the deformation of a reduced number of genuine samples. The technique used here is based on convolution between deforming polynomials representing the deformations and the signals representing the horizontal and vertical moves of the pen, required for the reproduction of the original samples. The result of the convolution produces the deformation of those signals and, consequently, the deformation of the tracing obtained from them.

1. Introduction

In signatures off-line acquisition every dynamic information relevant to the writing process is lost, since the acquisition occurs after the writing is done.

We use in our work the off-line signature acquisition and consequently the tracing sequentialization is lost. However, this characteristic is of major importance for the development of the proposed signature deformation technique. The partial reconstruction of the normal writing tracing sequentialization starting from off-line samples is described by [DOE93]. Although the algorithms for re-connection of the normal writing tracing produce good results, they are not perfect. Thus, in spite of the feasibility of the adaptation of those algorithms to the signature writing, deeper studies for the development of automatic signature deformation tools will be required. These tools can be found at [PLAM89], [PLAM90], [RAND90] and [SABO90].

The steps required for the obtaining the sequence of points defining the signature tracing are thinning, shrinking, determination of characteristic points, determination of trace segments, reconnection and trace sequentialization.

A sample of input signature can be seen on Fig. 1.

Considering that the signature samples obtained by the off-line acquisition process have been previously filtered (that is, without noise) and thresholded, the stage of samples pre-processing consists of the re-construction of the sequentialization of the tracing.

Fig. 1: Sample of input signature

2. Important Concepts

2.1 - Thinning

To obtain a sequence of points which will allow the following of the trace it is necessary that each point, except extreme and trace-crossing points, have a predecessor and a successor. This restriction does not apply to the component points of the image obtained soon after the signature acquisition. Thus, it is necessary to select the component points of the signature skeleton. The thinning algorithm produces a trace representation complying with that restriction. Each trace segment of the thinned image is exactly 1 pixel thick. The choice of the thinning algorithm is of major importance, since the performance of the following algorithm depends directly on the quality of the resulting skeleton. Fig. 2 shows the thinned image corresponding to the sample image and the improper **"points"** originated by most thinning processes.

2.2 - Shrinking

The improper **"points"** generated by the thinning process are normally 3 pixels long. The controlled application of three interactions of the thinning algorithm erases three pixels from each segment ending, including those that form a normal tracing. The loss of three or four pixels of normal segments is only relevant if the segment is too small. In these cases, the segment may be completely erased.

The disappearance of small segments does not produce significant losses in the signatures modelling, normally composed by lots of points. However, these segments can define characteristic points determining the beginning or ending of real segments and, in these cases, should not be eliminated. Fig. 3 shows the thinned image without the improper points that result from the thinning process.

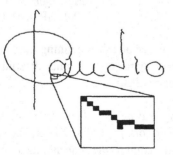

Fig. 2 Thinned image Fig. 1   **Fig. 3 Result thinning process.**

2.3 - Characteristic Points

After the thinning, points in the image that delimit the trace segments can be observed. These characteristic points are either extreme points or points where trace crossing of two or more traces occur, as shown in Fig. 4.

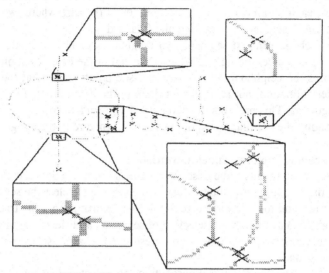

Fig. 4. Characteristics points.

2.4 - Trace Segments

A trace segment is a nonempty sequence of points between two characteristic points either equal or different. The sequence of points which define a trace segment do not necessarily represent the sequence of the tracing. At this stage of the pre-processing, only the set of points defining the segment and the neighbourhood of each of these points are determined.

The algorithm for extraction of trace segments considers that there are not two or more pixels associated to the same characteristic points, as shown in Fig. 5(a). The result of the removal of the undesirable pixel is shown in Fig. 5(b).

The elimination of these and other similar pixels does not allow the characteristic points associated to the trace crossing to have more than one pixel. As a consequence, in the crossing of two or more traces, only one pixel is defined.

Fig. 5 . The result of the removal of the undesirable pixel

Considering the restriction described on the previous paragraph, the algorithm for extraction of segments analyses the neighbourhood of each of the characteristic points. For each non analysed neighbour of each characteristic point, one segment is initiated. The subsequent point of the sequence is determined by analysis of the neighbours of the last point included in the sequence. In this case the last point and its two closest neighbours are not considered, in compliance with the trace direction. This procedure is repeated until there is not neighbour of the current point. In the event of reaching an ending or a crossing ,or in the case where two or more neighbours occur, a segment is identified and finalised.

For each determined segment two neighbours of characteristic points are analysed: one at the beginning and other at the end of the determination. The algorithm memorises the neighbours of the characteristic points reached at the end of the segment determination, not re-analysing them in the determination of the beginning of new segments. This prevents the duplicity of segments to occur.
Fig. 6 shows the decomposition of the sample signature in trace segments.

2.5 - Reconnection and sequentialization

In the experiments to be described, both the reconnection and the sequentialization of the tracing are manually done. It is possible to select the segments that compose a trace and to assign a course to the trace, starting from the trace segments and the original image. Both the selection of segments and the assignment of course are based on heuristics, but produce good results when a pre-selection of the signature samples is made.

Once the segments that compose a trace are selected, they may be put together, if necessary, in the endings corresponding to the points that belong to equal crossings, thus forming a unique sequence of points corresponding to the whole trace. After that, this sequence, unique for each trace, may have to be inverted to reflect the tracing sequentialization.

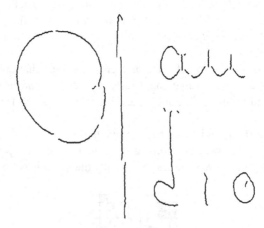

Fig. 6 . The decomposition of the sample signature in trace segments

It may occur that the same segment belongs to two or more traces. This is due to the superposition of traces or to the appearance of the "false" segments, previously discussed. This does not cause problems in the manual reconnection and sequentialization, since they require human decisions. Fig. 7 shows the result of the segments composition.

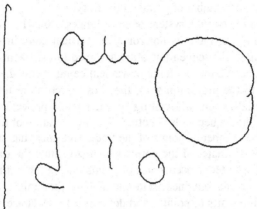

Fig. 7 . The result of the segments composition.

2.6 - Traces Mass Centres

The trace mass centre (x_o, y_o) is defined by the equation:

$$\left(x_0, y_0\right) = \left(\frac{1}{n}\sum_{i=1}^{n}x_i, \frac{1}{n}\sum_{i=1}^{n}y_i\right)$$

Fig. 8 shows the mass centres of the signature shown on Fig. 1.

After finding the mass centre (x_o, y_o) of a trace, all of its points must be translated to $\left(x'_i, y'_i\right) = \left(x_i - x_o, y_i - y_o\right)$, which permits the individual processing of traces, avoiding the positional dependency between traces.

Fig. 8 . The mass centres of the signature shown on Fig. 1.

2.7 - Reconstruction of Tracing Movements

The time unit used as a system of co-ordinates in our work, has a value equal to the value of the distance, either horizontal or vertical, between to consecutive points of the trace, in accordance with the type of movement. This time system of co-ordinates represents the shapes of the two movements, which will have a duration, in time units, equal to the number of points in the trace.

After definition of the time system of co-ordinates, from the first point of the sequence representing the trace, its horizontal or vertical co-ordinates are ordinately projected according to the sequentialization assigned to the tracing, in an vertical or horizontal axis, in accordance with the movement being obtained.

Perpendicular to the projection axes, the axes representing time keep the information of sequentialization, associating to each point projection, the position in which it occurs in the sequence. It is noticed that the duration of the tracing is equal, in time units, to the number of points of the trace, and, thus, has the only purpose of maintaining the co-ordinates of the points which constitute the trace duly sorted, in compliance with the order of occurrence of points during the writing process. Fig. 9 illustrates the movements obtained from one trace of the sample signature. It is noticed that the tracing starts at point a and describes the sequence $< a, b, j, c, k, d, l, e, m, f, n, g, o, h, p, i >$.

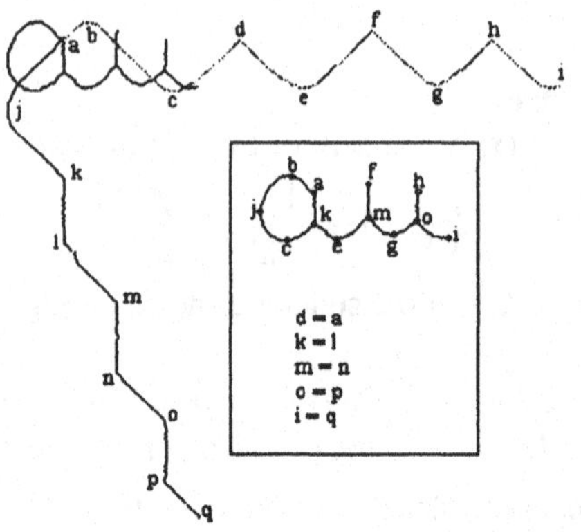

Fig. 9. The movements obtained from one trace of the sample signature.

Due to the choice of the time system of co-ordinates, the movements can be easily represented. As the time simply reflects the position of points within the trace, the movements can be represented as sequences of co-ordinates with the same order of the sequence of points that define the trace.

If the sequence $\langle (x'_0, y'_0), (x'_1, y'_1), ..., (x'_{n-1}, y'_{n-1}) \rangle$ represents a trace, then the sequences $\langle x'_0, x'_1, ..., x'_{n-1} \rangle$ and $\langle y'_0, y'_1, ..., y'_{n-1} \rangle$ represent, respectively, the forms of horizontal and vertical movements associated to it.

The sequences of co-ordinates are interpreted as polynomial of order n-1, where n is the number of points of the trace. For example, the polynomial obtained from $\langle x'_0, x'_1, ..., x'_{n-1} \rangle$ and $\langle y'_0, y'_1, ..., y'_{n-1} \rangle$ are $x'_0 t^0 + x'_1 t^1 + ... + x'_{n-1} t^{n-1}$ and $y'_0 t^0 + y'_1 t^1 + ... + y'_{n-1} t^{n-1}$, respectively.

3 - Deformations of Signatures

The convolution describes the action of an observation instrument when it produces a weighted average of a range of some physical quantity. As the form of the weighting function does not change significantly, the observed quantity is a value of the convolution of the distribution of the desired quantity with the weighting function, instead of the value itself. In this work, the movements representing the traces of the signatures will be considered as physical quantity and after this, it will be described how simple pondering functions modify those traces.

3.1 – Convolution

The convolution of a function $f(x)$ that can be integrated and of a sample function $g(x)$ called instrument, is given by :

$$h(x) = \int_{-\infty}^{+\infty} f(u)g(x-u)du$$

It is important to notice that the convolution is a function of x. However, to determine $h(x)$ in the point x_1, it is needed to know $f(x)$ for all x.

The relationship between the general characteristics of $h(x)$ and $f(x)$ is illustrated in Fig. 10, where it can be observed that $h(x)$ is smoother than $f(x)$. A proper choice of $g(x)$ can modify, either much or little, the characteristics of $f(x)$.

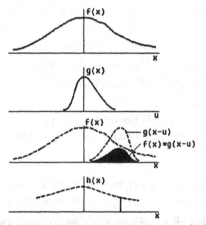

Fig. 10 . The relationship between the general characteristics of $h(x)$ and $f(x)$.

3.2 - Serial Product

The polynomial representing the movements obtained from the pre-processing of the signature images are discrete. Thus, $f(x)$ cannot be evaluated for every x. however, the approximate convolution can be obtained in the form of a serial product. Considering the polynomial

$$A = a_0 + a_1 x + a_2 x^2 + a_3 x^3 + ...$$
$$B = b_0 + b_1 x + b_2 x^2 + b_3 x^3 + ...$$

Thus, the serial product $A*B$ is given by:

$$A*B = \sum_{i=0}^{n} c_i x^i \text{ , where } c_i = \sum_{j=0}^{i} a_j b_{i-j} \text{ .}$$

The product has an important connection with the convolution. Suppose that two-function f and g are given, and it is necessary to numerically calculate its convolution. Then if f and g are the sequences of values $<f_0, f_1, f_2, f_3, ... , f_m>$ and $< g_0, g_1, g_2, g_3, ... , g_n >$ in intervals of same short duration w, then the integral of approximate convolution $f*g$ can be obtained by adding products of corresponding values of f and g.

3.3 - Uniform Deformations through Degree Instruments

A uniform deformation of signature is that in which the deformation is constant for all the points of the signature. In the discussions that follows, the following conventions are accepted:

- $f_V(t)$ and $f_H(t)$ denote, respectively, the polynomials representing the vertical and horizontal sequences of co-ordinates, in time t, described by the pen during the writing of a trace T;
- $g_V(t)$ and $g_H(t)$ are the instruments (or sampling functions) to be used for the deformation of the forms of vertical and horizontal movements, respectively. Such instruments will be generically called deforming polynomials;
- $h_V(t)$ and $h_H(t)$, with $h_V(t) = f_V(t) * g_V(t)$ and $h_H(t) = f_H(t) * g_H(t)$, denote, respectively, the polynomials representing the vertical and horizontal co-ordinates, in time t, that the pen must describe to generate the deformed trace T;
- $f(t)$ generically denote $f_V(t)$ or $f_H(t)$, and $g(t)$, $g_V(t)$ or $g_H(t)$;
- f_{t_1} and g_{t_1} are simplifications of $f_V(t_1)$, or $f_H(t_1)$, and $g_V(t_1)$, or $g_H(t_1)$, depending on the context; and
- the width w of the intervals between the acquisition of two consecutive co-ordinates, horizontal or vertical, is equal to the time unit.

3.3.1 - Scale

Considering the deforming polynomial of order 0, $g(t)=g_0$, where g_0 is constant. The serial product between $f(t)$ and $g(t)$ is the sequence $h(t)=<f_0*g_0, f_1*g_0, f_2*g_0, ..., f_m*g_0>$ which can also be written as $h(t)=<f_0, f_1, f_2, ..., f_m>*g_0$ or $h(t)=g_0*f(t)$. It is clearly noticed that, in these conditions, the reduced equation for $h(t)$ produces a uniform modification of the amplitude of $f(t)$. Applying this thinking more specifically to $f_V(t)$ and $f_H(t)$, we reach $h_V(t)=f_V(t)*g_V$ and $h_H(t)=f_H(t)*g_H$, where g_V and g_H are constant polynomials of order 0, deforming respectively and uniformly, the vertical and horizontal amplitudes of the original tracing.

If g_V and g_H have the same value, then the horizontal/vertical proportionality of the tracing is maintained; otherwise, the tracing is flat or stretched in one direction relative to the other, depending on the values of g_V and g_H. If g_V or g_H or both are negative, the trace is reflected and possibly scaled in one or in both directions.

The equation $(x'',y'') = (x' * g_V, y' * g_H)$ established the relationship between points of the deformed trace (x'', y'') and of the original (x', y'). Fig. 11 illustrates some examples of uniform scale deformations.

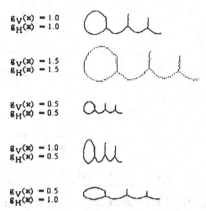

Fig. 11 . **Examples of uniform scale deformations.**

3.3.2 - Rotation

The rotation of a point $Pt'=(x', y')$, α_o relative to the origin (mass centre), moves this point to $Pt''=(x'', y'')$, in a way that the distance between $Pt''=(x'', y'')$ and the origin is equal to the distance between Pt'' and the origin. Before the rotation, the co-ordinates (x', y') define an angle α_1 between the axis x and the straight line passing through the origin and the point Pt'. After the rotation, the co-ordinates (x'', y'') define the angle $\alpha_2=\alpha_1+\alpha$ between the axis x and the straight line passing through the origin and the point Pt''.

In the serial $h_V(t)=f_V(t)*g_V(t)$ and $h_H(t) =f_H(t)*g_H(t)$, the pair $<f_H(t), f_V(t)>$ represents the co-ordinates of the point that occurs in time t before the rotation and, $<h_H(t), h_V(t)>$, the co-ordinates of the same point after the rotation. Thus, the co-ordinates of the rotated points $<h_H(t), h_V(t)>$ should be given by:

$$h_H(t) = f_H(t) * \cos\alpha_H + f_V(t) * \sin\alpha_H$$
$$h_V(t) = -f_H(t) * \sin\alpha_V + f_V(t) * \cos\alpha_V$$

where the rotation angles α_H and α_V has been differed to allow more flexibility of deformation.

Thus the deforming polynomials given by:

$$g_H(t) = \frac{f_H(t)*\cos\alpha_H + f_V(t)*\operatorname{sen}\alpha_H}{f_H(t)}$$
$$g_V(t) = \frac{-f_H(t)*\operatorname{sen}\alpha_V + f_V(t)*\cos\alpha_V}{f_V(t)}$$

Fig. 12 illustrates uniform deformations of rotation.

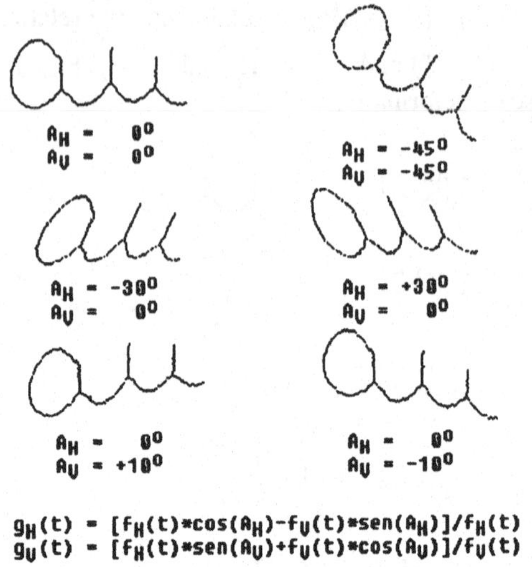

$$g_H(t) = [f_H(t)*\cos(A_H)-f_U(t)*\text{sen}(A_H)]/f_H(t)$$
$$g_U(t) = [f_H(t)*\text{sen}(A_U)+f_U(t)*\cos(A_U)]/f_U(t)$$

Fig. 12 : Uniform deformations of rotation

3.3.3 - Scale and rotation

The scale and rotation polynomials can be combined to represent simultaneous deformations of scale and rotation. g_{Ve} and g_{He} being the scale deforming polynomials, and g_{Vr} and g_{Hr} the rotation deforming polynomials, then the polynomials $g_V = g_{Ve}*g_{Vr}$ and $g_H = g_{He}*g_{Hr}$ represent respectively, vertical and horizontal deformations both of scale and rotation. The equations

$$g_H(t) = \frac{f_H(t) * \cos\alpha_H + f_V(t) * \sin\alpha_H}{f_H(t)} * g_H$$

$$g_V(t) = \frac{-f_H(t) * \sin\alpha_V + f_V(t) * \cos\alpha_V}{f_V(t)} * g_V$$

represent the deforming polynomials both of scale (constants g_V and g_H) and rotation (constant angles α_H and α_V).

3.4 - Variable deformations through degree 0 instruments

The deformations are not uniform through all the tracing. Short variations of the parameters used for deformation of scale and/or rotation occur as the tracing evolves. Other more intense variations, practically observable, occur at the beginning and at the ending of the tracing. Thus, in the relationships $h_V(t)=f_V(t)*g_V(t)$ and $h_H(t) = f_H(t)*g_H(t)$, between the deformed trace $h(t)$ and the original $f(t)$ the coefficients of the order 0 polynomials $g_V(t)$ and $g_H(t)$, are not constant. However, the variations of the coefficients are not totally random, as some could imagine. The strong correlation between points that exists in the original trace should be maintained in the deformed trace.

In this way, there should be a strong correlation between the values that $g_V(t)$ and $g_H(t)$ can assume, that is, considering the scale as an example, it is not allowed to utilise a scale factor for a point in the sequence defining the trace and, right after, do it for the subsequent point, another scale factor different from the previous one. This would cause a discontinuity of the deformed trace. The same thinking should be used in the definition of variation of the rotation angles.

The theory previously developed is extended for the case of non-constant coefficients and the conventions accepted are those previously described, plus the ones that follow:

- $g_H(t)$ and $g_V(t)$ are polynomials of order 0, deforms of horizontal and vertical scales, respectively, represented by functions of time;

- $$g_H(t) = \frac{f_H(t) * \cos[\alpha_H(t)] + f_V(t) * \sin[\alpha_H(t)]}{f_H(t)} \quad \text{and}$$

$$g_V(t) = \frac{-f_H(t) * \sin[\alpha_V(t)] + f_V(t) * \cos[\alpha_V(t)]}{f_V(t)} \quad \text{are polynomials of order 0,}$$

deforms of vertical and horizontal rotations, respectively, where $\alpha_H(t)$ and $\alpha_V(t)$ are functions of time representing the variations of the rotation angles; and

- D represents the duration of the tracing

3.4.1 - Random Variations of the Coefficients of the Deforming Polynomials

The deformations of traces through a deforming polynomial of order 0, with random coefficient, produces a trace with characteristics totally different from those of the original trace, because the strong relationship between close points is lost. Thus, the coefficient of a deforming polynomial of order 0 does not have a random variation, but varies according to some standard. This represents well what happens during the signatures writing. As the deforming polynomials represent disturbs occurring during writing, their coefficients cannot vary abruptly, because the disturbances does not vary abruptly.

3.4.2 - Standards of Variation of the Deforming Polynomials Coefficients

Considering that the more perceptible deformations occur in the trace extremes, a good representation of the variation of the deforming polynomials coefficients is through exponential.

Remembering that D represents the tracing duration,

Fig. 13 illustrates the exponential that can be used to define the intensity of the deforming polynomials coefficients at each time.

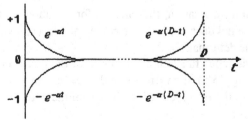

Fig. 13 . Deforming polynomials coefficients at each time

It is noticed that if $t=0$, at the beginning of the tracing, $e^{-\alpha t}=1$ and $e^{-\alpha(D-t)}$ tends to zero and, if $t=D$, end of the tracing, $e^{-\alpha t}$ tends to zero and $e^{-\alpha(D-t)}=1$. The constant α, positive, represents the speed with witch the exponential $e^{-\alpha t}$ and $-e^{-\alpha(D-t)}$ tend to zero, as well as the speed with witch the exponential $e^{-\alpha(D-t)}$ and $-e^{-\alpha t}$ tend to 1. The bigger the value of α, the faster the tendencies occur. Practically, reasonable values of α are less than 0.1.

The exponential previously described produce an undesirable collateral effect, in case they are utilised in the scale deforming polynomials.

It is noticed, for instance, that if the coefficient in a scale deforming polynomial of order 0 is given by $e^{-\alpha(D-t)}$, then the scale factors utilised for the deformation of the first points of the trace, depending on α and D, tend to zero. This would concentrate these points in a vertical axis that passes through the mass centre of the trace. That is not what is intended. The use of exponential functions that satisfy the limits

$$\lim_{t \to D} f_1(\pm e^{-\alpha t}) = 1 \text{ and } \lim_{t \to 0} f_2\left(\pm e^{-\alpha(D-t)}\right) = 1,$$

minimise those collateral effects present in the polynomials of order 0 that deform, in scale, the points of a trace.

For example, the functions : $f_1(e^{-\alpha t}) = \left(\dfrac{2+e^{-\alpha \cdot t}}{2}\right)$ and $f_2\left(e^{-\alpha(D-t)}\right) = \left(\dfrac{2+e^{-\alpha \cdot (D-t)}}{2}\right)$

produce interesting results, as shown in Fig. 14.

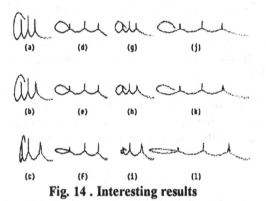

Fig. 14 . Interesting results

3.4.3 - Rotation

In the non-uniform rotation, the function for variation of the coefficients of the deforming polynomial does not have the collateral effects previously described in the non-uniform scale deformation.

Thus, any continuous function can be used for the variation of those coefficients. For instance, in Fig. 15(a), the variation of the rotation angle is uniform and is of 0°, in the beginning of the trace, for 30°, at the end of the trace.

3.4.4 - Scale and Rotation

The deformations of non-uniform scale and rotation can be obtained starting from just one deforming polynomial. Fig. 16 illustrates some combined deformations of non-uniform scale and rotation.

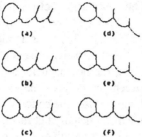

Fig. 15 . The variation of the rotation angle

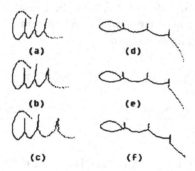

Fig. 16 . Some combined deformations of non-uniform scale and rotation.

3.5 - Deformations through instruments of Degree N, N>0.

In the deforming polynomials of order 0, the deformation of a point depends only on the original position of the point itself and of the coefficient, which represents the deforming polynomial. The correlation between the deformed point and its neighbours must be assured by the pattern of deformation of the deforming polynomial. In the deforming polynomials of order greater than 0, the deformation of a point also depends on the original position of its neighbours. Thus, the correlation between close points is more easily obtained, as may be seen on Fig. 17. The deforming polynomial used for deformation of the horizontal scale of the trace of the letters AU at Fig. 17, has order 6 and it is given by:

$$g_H(t) = \frac{c_6 t^6 + c_5 t^5 + c_4 t^4 + c_3 t^3 + c_2 t^2 + c_1 t^1 + c_0}{c_6 + c_5 + c_4 + 2 + c_2 + c_1 + c_0}$$

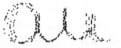

Fig. 17 . The correlation between close points.

where $c_0 = c_6 = 1$, $c_1 = c_5 = 2$, $c_2 = c_4 = 3$, e, c_3 is an aleatory value between 0 and 4. It is clearly noticed in Fig. 17 that, although the coefficient $c3$ tends to randomly distribute the points of the trace, the homogeneity of the other coefficients tends to maintain the correlation between the points. The choice of this polynomial, more specifically, of coefficient $c3$, emphasises the concept being discussed. Actually, possibly, none of the coefficients will be aleatory but functions that can be differentiated.

Continuing to talk about the previous polynomial, representing it as a sequence of values, it is obtained <1/14, 2/14, 3/14, c_3/14, 3/14, 2/14, 1/14> where $c3$ varies from point to point according to its definition. Considering that (x_{t0}, y_{t0}) are the co-ordinates of the point $Pt'(t_0)$ of a trace T, then, the utilisation of that deforming polynomial for the deformation of the trace T makes the point $Pt'(t_0)$ to be moved for a point between the co-ordinates given by:

$$\left(\frac{x_{t-3}*1 + x_{t-2}*2 + x_{t-1}*3 + x_{t0}*0 + x_{t1}*3 + x_{t2}*2 + x_{t3}*1}{14}, y_{t0} \right)$$

$$\left(\frac{x_{t-3}*1 + x_{t-2}*2 + x_{t-1}*3 + x_{t0}*4 + x_{t1}*3 + x_{t2}*2 + x_{t3}*1}{14}, y_{t0} \right)$$

where the points $Pt'(t_{t-3})=(x_{t-3}, y_{t-3})$, $Pt'(t_{t-2})=(x_{t-2}, y_{t-2})$, $Pt'(t_{t-1})=(x_{t-1}, y_{t-1})$, $Pt'(t_{t1})=(x_{t1}, y_{t1})$, $Pt'(t_{t2})=(x_{t2}, y_{t2})$ e, $Pt'(t_{t3})=(x_{t3}, y_{t3})$ are the closest neighbours of $Pt'(t_0)$. The previous equations emphasise the dependency that the deformation of a point has relative to the position of its n neighbours , where n is the order of the deforming polynomial, and with what intensity these dependencies occur, this given by the coefficients of the terms of the deforming polynomial relative the neighbour points.

An example of the use of deforming polynomials of order bigger than 0 is given below:

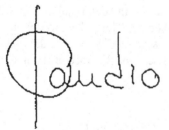

Fig. 18 . Original signature

Fig. 19 . Deformations signature

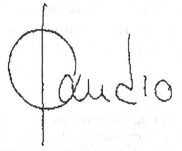

Fig. 20 . Deformations signature   **Fig. 21 . Deformations signature**

Fig. 19, Fig. 20 and Fig. 21 show three deformations of the original signature, on Fig. 18. The deforming signatures were obtained starting from the original signature through deforming polynomials of order 5.

The polynomials for generation of the three deformations produce non-uniform scale deformations in both directions, horizontal and vertical, and are given by:

$$h_H(t) = 0.01 * c_{H_1}(t) * t + 0.05 * c_{H_2}(t) * t^2 + \left[1 + 0.1 * c_{H_3}(t)\right] * t^3 + 0.05 * c_{H_4}(t) * t^4 + 0.01 * c_{H_5}(t) * t^5$$

$$h_V(t) = 0.01 * c_{V_1}(t) * t + 0.05 * c_{V_2}(t) * t^2 + \left[1 + 0.1 * c_{V_3}(t)\right] * t^3 + 0.05 * c_{V_4}(t) * t^4 + 0.01 * c_{V_5}(t) * t^5$$

where each term $c_{Hi}(t)$ and $c_{Vi}(t)$ is a differentiable function. For each trace of the original signature it was used an independent pair of deforming polynomials, that is, dependencies between deformation of different traces were not considered.

The functions $c_{Hi}(t)$ and $c_{Vi}(t$ were obtained through the Gregory-Newton's inter-polating polynomial, starting from 6 aleatory values between -1 and +1, homogene-ously distributed in the duration time of the corresponding traces. Fig. 22 illustrates the interpolation of the values 0.36, -0.19, 0.80, -0.56, 0.84 and, -0.82 in the corre-sponding times, 0, 62, 124, 186, 248 and 310, being D=310. It can be seen on Figure 22 that, due to the interpolation $f(t)$ is known for every t complying with $0 \le t \le D$.

As the values of $c_{Hi}(t)$ and $c_{Vi}(t)$ should be known at any time between the begin-ning and the end of the tracing, the distribution of the aleatory points in the time of the duration of all the tracing is justified.

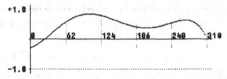

Fig. 22 . The interpolation of the values 0.36, -0.19, 0.80, -0.56, 0.84 and, -0.82

The limits in which the values for interpolation can be randomly selected, -1 and +1, allow an easy manipulation of the intensities with which each term of the de-forming polynomial contributes for the total deformation. For instance, the constants 0.01, 0.05 and 0.1 when used in the interpolations, limit the contributions of each term of the deforming polynomial. Finally, the addition of the constant 1 to the coefficient $1 + 0.1*c_{H_3}(t)$ of t^3 defines a stronger tendency to the maintenance of the tracing. Isolately interpreting this term, it can be said that the constant 1 represents the trace without deformation and $0.1*c_{H_3}(t)$, with $-1 \leq c_{H_3}(t) \leq +1$, a tendency of about 10% to deformation.

4. Conclusion

In a first important observation, we can verify that, if each variable interfering in the process of the writing of the signature traces is represented under the form of a coefficient of a deforming polynomial, then, this polynomial has the power of weighting the interference's of each of those variables, in order to produce deformations, at least, very close to those which occur in real situations. It is also important to observe that as there are variables tending to the deformation of the tracing, there are also variables which tend to keep the tracing - these, inclusively, having a bigger contribution than the former. Thus, for any interpretation of the variables interfering in the process of writing of traces, the use of an adequate deforming polynomial, or the composition of various polynomials, as previously described, has the power of capturing any deformation of tracing wanted, from the model of traces of an original signature.

The results so far obtained are promising and can be used for future works with much reliability.

Bibliographic References

[DOE93] – Doermann, D. S. & Rosenfeld, A., "The Interpretation and Reconstruction of Interfering Strokes", In Frontiers in Handwriting Recognition III, pages 41-50, CEDAR, SUNY Buffalo, New York, USA, May 25-27, 1993.

[PLAM89] – Plamondon R, Lorette G – "Automatic Signature Verification and Writer Identification – the State of the Art ", Pattern Recognition, vol. 22, n° 2, pag. 107 a 131, 1989.

[PLAM90] – Plamondon R, Lorette G, Sabourin R – "Automatic Processing of Signatures Images : Static Techniques and Methods ", Handwritten Pattern Recognition, 1990.

[RAND90] – Randolph D, Krishnan G – "Off-Line Machine Recognition of Forgeries" – Machine Vision Systems Integration in Industry, vol. 1386, pp. 225 a 264, 1990.

[SABO90] – Sabourin R, Plamondon R, Lorette G – "Off-Line Identification with Hand-written Signature Images : Survey and Perspectives", SSPR, pp. 377-391, 1990.

An Application of the Sequential Dynamic Programming Matching Method to Off-line Signature Verification

Mitsu Yoshimura[1] and Isao Yoshimura[2]

[1] Chubu University, Matsumoto-cho, Kasugai-shi, Aichi 487, Japan
[2] Science University of Tokyo, Kagurazaka, Shinjuku-ku, Tokyo 162, Japan

Abstract. This paper proposes a method (SDP method) of off-line signature verification incorporating a sequential application of the dynamic programming matching method in its preprocessing stage. Based on the frequencies of black pixels in each signature projected on the x-axis, the SDP method determines an optimum segmentation of any signature into the same number of components as the representative reference signature. The SDP method measures the dissimilarity between the suspect signature and the representative signature, and accept the suspect signature as genuine without any further scrutiny, if the dissimilarity is below a threshold. When the suspect signature is not accepted, the SDP method subsequently applies a verification procedure proposed by Yoshimura and Yoshimura(1997) to it. This paper evaluates also the effectiveness of the SDP method through an experiment using a database of 800 Japanese signatures, half of which were obtained on the area of countersignature on sheets similar to a traveler's check, whereas the remaining were on the area of authentic signature. The experiment used the countersignatures as suspect signatures to obtain estimates of mis-verification rate, taking five signatures alternatively from the twenty authentic signatures as the reference signatures. It achieved values as low as 12.9% in an average mis-verification rate.

1 Introduction

Recently being interested in verification of Japanese signatures, we focus our attention, in this paper, on off-line verification of Japanese signatures, although the proposal need not be limited to it.

In off-line signature verification, the dissimilarity of two signature images, e.g., those of a suspect and an authentic signature are measured and the choice of a suitable similarity measure is important to decrease error rates in signature verification. Concerning this, Yoshimura and Yoshimura [11, 12] remarked that Japanese are likely to sign their name, which are generally composed of three, four or five characters, keeping the component characters apart as are seen in Fig. 1. The segmenting of signatures into their components and measuring dissimilarity between two signatures based on the achieved segmentation is, therefore, quite useful to decrease mis-verification rates, as is evidenced in their above-mentioned papers.

To realize the segmentation of signatures in their experiments, they adopted a method which identified the boundaries of components in a signature as the places where the frequency of black pixels projected on the x-axis was almost vanished. Although the adoption of this method achieved a certain amount of decrease in mis-verification rates, the results were not satisfactory, leaving much room for improvement. It prompted us to devise a new method of segmentation.

The issue to be addressed in this paper is to segment any signature into the same number of components as the reference signatures. Although two-dimensional matching techniques are in frequent use to address this issue, one-dimensional techniques are also useful, if important features of signature images are converted to a series of one-dimensional values (cf. [6, 8, 9]) such as the frequency distribution of black pixels projected on the x-axis.

One-dimensional techniques of segmentation feasible in the situation similar to the above case have been developed in the field of speech recognition [2, 5]. In fact, a speech is necessarily realized as a time series of sound without any loss of information and the implication of speech can be recognized exactly only when it is matched with some reference words which are a set of segmented sound series. We surveyed them and had a particular interest in those extensively developed and investigated by Oka and his colleagues under the name "continuous dynamic programming matching" [1, 2, 3, 4, 7], convincing themselves that a sequential application of a dynamic programming (DP) matching method was effective also for off-line verification of Japanese signatures. They, therefore, devised a method, which is referred to as the SDP method in the following, incorporating it as a tool in the preprocessing stage of the procedure.

In what follows, this paper briefly explains the devised method in Section 2 and presents, in Section 3, an experimental evidence for the effectiveness of the devised method, subsequently giving discussions in Section 4.

2 SDP Method

2.1 Input Signature

The SDP method assumes that any signature is input in the verification system as a binary pattern $\{f(x,y); 1 \leq x \leq n_x, 1 \leq y \leq n_y\}$, where $f(x,y) = 1$ if the grid (x,y) is on the signature, and $= 0$ otherwise, taking $n_x \times n_y$ as the size of the meshed area. It also assumes that the system keeps a number, say a, of authentic signatures as the reference for verification, which are referred to as the reference signatures and a is set as five in later experiments. In the SDP method, properly chosen one signature in the reference signatures is used as the representative reference signature and it is segmented into a number, say b, of components through a suitable segmentation procedure, e.g., a visual inspection in advance of the application of the method. The parameters related to the $n-$th component ($n = 1, 2, \ldots, b$) of the representative reference signature such as the width s_n, the starting point x_{hn} and the marginal distribution $\{h_n(x); 1 \leq x \leq s_n\}$ of black pixels projected on the $x-$axis, i.e.,

$$h_n(x) = \sum_{y=1}^{n_y} f(x_{hn} + x, y); \qquad 1 \le x \le s_n \tag{1}$$

are referred to in the segmentation of any input signature with frequency distribution $\{g(x); x_1 \le x \le x_2\}$ of black pixels projected on the $x-$axis, where x_1 and x_2 are the least and the greatest $x-$coordinate such that $g(x) > 0$.

2.2 Segmentation Procedure

When a signature is input in the verification system, the SDP method tries to segment it, in its first step, into the same number of components as the representative reference signature through a sequential application of a DP algorithm, which will be shown in Appendix. Briefly, it first seeks for the best warping function $\tau(t)$ to match the coordinate t of $g(t)$ to the coordinate τ of the first component $h_1(\tau)$ of the representative reference signature so that they achieve the best fit within the range depicted in Fig. 2, where t and τ are dummy variables to explain the relationship. The warping function thus obtained identifies the endpoints $x_{11}(= n1)$ and x_{12} of the first component of the input signature, whereas it is adjusted by the adjusting algorithm explained later in Appendix. It next seeks for the best warping function for the next component, the third one, and so on, sequentially applying the same algorithm until the input signature is segmented into b components. Although it may, eventually, be forced to terminate the procedure before the b components is obtained because the remained part of the input signature to be contrasted to the representative reference signature becomes empty, it need never be considered as a serious problem because the input signature is successfully judged as a forged signature in the SDP. Note that the algorithm explained in Appendix automatically calculates a measure, say $FD(h, g)$, of dissimilarity of two signatures h and g, which is referred to as the frequency difference in the following.

Steps of the segmentation procedure is visualized in Fig. 3. The top figure of Fig. 3 is an example of the representative reference signature used in the experiment in this paper and the second one is an example of input signatures of the same autograph as the reference signature. The number of black pixels is projected on the x-axis, which makes a frequency distribution $g(x)$ or $h_n(x)$ of black pixels shown on the x-axis of each figure. By placing the frequency distribution of the top figure on the y-axis and that of the second one on the x-axis of the third figure of Fig. 3, the absolute difference between them is shown in the third figure as a two-dimensional gray pattern, where the steps of the pattern from the bottom to the top corresponds to the segmentation of the reference signature. In other words, the horizontal coordinate of the starting point of the first step of the figure is x_{11}, the second step is x_{21}, and so on. Note that the values in the upper left area are omitted from displaying because they are not used in the SDP method. The fourth figure of Fig. 3 is a gray pattern of the dissimilarity FD for various right endpoint (x, y) of each segmentation of the input signature.

(a)　　　　　　　**(b)**　　　　　　　**(c)**

Fig.1　Examples of signatures in the database
(a) Authentic signatures, (b) Genuine signatures, (c) Forged signatures.

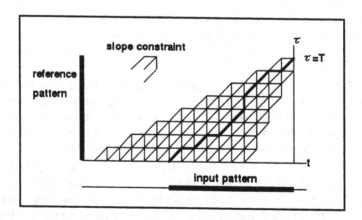

Fig.2　Illustration of the warping function
Thick line in the figure represents the warping function $\tau(t)$ which is optimum
with respect to a dissimilarity measure.

On this figure, the least dark position on the horizontal line on each step should be regarded as the most suitable point of segmentation of the input signature. The curve on the fifth figure of Fig. 3 shows the darkness corresponding to each horizontal line in the fourth figure, the lowest place of which corresponds to the most plausible candidate of the boundary of segmentation of the input signature as is shown in the bottom figure of Fig. 3.

2.3 Verification Procedure

The SDP is composed of two stage of verification, i.e., the first stage which uses the FD and the second stage which applies a pattern matching method to suspect signature not accepted in the first stage. In the first stage, it accept the suspect signature as genuine if the FD is below a threshold which is set as the median of the FD's between the representative reference signature and other reference signatures multiplied by a threshold coefficient, say C_{c1}, determined in advance. In the second stage, it measures a component-wise dissimilarity measure, say $D(h_n, g_n); n = 1, 2, \ldots, b$, obtained by a well known pattern matching method (see e.g. [11, 12]), where h_n and g_n symbolically represent the n-th component of the reference signatures and the suspect signature. In fact, the dissimilarity $D(h_n, g_n)$ between the reference signatures and the suspect signature is measured through a multivariate analysis of the reference signatures. The overall dissimilarity $D(h, g)$ between the reference signatures and the suspect signature is obtained by summing up $D(h_n, g_n)$ with respect to b components. The SDP in the second stage verifies the suspect signature if $D(h, g)$ is below a threshold, and rejects it as forgery otherwise, where the threshold is set as the median of the dissimilarity among representative signatures multiplied by a threshold coefficient, say C_{c2}, determined in advance.

3 Verification Experiment

3.1 Objective

Since the effectiveness of the proposed method should be examined experimentally, we conducted a verification experiment using a database once provided by us.

The ideas newly taken into the proposed method are essentially summarized in two points; one to incorporate the DP matching method to realize the optimum segmentation of signatures and the other to use the achieved dissimilarity FD to verify suspect signatures in the first stage. Consequently, the experiment was designed considering these two points. Concerning the first point, the effectiveness of the new procedure can be checked by comparing the new segmentation method with the minimal frequency method incorporated in the method by Yoshimuras[12] (previous method). In other words, the comparison of two methods, the one which replaces only the segmentation procedure of the previous method with the one incorporated in the proposed method and the other which does not, must be made as a part of the experiment. The second point can be checked through a comparison experiment for the proposed method and the previous method, subtracting the effect of the first point from the effectiveness realized in the latter part of the experiment.

3.2 Material

The above-mentioned signature database is composed of 800 binary signatures, each of which is transformed from a signature image written on a sheet shown in Fig. 4, where the sheet is almost the same as a traveler's check [10] and the size $n_x \times n_y$ is set as 608×144. Examples of signatures in the database have been already shown in Fig. 1. The size Half of the 800 signatures are composed of twenty repeated writings of twenty autographs for twenty people, which were written on the upper area of the sheet as authentic signatures, while the remaining 400 signatures are composed of ten forged and ten genuine writings of each autograph, which were written on the lower area of the sheet as countersignatures. Forged signatures were obtained after some practices of imitation looking at corresponding authentic signatures.

3.3 Design

Since the ability of any verification method is reasonably evaluated by mis-verification rates, they were estimated in the experiment which regarded authentic signatures as the reference signatures and countersignatures as suspect signatures. More precisely, the twenty authentic signatures of the same autograph in the database were split into four groups with $a = 5$ reference signatures and the authenticity of ten genuine and ten forged countersignatures of the corresponding autograph were verified for each group of reference, subsequently leading the average error rate in the verification as a measure of the ability of each method.

In the experiment, the three method were compared; one is the proposed method, the second (Hybrid) is the one which uses the newly proposed segmentation method leaving the verification stage unchanged from the previous method, and the last is the previous method.

Although the threshold coefficients C_{c1} and C_{c2} should, in principle, be determined in advance and we had, through past experiences, an insight that $C_{c1} \neq 1.0$ $C_{c2} \neq 1.5$ are reasonable, various values were tried in the experiment for the sake of sensitivity analysis. Note that there are two type of errors with trade-off relationships, i.e., the Type I error which rejects a genuine signature as forged and the Type II error which accepts a forged one as genuine, and the choice of the coefficients greatly affects the balance of them.

3.4 Results

The summarized result of the experiment is shown in Table 1 as mis-verification rates, where the "Average" represents a simple arithmetic mean of two types of errors.

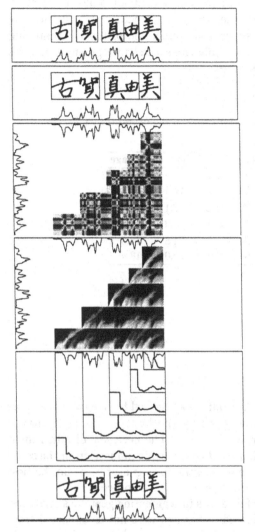

Fig.4 Example of original signature images

Fig.5 Example of erroneously segmented signatures

Fig.3 Illustration of the application of the SDP method.

The top is a representative reference signature with the frequency distribution of black pixels projected on the x-axis beneath the image.

The second is an input signature to be matched with the reference signature.

The third is the gray pattern of absolute differences between the two frequency distributions; one is for the reference signature placed on the y-axis and the other is for the input signature placed on the x-axis. The values in the upper left area are omitted because they are not used.

The degree of darkness in the fourth represents the dissimilarity values taken on the endpoint of segmentation.

The least dark point on the horizontal line of each step corresponds to the endpoint of an achieved segmentation.

The fifth shows the standardized darkness of the fourth figure on the y-axis.

In the bottom, the result of segmentation is shown.

In average, the newly proposed method achieved about 3.5% decrease in mis-verification rates compared with the previous method, which implies the effectiveness of the adoption of the sequential application of the DP matching method in its first stage of verification, while the success in segmentation does not much affect the decrease of error rates.

Table 1. Mis-verification rates obtained in the experiment (%)

Method	C_{c1}	C_{c2}	Type I	Type II	Average
Proposed	1.1	1.4	10.9	14.9	12.9
	1.1	1.5	8.5	17.4	12.9
Hybrid	–	1.7	15.6	17.1	16.4
	–	1.9	9.6	15.9	17.8
Previous	–	1.6	18.0	14.5	16.3
	–	1.9	7.9	29.0	18.4

4 Discussions

4.1 Adjustment Algorithm

The adjustment algorithm shown in Appendix was devised to give more plausible boundaries of components than those given by the best, in the sense that it achieves the least dissimilarity, warping function. The necessity of adjustment comes from the instability of the tail area of components induced from the sense of ordinary people that the tails are not essential for signatures, while the best warping function does not take this information into the procedure.

The adjustment achieved successful results in segmenting genuine signatures, whereas realized segmentation for forged signatures were not as expected from visual inspection. Example of mis-segmented images are shown in Fig. 5, all of which were forgery and were rejected probably due to the improper segmentation. We believe that the adjustment provided the SDP an advantageous feature to segment forged signatures improperly.

Note that some constants such as 0.7, 12 and so on in the algorithm were determined through a preliminary trial and, therefore, are not regarded as exact constants. Other values can be set to obtain similar results of segmentation.

4.2 Implication of the Result

The signatures regarded as suspect signatures in the experiment to estimate mis-verification rates were written under slightly different conditions with respect to the area, size and background image from the reference signatures [10]. If the conditions could be the same between the suspect and the reference signatures, better results would be expected. On the contrary, if the writing conditions such as the tools for writing are different between them, the result might be poorer. Likewise, if the database were prepared from signatures on a sheet without background, a better performance could be expected, because then preprocessing to obtain binary patterns would be simpler and, therefore, produce clearer images.

The most prominent finding from the experiment is that the use of the frequency difference FD obtained under a sequential application of the DP matching method in its first stage [12], leaving the succeeding stage unchanged, much improved the performance of the method.

4.3 Representative

The error rates in Table 1 were obtained when the first authentic signature among randomly ordered reference signatures was used as the representative reference signature. If a more appropriate one could be selected as the representative signature using, e.g., a clustering method [8], a better performance would be achieved.

4.4 Number of Reference Signatures

In the experiment, the number a was set as five. We think this number is reasonable in practical use if we have multiple sheets of traveler's checks in hand.

4.5 Threshold Coefficient

The determination of the threshold is always the most difficult issue to be addressed. In the SDP method the determination of two threshold coefficients, C_{c1} and C_{c2}, which were not mutually independent to change the ability of the method, were required. We considered that C_{c1} should be set as the value around the median of the frequency differences between the representative and the remaining reference signatures, because it was used in the preliminary stage. In fact, it was set as the median multiplied by a threshold coefficient 1.1 in the experiment. In contrast, the threshold coefficient C_{c2} was set as a greater value than 1.0; the plausible value was around 1.5 according to our past experiences. The values shown in Table 1, which were values achieved a fairly good results, are almost the same as anticipated.

Note that the choice of the threshold should be dependent on the strategy to balance two type of errors.

4.6 Applicability

Although this paper took only Japanese signatures into consideration, the proposed method can be applied for other cases such as Chinese or Korean signatures, where signatures are usually written in a segmented manner. The number b of components may be dependent on the nationality of signature as well as the personal tendency of it.

Acknowledgment   This study was supported in part by the Grant-in-Aid for Scientific Research (C08680415) of the Ministry of Education, Science, Sports and Culture of Japan.

References

1. Hayamizu, S. and Oka, R.: Experimental studies on the connected words recognition using continuous dynamic programming. Trans. IEICE J67-D, 6 (1984) 677-684.
2. Oka, R.: Continuous words recognition by use of continuous dynamic programming for pattern matching (in Japanese). Trans. Committee Speech Research; The Acoustical Soc. of Jpn., S78-20, (1978) 145-152.
3. Oka, R. and Hayamizu, S.: Vector continuous dynamic programming algorithm utilized for sentence recognition of continuous speech (in Japanese). Trans. Committee Speech Research; The Acoustical Soc. Jpn., S82-71, (1982) 561-568.
4. Oka, R. : Phonetic recognition of each frame by partial matching method based on continuous dynamic programming (in Japanese). Trans. IEICE J70-D, (1987b) 917-924.
5. Rabiner, L.R. and Schimdt, C.E.: A connected digit recognizer based on dynamic time warping and isolated digit templates. Proc. IEEE Int. Conf. Acoust. Speech Signal Proces., (1980) 194-198.
6. Sato, Y. and Kogure, K.: On-line signature verification based on shape, motion, and writing pressure. Proc. 6-th Int. Conf. Patt. recog., (1982) 823-826.
7. Takahashi, K., Seki, S., Kojima, H. and Oka, R.: Spotting recognition of human gestures form time-varying images (in Japanese). Trans. IEICE Jpn. J77-DII, 8 (1994) 1552-1561.
8. Yoshimura, M., Kato, Y., Matsuda, S. and Yoshimura, I.: On-line signature verification incorporating the direction of pen movement. Trans. IEICE E74, 7 (1991) 2083-2092.
9. Yoshimura, I., Yoshimura, M., Hosaka, K., and Yokoi, H.: Experiment of person verification based on on-line writing of one's address. Proc. 6-th Int. Conf. Handwriting Drawing, (1993)138-140.
10. Yoshimura, I. and Yoshimura, M.: Off-line verification of Japanese signatures after elimination of background patterns. Int. Jour. Patt. Recog. Artificial Intelligence, 8 (1994) 69-78.
11. Yoshimura, I., Yoshimura, M. and Tsukamoto, T.: Investigation of an automatic verification system for Japanese counter-signatures. Proc. 7-th Biennial Conf. Int. Graphonomics soc., (1995) 86-87.
12. Yoshimura, M. and Yoshimura, I.: Investigation of a verification system for Japanese countersignatures on traveler's cheques (in Japanese). Trans. IEICE J80DII (1997) 1764-1773.

APPENDIX

1. DP Algorithm

The SDP method searches for the endpoints of n–th component of the input signature referring to the segmentation of the representative reference signature sequentially from $n = 1$ to $n = b$. It first uses the DP Algorithm shown below, subsequently adjusting the endpoints to get better separation of component characters using the Adjusting Algorithm shown in the next section.

1. Calculate the absolute difference $q_n(x, y)$ of $h_n(y)$ and $g(x)$ for any (x, y) using the following formula:

$$q_n(x, y) = |h_n(y) - g(x)| \qquad (x_{n1} \leq x \leq n_2, \ 1 \leq y \leq s_n), \qquad (2)$$

where $x_{n1} = x_1$ for $n = 1$ and x_{n1} is sequentially determined for n greater than 1 through the application of the Adjusting Algorithm.

2. Calculate the accumulated value $p_n(x, s_n)$ of $q_n(x, y)$ from $y = 1$ to $y = s_n$ for x in the range of (x_{n1}, x_2) using the following formula:

$$p_n(x_{n1}, y) = p_n(x_{n1} - 1, y) = M \qquad (1 \leq y \leq s_n) \qquad (3)$$

$$p_n(x, 1) = 2q_n(x, 1) \qquad (4)$$

$$p_n(x, 2) = \min \begin{cases} p_n(x - 2, 1) + 2q_n(x - 1, 2) + q_n(x, 2) \\ p_n(x - 1, 1) + 2q_n(x, 2) \\ p_n(x, 1) + q_n(x, 2) \end{cases} \qquad (5)$$

$$p_n(x, y) = \min \begin{cases} p_n(x - 2, y - 1) + 2q_n(x - 1, y) + q_n(x, y) \\ p_n(x - 1, y - 1) + 2q_n(x, y) \\ p_n(x - 1, y - 2) + 2q_n(x, y - 1) + q_n(x, y) \end{cases} \qquad (3 \leq y \leq s_n),$$
$$\qquad (6)$$

where M is an arbitrary large number such as 999999.

3. Calculate a normalizing constant $c_n(x, y)$ to normalize $p_n(x, y)$ with respect to the path length of the warping function using the following formula:

$$c_n(x_{n1}, y) = c_n(x_{n1} - 1, y) = 0 \quad (1 \leq y \leq s_n) \qquad (7)$$

$$c_n(x, 1) = 2 \qquad (8)$$

$$c_n(x, 2) = \begin{cases} c_n(x - 2, 1) + 3 \\ c_n(x - 1, 1) + 2 \\ c_n(x, 1) + 1 \end{cases} \qquad (9)$$

$$c_n(x, y) = \begin{cases} c_n(x - 2, y - 1) + 3 \\ c_n(x - 1, y - 1) + 2 \\ c_n(x - 1, y - 2) + 3 \end{cases} \qquad (3 \leq y \leq s_n) \qquad (10)$$

4. Calculate the dissimilarity $a_n(x)$ of the n–th component of the two signatures for $x_{n1} \leq x \leq x_2$ using the following formula:

$$a_n(x) = \frac{p_n(x, s_n)}{c_n(x, s_n)}; \quad (x_{n1} < x \leq x_2) \tag{11}$$

2. Adjustment Algorithm for Endpoints

The DP Algorithm suggests us the best warping function to specify the right endpoint x_{n2} of the n–th component of the input signature as the x–coordinate which gives the least dissimilarity $a_n(x)$. However the point is not sufficiently reasonable as the endpoint because there are often multiple points to give similar values to the least dissimilarity. The SDP, therefore, adjusts the endpoint using the following algorithm:

1. If $n = b$ then set x_2 as x_{n2}, otherwise set the least x such that $x_{n1} + 0.7s_n < x$, $12 < a_n(x)$ and $12 < a_n(x+1)$ as x_{n2}. If such x does not exist, search for the range of x where $a_n(x)$ is monotone decreasing within the range $x_{n1} + 0.7s_n < x \leq x_{n1} + 2s_n$ and set the least x as x_{n2}.

2. If $n < b$, $g(x_{n2}) = 0$, set the least x and the greatest x such that $g(x) = 0$ around x_{n2} as x_{n2} and $x_{(n+1)1}$, respectively. If $g(x_{n2}) > 0$, identify the range of x such that $g(x) = 0$ within the range $x_{n2} - 20 \leq x \leq x_{n2} + 30$ and set the x nearest to x_{n2} within this range as x_{n2} and the least x such that $g(x) > 0$ and $x_{n2} < x$ in the same range as $x_{(n+1)1}$. If these procedure does not succeed in getting adjusted values, then do not change x_{n2} and set $x_{n2} + 1$ as $x_{(n+1)1}$.

3. If the above procedure does not get adjusted values of x_{n2} and $x_{(n+1)1}$, then terminate the processing and calculate, if possible, $a_n(x_n)$ and the frequency difference $FD(h, g)$ as the sum of $a_n(x_n)$ from $n = 1$ through b.

A System for Automatic Form Reading

Thomas W. Rauber, Valério B. de Souza and Silvana Rossetto

Universidade Federal do Espírito Santo, Centro Tecnológico,
Departamento de Informática, Av. F. Ferrari, 29065-900 Vitória - ES, Brazil,
Phone:(+55)(27)3352654 Fax: 3352850 E-mail: thomas@inf.ufes.br

Abstract. The objective of this paper is to present the research activities of our group with respect to the development of an automatic form reading system. We aim to automate the registration of students by autonomously recognizing the handwritten information on application forms. An image of a digitized form is acquired, the relevant areas of the image are segmented and passed to a character recognition system. The complexity of the recognition is limited to isolated handwritten characters. We propose a prototype oriented method for the classification of the characters that seems to promise better results than classical feature-based techniques. The representation method of the characters are parametric curves. We try to adapt a prototype to a pattern in order to measure its deformation from the original prototype. Results are limited to conceptual proposals for the form reading system.

1 Introduction

One of the most challenging research topics in the area of intelligent systems is the simulation of the human capacity of perception. One example from this field is the recognition of handwritten characters. Existing systems that analyze printed text (Optical Character Recognition- OCR) have already reached a level of precision that permits to use them in everyday tasks, for instance to scan a document and extract the printed text in a machine-readable form.

On the other hand the automatic recognition of handwritten characters is a much greater obstacle for a reading system that wants to reach a satisfactory level. Nowadays there is not such a thing as a general purpose system, able to read a continuous handwritten text without exhibiting an elevated rate of error or rejection. Even with the restriction of only permitting isolated characters the best results did not yet reach a level of satisfactory quality.

The key for the construction of a successful system for handwritten character recognition is in the conception of the features that describe the involved patterns. The tendency relative to the feature model is quite clear. The traditional methods of preprocessing and feature extraction starting from the bitmap pattern of the character are being replaced by more analytic models. The focus of our research is to find a robust analytic model, able to cover the variability of the characters and at the same time keeping a good differentiation between the classes of the characters. We will use a prototype-based recognition system. This means that each character class is represented by several prototypes. The objective is to measure a distance of an unknown pattern from the prototypes. The similarity between the prototype and the pattern will decide the class membership.

Even an optimal classifier of isolated characters will be unable to provide an error free classification. Therefor the use of context is essential in document analysis. In some cases the context provides quite strong restrictions, for instance the permission of only using a limited number of street names in an address. This can considerably lower the error rates.

Our objective is to build an automatic form reader. After the acquisition of the form a preprocessing step will extract the boxes with isolated characters. The kernel of the system is the classifier of the isolated characters. Postprocessing includes the insertion of the recognized information into a database.

The field of application of the automatic form reader is a registration system for the annual selection tests at our university. The candidate provides his personal data on a form which is usually digitized manually. The global objective of the automatic form reader is to provide better acquisition rates than the manual data acquisition.

2 Research in Handwritten Character Recognition

The recognition of handwritten characters is still a wide-open field for research. There do not exist general purpose algorithms or systems that are able to solve this problem without errors. In recent research work many different proposals have been made with respect to the architectural concepts and paradigms of analysis. Together with the introduction of inexpensive digital computing power and the improvement of some algorithms the interest in the field of handwritten document analysis began to increase. The state of the art for *online* handwritten character recognition was reviewed for instance by Tappert et al. [26]. In the case when the temporal information of the pattern is absent we are dealing with *offline* recognition which generally makes the identification process more difficult. For a synthesis of results in the area of offline recognition see for instance Senior [23]. Normally the handwritten character recognition system is integrated into a more complex architecture of several modules, being part of a document analysis system.

The actually achieved results from the recognition of continuous writing (word, sentences) do not permit a totally automatic reading. The error rate is too high for a viable use. One restriction is the permission of only isolated characters. In many applications there is in fact the possibility to obtain isolated characters, for instance in forms where in predefined boxes only one character may be written at a time. In our research effort we will concentrate only on the recognition of isolated characters. In Revow et al. [22] one can find relevant references to research related to isolated handwritten recognition, for instance Impedovo [14] or Suen et al. [25]. Breuel [3] gives an overview from the standpoint of artificial neural networks for character classification.

2.1 From Pixels to Features

The most important aspect for the description of the patterns is the feature model. The feature can be binary or non-binary (ordered and non-ordered). A good feature model can compensate for a less sophisticated classifier. The nature of the features can vary considerably. Fourier descriptors were used in Impedovo et al. [12] and Impedovo [13]. Statistical moments of the pixels appear as the only feature model in Li and Xu [17] and Teh and Chin [27]. The Log-Polar transform was applied in the area of character recognition by Reber and Lyman [21] and Wechsler and Zimmerman [28]. The shadow masks of Burr [4], [5] transform a character into a kind of inverted LCD display.

Usually the features are calculated after some preprocessing has been performed that aims at improving the classification results. Noise and dropouts have to be removed. The character has to be normalized in relation to translation, slant, scale and rotation before the features are calculated.

The classifier that finally categorizes the characters based on the features is of quite irrelevant nature. One can use an artificial neural network, e.g. Haykin [10] where two models are recommended by practitioners, namely the Multilayer Perceptron and Radial Basis Function Networks (Kressel and Schürmann [15]). A classifier based on traditional statistical methods can equally be used, like for instance a Bayes classifier based on multivariate normal distributions (Duda and Hart [6]). The comparative results of the "StatLog" project (Michie et al. [19]) suggest that the choice of the classifier has little impact for a fixed set of features, regardless if the classifier is from the area of artificial neural nets or from more traditional statistical techniques.

Many recognition systems pass the bitmap matrix directly to the classifier which is often a model from the area of artificial neural networks, for instance Hecht-Nielsen [11], Blackwell et al. [1], Le Cun et al. [16], Boser et al. [2] and Simard et al. [24]. The expectation is that complex classifier structures compensate for the absence of an explicit feature extraction procedure. This assertion is especially true for the neocognitron of Fukushima [7].

A step quite often used as preprocessing is the normalization of the raw bitmap relative to similarity transformations (translation, scale, rotation) and slant. Then the values of the normalized bitmap are directly used as features. This can be done by scanning a typically $N \times N$ pattern, scanning the pattern line by line or column by column and generating a N^2-dimensional feature vector. A typical dimensionality reduction step is Principal Component Analysis (PCA). The highly correlated pixel values are mapped to the first $M \ll N^2$ dimensions which eliminate the linear correlation among the values. Finally the Principal Components are used as the final input to the classifier. This strategy was e.g. used by Kressel and Schürmann [15]. Breuel [3] states that many different strategies based on similar normalizing and feature extraction schemata, all lead to similar classification performances. This is due to the often intrinsic equivalence of the feature models.

2.2 Prototype-based Character Models

Actually one can observe a tendency in the area of isolated handwritten characters which is due to the unsatisfactory results of those methods which analyze the pure bitmap of the pattern. The derived features often suffer a loss of discriminatory information because eventually a certain heuristic destroys valuable pieces of the pattern structure. Instead of these features, analytical prototypes are used to represent the character classes. The classification is made via a similarity measure between a prototype and the unknown pattern. This strategy was e.g. pursued in Williams et al. [29] with elastic deformable prototype models. Revow et al. [22] present a modeling of the characters by B-splines. Here the prototype is gradually adapted to the pattern and its deformation (the departure from the "home position") is measured. Similarly Hastie and Tibshirani [9] propose a polygon-based representation for the prototypes. The line segments were adapted until a match between the pattern and the prototype was reached.

2.3 Classification

In the feature-based approach to character description the features are fed to the classifier. Traditionally statistical models were used for the classification. The motivation for the use of artificial neural networks is their capacity to learn from training data. The underlying probability density functions of the classes can be quite complex, but still the net can adapt to the data well. The user of the classifier has not to possess a profound background in multivariate statistics. Alternative classifier models which do appear of more academic interest are fuzzy sets, for instance the FOHDEL/FOHRES system of Malavija [18].

2.4 Document Analysis

As it was mentioned before the character classification is an essential part in a more global context of automatic document analysis. If the classification results are not very reliable the other parts of the system cannot compensate much for this deficiency. Our interest is in the area of automatic form reading. The integration of various components into a complete form reading system is pursued in the OSCAR project [20]. A general framework that we use as a basis for our development work is the NIST reference character recognition system (Garris et al. [8]).

It can be observed that the use of context information is indispensable in automatic form reading. Ambiguities can often be removed for the classification if the knowledge about permitted sequences of characters is available. Hidden-Markov- Models were used for this purpose as well as n-gram techniques.

Generally speaking one can state that the topic of automatic form processing is a very interesting field of research which leaves many entry points for an improvement of the performance and robustness of such a system.

3 Layout of the Form Reading System

The objective of our research work is the construction of an automatic form reading system. The input of the system are forms of the type that is shown in Fig. 1. The manually filled in parts consist of typical alphanumeric data, like name and address. Besides there exist binary tick boxes where a marker has to be set. The system has the task to acquire the digitized images of the forms, extract the respective fields and insert the recognized data into a record of a data base.

3.1 Description of the System

In the considered forms the space where the characters have to be inserted is well limited. One can also observe that the characters have to be written in handprinted form. These two facts facilitate the recognition task. First because the use of handprinted text reduces the number of possible training patterns. Secondly, the segmentation of the characters within a word is strongly simplified because the characters have to be filled into predefined boxes and therefor can be encountered at positions known a priori.

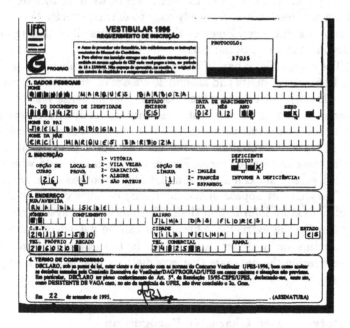

Fig. 1. A student registration form.

We are developing a software environment that permits to specify a generic form template via a graphical interface. Starting from this specification, other modules which define the kernel of the system will segment the characters, recognize them and insert them into a database. All these modules together define the automatic form reader.

The kernel of the system will be a module that analyzes the document in the following steps:

1. Preprocessing of the document image (noise elimination, contrast adjustment, etc.)

2. Segmentation of the interesting fields, extraction of the isolated character

3. Recognition of the character together with a degree of similarity to each of the possible classes.

4. Use of context information via a dictionary in order to correct erroneously classified characters

5. Insertion of the data records into a database

ad 1.) After the form has been acquired by the digitizing device (usually a scanner) the objective is to extract information starting from the digital image and the a priori knowledge about the structure of the form. The initial image contains noise due to imperfections of the digitalization process. Besides the image is normally a multi-value grey-level signal. A low-pass filtering, followed by a simple thresholding transforms the multi-level image into a binary image.

ad 2.) The relative position of the areas where we have to look for a character is known. Therefor it is relatively easy to extract exactly the set of pixels that compose the character. However there is the possibility that the character extends beyond the limits of the box.

ad 3.) The most difficult part of the system is the recognition of the character. There are 10 classes of digits and around 30 classes of characters (if only capital letters are allowed). Each pattern is subjected to an immense variety of different writing styles. Under these circumstances only an analytical character model that is quite robust has chances to reduce the error to acceptable rates.

ad 4.) A limited vocabulary for the allowed words permits the successive reduction of the classification error. For instance in a text field which contains the name of a person one can exclude implausible hypotheses, i.e. names that do not exist. They do not appear in the dictionary of allowed names. Hence with the help of contextual information the performance of the system improves.

ad 5.) Finally the extracted information can be passed to a data processing infrastructure on a higher logical level. In the example of the automatic student registration the database that was digitized manually is now generated automatically.

3.2   Prototype-based Character Recognition

As it was mentioned before the character classifier is the most important part of the automatic form reader. If its performance is good the system can reach a relatively high level of autonomy. The general strategy for our classifier system is:

- Each character is represented by a small number of prototypes.
- Each prototype is represented by a set of analytical parametric curves.
- The free parameters of the system are the coefficients of the parametric curves.
- An iterative adaptation algorithm tries to correlate the prototype to the bitmap of the unknown pattern.
- The deviation of the free parameters from their original value serves as a measure of distance between the prototype and the unknown pattern.
- We try to avoid the introduction of heuristics. Only then the general purpose character of the method can be preserved.

Presently our research tries to model a character prototype as a set of parametric curves for each of the two components x and y as $z(t) = (x(t), y(t))$. In Fig. 2 three examples are given which try to model prototype patterns of digits. The pattern "3" for instance is composed of two curves of order 2. The upper stroke is defined as: $z(t) = (-2(t - 0.5)^2, -0.5t) = (-0.5 + 2t - 2t^2, -0.5t)$

We realize that we can represent this curve by the coefficients $w_{0x} = -0.5$, $w_{1x} = 2$, $w_{2x} = 2$, $w_{0y} = 0$, $w_{1y} = -0.5$.

We try to match the unknown bitmap pattern with the prototype pattern in an iterative procedure. This will generate a sequence of modified prototype patterns at step k which have weights $\{w_{ix}^{(k)}\}$ and $\{w_{iy}^{(k)}\}$. The final weights $\{w_{i.}^{(n)}\}$ at step n will be compared to the initial weights $\{w_{i.}^{(0)}\}$ which allows to measure a distance between a modified prototype and its initial representation.

Surely there remain many open questions, for instance how the prototypes are modified iteratively and how they are normalized after the adaptation has been completed. Our current research activities do try out several alternatives, but do not yet allow a presentation of a workable prototype.

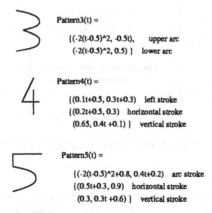

Pattern3(t) =

  {(-2(t-0.5)^2, -0.5t),   upper arc
  (-2(t-0.5)^2, 0.5) }   lower arc

Pattern4(t) =

  {(0.1t+0.5, 0.3t+0.3)   left stroke
  {(0.2t+0.5, 0.3)   horizontal stroke
  (0.65, 0.4t +0.1) }   vertical stroke

Pattern5(t) =

  {(-2(t-0.5)^2+0.8, 0.4t+0.2)   arc stroke
  {(0.5t+0.3, 0.9)   horizontal stroke
  (0.3, 0.3t +0.6) }   vertical stroke

Fig. 2. Digits composed of parametric curves.

3.3 The NIST Form-Based Handprint Recognition System Framework

We use a publicly available form-based character recognition system as a framework for the integration of our work. This system is the NIST reference character recognition system (Garris et al. [8]). The software supports several processing modules that are needed to do forms processing. An image processing part is responsible for the extraction of lines of handwritten text. Afterwards single characters are isolated. The pixels of the patterns are inserted into a vector from which the Principal Components are extracted. These Principal Components are used as the input features of a Multilayer Perceptron classifier. The network is trained to recognize single isolated characters. The classification results is a string of sequences of candidate characters. The use of a dictionary permits to check the plausibility of a character combination. The use of context via the dictionary improves the quality of the classification considerably. Finally the complete result of the classification are character strings associated to positions on the form.

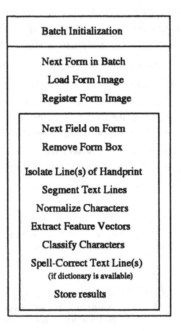

Fig. 3. Organization of the functional components within HSFSYS2, from [8].

The layout of the latest version of the software system HSFSYS2 can be observed in Fig. 3. Forms are continuously processed and the results are stored. There is a template information which contains the layout of the form that helps to identify to which field the extracted information belongs. Some of the modules are quite useful for our purposes, especially the isolation of lines of handprint and the postprocessing with the help of the dictionary.

Other parts of the system will be modified, especially the isolated character classification part. As it was mentioned before the feature model that is used in the HSFSYS2 system are the Principal Components of the vectorized images of the patterns. The usual normalization steps are performed, similarity and slant correction. In order to integrate our prototype-based character recognizer we have to modify the modules "Normalize Characters", "Extract Feature Vectors" and "Classify Characters" in Fig. 3.

4 Application Area

Each year an application of high-school students and professionals takes place in order to occupy the available vacancies at our university (UFES). There were a total of 2420 vacancies in 1995 and were disputed by 20282 candidates. Each of these candidates filled in a form with his personal data and the course of his choice, besides some additional data. In the absolute majority of the cases the form was filled in by hand, seldomly by a typewriter. Each of the forms had to be digitized manually in order to provide the information contained in the forms for the organizing committee.

The process of digitization involved a total of 8 operators, which worked in 6-hour shifts during 23 days. This data acquisition process was subjected to errors. Therefor key data of the forms such as the name of the candidate, the opted course, zip code etc. were redigitized. Otherwise the possibility of an erroneous exclusion of a candidate could have been occurred. With the repetition of the data acquisition process it was possible to detect typing errors and correct them. During this phase the fields in the forms that had been incorrectly digitized were registered via counters in the acquisition software. In order to give an idea of the score: "Name" was 3813 times typed incorrectly considering a total of 20093 inscription forms.

A general purpose scanner can acquire a digital image of a form in less than 15 seconds, including the preparation time to place the document onto the scanning surface. A single operator could feed a total of 20000 forms in 13 days to the automatic form reading systems which would consist of the scanner, a general purpose microcomputer plus the recognition system software. We suppose that in the interval between two forms the information would have be extracted completely, i.e. in real-time. Resuming, we could say that an automatic form reading system could make the process of data acquisition faster and more reliable if the respective system were robust enough.

The potential application areas of an automatic form reading system are quite diverse. It could be used in areas where actually data acquisition is done manually, for instance in the public administration, in banks and other organization that use forms to collect information from their clients.

5   Conclusions and Future Work

We must emphasize again that our research work is just at the beginning. We have identified the objectives of the work and which part of an automatic form reader are important to us. We want to focus our attention to the recognition system of the isolated handwritten character since it is the key element to a successful system. We departed from the traditional features that were calculated from a bitmap of the pattern image. We propose the use of prototypes which are compared to the unknown pattern. Parametric curves are the basis of the representation of the prototypes. We aim at developing a fast iterative procedure which adapts the prototypes to the pattern. The modification of the prototype serves as a similarity measurement which will finally decide the character class. As a framework we use a publicly available software environment, the HSFSYS2. We will adapt the software in order to serve our needs, especially in the part of the recognition algorithms.

We can expect interesting conclusions from our current work. We hope that the way that we approach the pattern classification will deliver very good results.

References

1. Blackwell, K. T., Vogl, T. P., Hyman, S. D., Barbour, G. S., Alkon, D. L.: A new approach to hand-written character recognition. Pattern Recognition 25 (1992) 655–666
2. Boser, B., Guyon, I., Vapnik, I.: A training algorithm for optimal margin classifiers. in Proc. of COLT II, Philadelphia, USA (1992)
3. Breuel, T. M.: Handwritten character recognition using neural networks. Handbook of Neural Computation, IOP Publishing Ltd. and Oxford University Press (1997)
4. Burr, D. J.: A neural network digit recognizer. Proc. IEEE Conf. Systems, Man, and Cybernetics, Atlanta, GA (1986) 1621–1625
5. Burr, D. J.: Experiments on neural network recognition of spoken and written text. IEEE Trans. on Acoustics, Speech, and Signal Processing 36 (1988) 1162–1168
6. Duda, R. O., Hart, P. E.: Pattern Classification and Scene Analysis. J. Wiley & Sons, New York (1973)
7. Fukushima, K.: Neocognitron: A self-organizing neural network model for a mechanism of pattern recognition unaffected by shift in position. Biological Cybernetics 36 (1980) 193–202
8. Garris M. D., Blue J. L., Candela G. T., Grother P. J., Janet S. A., Wilson C. L.: NIST Form-based handprint recognition system. TR NISTTR 5959, U.S. Dept. of Commerce, NIST, Gaitherburg, MD 20899-001 (1997)
9. Hastie, T., Tibshirani, R.: Handwritten digit recognition via deformable prototypes. AT&T Bell Technical report (1994),
   URL: ftp://playfair.stanford.edu/reports/hastie/zip.ps.Z
10. Haykin, S.: Neural Networks - A Comprehensive Foundation. Macmillan College Publ. Comp. (1994)
11. Hecht-Nielsen, R.: Neurocomputing, Addison-Wesley. Reading, MA (1990)
12. Impedovo, S., Marangelli, B., Fanelli, A. M.: A Fourier descriptor set for recognizing nonstylized numerals. IEEE Trans. on Systems, Man, and Cybernetics 8 (1978) 640–645

13. Impedovo, S.: Plane curve classification through Fourier descriptors: An application to Arabic hand-written numeral recognition. Proc. 7th IAPR Int. Conf. on Pattern Recognition (1984) 1069-1072

14. Impedovo, S.: Fundamentals in Handwriting Recognition. Springer Verlag (1994)

15. Kressel U., Schürmann J.: Pattern classification techniques based on function approximation. in Handbook on Optical Character Recognition and Document Analysis, H. Bunke and P.S.P. Wang (Eds.), World Scientific Publishing Company (1996)

16. Le Cun, Y., Boser, B., Denker, J. S., Henderson, D., Howard, A., Hubbard, W., Jackel, L.: Handwritten digit recognition with a back-propagation network. In Touretzky, ed., Advances in Neural Information Processing Systems 2 (1990) Morgan Kaufman, Denver

17. Li, R. Y., Xu, M.: Character recognition using a fast neural-net classifier. Pattern Recognition Letters 13 (1992) 369-374

18. Malavija, A.: FOHRES: Fuzzy online handwriting recognition system. 2.nd Int. Conf. on Fuzzy Set Theory, Durham (1993)

19. Michie, D., Spiegelhalter, D. J., Taylor, C. F.: Machine Learning, Neural & Statistical Classification. Ellis Horwood (1994)

20. Newman, R.: The OSCAR project: An offline script and character recognition toolset. URL: http://hcslx1.essex.ac.uk/oscar/oscar.html

21. Reber, W. L., Lyman, J.: An artificial neural system design for rotation and scale invariant pattern recognition. in Proceedings of IEEE First International Conf. on Neural Networks, San Diego, CA, IV (1987) 277-283

22. Revow, M., Williams, C. K. I., Hinton, G. E.: Using generative models for handwritten digit recognition. IEEE Trans. on Pattern Analysis and Machine Intelligence 18 (1996) 592-606

23. Senior, A. W.: Off-line handwriting recognition: a review and experiments. Technical report CUED/F-INFENG/TR 105, Cambridge University Engineering Dept. Cambridge, England (1992)

24. Simard, P. Y., Le Cun, Denker, J. S.: Efficient pattern recognition using a new transformation distance. in Advances in Neural Information Processing Systems 2 (1993) Morgan Kaufman, San Mateo, USA

25. Suen, C. Y., Nadal, C., Legault, R., Mai T. A., Lam, L.: Computer recognition of unconstrained handwritten numerals. Proc. of the IEEE 80 (1992) 1162-1180

26. Tappert, C. C., Suen, C. Y., Wakahara, T.: The State of the Art in On-Line Handwriting Recognition. IEEE Trans. on Pattern Analysis and Machine Intelligence 12 (1990) 787-808

27. Teh, C.-H., Chin, R. T.: On image analysis by the methods of moments. IEEE Trans. on Pattern Analysis and Machine Intelligence 10 (1988) 496-513

28. Wechsler, H., Zimmerman, G. L.: 2-D Invariant Object Recognition Using Distributed Associative Memory. IEEE Trans. on Pattern Analysis and Machine Intelligence 10 (1988) 811-821

29. Williams, C. K. I., Revow, M. D., Hinton, G. E.: Using a neural net to instantiate a deformable model. Advances in Neural Information Processing Systems 7 (1995) 965-972, The MIT Press

Automatic Extraction of Filled-in Information from Bankchecks Based on Prior Knowledge About Layout Structure

Alessandro L. Koerich[1] and Luan Ling Lee[2]

[1] Federal University of Santa Catarina, Electronic Instrumentation Laboratory,
CP 476, 88040–900, Florianopolis, SC, Brazil
[2] State University of Campinas, Department of Communication,
CP 6101, 13083–970, Campinas, SP, Brazil

Abstract. This paper presents a technique for extracting the filled–in information from bankchecks based on prior knowledge about their layout structure. We have analyzed the bankcheck characteristics and proposed a model that can be used to locate and extract the filled–in information applicable to any bankcheck. The model is based on prior knowledge about the check layout structure and on the identification of the check by reading the information stored in the MICR line. To eliminate the redundant information from a bankcheck image, such as the background pattern, the printed lines and the printed characters, we perform as follows. First of all we subtract the digitized check image from the check's background pattern image which is previously stored in the recognition system. Then the areas where the filled–in information is supposed to appear are extracted through a template. The elimination of the baselines in the image is based on projection profiles, while the printed characters are eliminated through a subtraction operation. Experimental results from testing Brazilian bankchecks show that the proposed method is capable of extracting the filled–in items from bankchecks achieving accuracy rates varying from 88.7% to 98.3%.

1 Introduction

Recent advances in information and communications have increased the needs and therefore interests in automated reading and processing of documents. The automation of bankcheck processing has been a problem of interest to financial institutions for many years [1]–[6]. Millions of handwritten or machine printed bankchecks have to be processed everyday. In spite of intensive research efforts, the degree of automation in processing bankchecks is very limited. When a check is processed only the information encoded in MICR line or in bar codes can be handled automatically. The filled–in information such as digit amounts, worded amounts, signatures, etc. is still manually handled. There are some proposed methods that can be applied to recognize the handwritten and machine printed items from bankchecks, however, these methods generally can not be applied directly to real documents because most of them deal with images in which only isolated elements exist, without considering the presence of background patterns [5]. The background pattern and the printed information presented on checks often disturb the recognition/verification process of filled–in items.

Two problems can be addressed in processing of bankchecks: the extraction of the information introduced by bank's customers and the recognition of these extracted information items. Our goal is to provide an approach for processing bankcheck images and to extract the filled–in information, allowing the independent processing of each item of information by recognition or verification systems. The proposed scheme for extracting the filled–in information from bankchecks is based on prior knowledge about their layout structure. We assume that a sample of all background patterns used by financial institutions are available in a database. Once a check is identified, the corresponding background pattern is select and used to eliminate the background pattern in the check. Next, using a template, the desired information items are extracted from the check image. The baselines are located and eliminated based on the information of projection profiles. An image containing printed characters strings is generated and used to eliminate the characters strings printed on the check. Finally, the different items of interest are segmented.

The organization of this paper is as follows. Section 2 describes the characteristics and the elements of Brazilian bankchecks. Section 3 describes the proposed modeling for bankchecks which allows the information extraction. The operations involved in extracting the filled–in information are detailed in Section 4. The experimental results are presented in Section 5. Finally, the conclusions are stated in the last section.

2 Characteristics of Checks

The machine understanding of bankchecks is an essential problem to the automation of financial institutions. One of the primary interests in document image understanding is that of using knowledge about the basic structure of documents to analyze and identify the different components of a document. This involves the syntactic and semantic interpretation of the various components of a document image, and requires domain knowledge about document features and characteristics. Layout analysis methods have been suggested for restricted types of documents such as bankchecks.

Brazilian bankchecks present a complex layout structure with variations in color, background patterns and stylistic characters or symbols, but still maintain a standard size measuring approximately 175 mm by 79 mm. The information presents on bankchecks is also standardized, such as position, size and contents of the related fields. Moreover, all checks issued by each Brazilian financial institution follow the same rules governed by Brazilian Central Bank. Fig. 1 shows a sample image of a Brazilian bankcheck digitized with 200 dpi spatial resolution and 256 gray levels.

324

Fig. 1. A sample of Brazilian bankcheck

A bankcheck image can be partitioned into 9 blocks as shown in Fig. 2. In the first top row (1), several strings of printed numerals and alphabets are found. Some of these strings are used to identify the bank, the agency, the account number, and the check serial number while the others are information just for internal banking processing, such as routing number, verification digits, etc. On the most right hand side of the first row (2), an area is reserved for digit amount filling. In the second row (3) there are two baselines for worded amount filling. The third row (4) is reserved for payee's name which can be optional during the filling. On the right hand side of the fourth row (6), we find an area reserved for the name of the city and the date. Note that some specific fields are reserved for day, month and year information. In general, the financial institution's name and logo, and the agency's name and address are printed on left hand side (5)(7) just below the fourth row. The area on the right hand side (8) just below the fourth row is reserved for handwritten signatures. There are printed lines functioning as guidelines for signature, worded amount, payee's name and date fields. Finally, in the bottom row (9) all identification numbers printed in the first row are encoded and printed with CMC–07 font by a MICR (Magnetic Ink Character Recognition) process.

(1) BANK NUMBER, AGENCY NUMBER, ACCOUNT NUMBER, CHECK SERIAL NUMBER AND INTERNAL PROCESSING NUMBER AREA		(2) DIGIT AMOUNT AREA
(3) WORDED AMOUNT AREA		
(4) PAYEE'S NAME AREA		
(5) FINANCIAL INSTITUTION'S LOGO AREA	(6) CITY AND DATE AREA	
(7) FINANCIAL INSTITUTION'S NAME AND ADDRESS AREA	(8) CUSTOMER'S SIGNATURE AND INFORMATION AREA	
(9) MICR CMC-7 LINE		

Fig. 2. Block division of a bankcheck

In spite of this standardization, some variations may occur. We must to account with small variations in size of printed characters, texture of background, location of each partitioned blocks, as well as other variables. Next we describe our model for processing Brazilian bankchecks taking into account every variation that can occur.

3 Bankcheck Modeling

In the previous section we have studied the structure of the bankcheck layout and performed an abstract division in terms of printed information and fields for data filling. Now we consider the overlapping of different kinds of information. The proposed bankcheck modeling is based on three primary elements of information: *background pattern, printed information and filled–in information. Background pattern* refers to the check's colored background pictures and drawings over which all other information elements are added. In our model we regard the financial institution's logo as part of background. *Printed information* is represented by printed characters strings, baselines, boxes, and symbols which specify the financial institution's name, identification numbers, baselines and also all information with respect to the agency and to the customer's personal information. *Filled–in information* refers to information in bankcheck which is introduced by bank's customers, such as the digit amount, the worded amount, the payee's name, the city and date, and the signature. Note that the filled–in information can be either handwritten or machine printed. Fig. 3 illustrates an abstract information division in these three layers.

Based on the well-defined characteristics of bankchecks and on the bankcheck information division, our goal is to extract the filled–in information from a scanned copy of every check, whatever the issuing financial institution. In addition our model also consider two essential aspects which make the information extraction no trivial: the variation of the position of the partitioned blocks in the check and the overlapping of different kinds of information.

Analyzing the proposed check division, either in blocks or in layers, we found that in the context of bankcheck information processing, the background pattern is redundant, as well as the printed information, except the information stored in MICR lines. This is due to the fact that, as mentioned before, it contains all needed information to identify a bankcheck. Moreover, the information introduced by bank's customers, i.e., the filled–in information is really important in bankcheck information processing.

4 Background Pattern and Printed Information Elimination

In a practical sense, the background pattern and the printed information only make the recognition of the filled–in information difficult. It is desirable that the

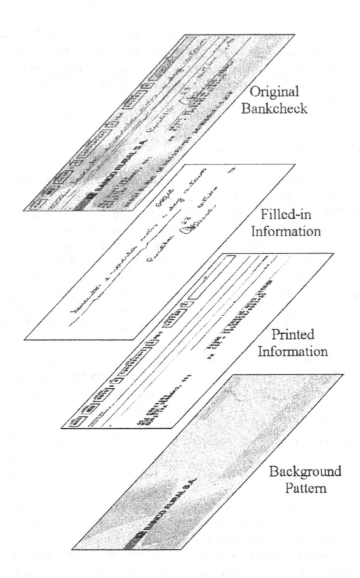

Original
Bankcheck

Filled-in
Information

Printed
Information

Background
Pattern

Fig. 3. Abstract layered division of information on bankchecks

background pattern and the printed information be eliminated from the check image, without losing any information of the interested parts, i.e., the filled–in information.

The process of extracting begins with the document insertion in a MICR scanner. The scanner reads the characters printed with magnetic ink and encoded using CMC–07 font style. With this reading, the bank and the customer are identified. Thus the related background pattern and parameters required to generate the template and an image which contains the printed characters can be select. The document is also digitized using an optical scanner generating a digitized bankcheck image. This image undergoes a position adjustment before any further processing. This image is introduced into a subtraction module that performs a subtraction operation over the correspondent background pattern selected from the database. Using the bank parameters stored in the database, we are able to generate a template which is used in the operation to extract the filled–in information from the bankcheck. The template image contains the positions of interested information items. The extracted information still has the presence of baselines and printed characters strings. The baselines are eliminated using an approach based on projection profiles, while the printed characters strings are eliminated through a subtraction operation. Next, we describe in details the operations involved in the elimination of the background pattern, baselines, and printed characters strings.

4.1 Position Adjustment

The dominant orientation of the baselines determines the skew angle of a document. A bankcheck generally have zero skew, that is, horizontally and vertically printed lines are parallel to the respective edges [6]. However, when a document is manually scanned, nonzero skew may be introduced, as well as shifts in vertical and in horizontal positions. Since frequently image analysis requires zero skew and zero shift, the correction of the image position should be performed. In fact, the position adjustment is fundamental for the successful extraction of filled–in information from bankchecks. We must perform skew angle correction, vertical and horizontal position adjustments.

Skew Angle Correction We have designed a method to evaluate and correct the skew angle of bankcheck images. Our approach is based on the information provided by projection profiles. First of all, the bankcheck image is divided in several vertical windows (strips) along horizontal axis. Each window occupies the whole vertical dimension of the bankcheck. Based on prior knowledge about the bankcheck layout structure, we can select an area around the second baseline that will be used as a reference to compute the skew angle of the bankcheck. For each window we select an area around the second baseline, and compute its horizontal projection profile, and determines the position where the maximum of horizontal projection profile occurs. Then we perform a vertical shift for each window until the maximum of the projection profile matches with that of the selected reference. If we analyze a bankcheck image we found that the maximum value that a horizontal projection profile can assume is the own baseline length.

Vertical Position Adjustment The adjustment of bankcheck vertical position is performed using the information obtained from the skew angle correction operation. We compute the horizontal projection profile around the second baseline and searching for the position where the maximum point of profile occurs. If this position is similar to the reference position present in the bankcheck model, the bankcheck is adjusted, otherwise, a shift is performed in digitized image until both positions be coincident.

Horizontal Position Adjustment The correction of bankcheck horizontal position is performed using the information provided by the vertical projection profile. We compute the vertical projection profile for a specific area in the top row of the bankcheck and searching for the position of the first block of printed information. This position is compared with the position imposed by the model. If two positions are similar, the bankcheck is adjusted; otherwise, a shift operation is performed in the bankcheck image until two positions coincide. After this operation, the bankcheck image is size normalized, and the other steps of processing can be applied.

4.2 Background Elimination

To recover the filled–in information from a bankcheck, recently Yoshimura et al. have suggested the subtraction of the image of an unused bankcheck from its filled–in bankcheck image version [7]. A main drawback of this method is that for each filled–in bankcheck, we have to maintain its unfilled image sample requiring, therefore, a large amount of memory space. In addition, we have found that thresholding techniques are unable to provide satisfactory results; specially when both the printed information and the filled–in information image present a similar gray level [8].

To solve this problem we propose an approach based on the method proposed by Yoshimura. To eliminate the background pattern on a bankcheck, we subtract the bankcheck image from the background pattern sample image. Such approach is perfectly plausible since Brazilian bankchecks issued by the same financial institution present a common background pattern. The subtraction operation is denoted by:

$$I_{WB}(x,y) = I_{CD}(x,y) - I_{CB}(x,y) \tag{1}$$

where $I_{CD}(x,y)$ is the digitized–position–adjusted bankcheck image, $I_{CB}(x,y)$ is a sample of background pattern and $I_{WB}(x,y)$ is the resulted image. Note that the selection of the correct bankcheck background pattern is based on previous knowledge about which financial institution has issued the check through MICR line reading.

4.3 Extraction of Filled-in Information

Humans have an impressive ability to extract information from documents when suitable training is given. This extraction process relies heavily on our knowledge of the domain and occurs at any levels of complexity and abstraction [9].

As mentioned before, the proposed extraction method relies on prior knowledge about bankcheck layout structure. Using such knowledge results in simple, elegant and efficient bankcheck decomposition approach. Due to layout standardization of Brazilian bankchecks, we design a template for extracting the desired information from every bankchecks. We use the block division, as shown in Fig.

2, to segment the information on bankchecks. The blocks are formed based on the following information introduced by the bank's customers: digit amount (2), worded amount (3), payee's name (4), city and date (6), and signature (8). The template is adjusted according to the financial institution which has issued the bankcheck. We construct a knowledge based rules for the adjust of the template based on identification of check through MICR line information [10]. The template, covers the positions where the desired information items appear, as shown in Fig. 4. The assigned areas represent the areas of the bankcheck that will be extracted. The different gray levels represent the different information items. The interested areas are extracted from bankcheck image, resulting in an image free of many printed elements, as shown in Fig. 5.

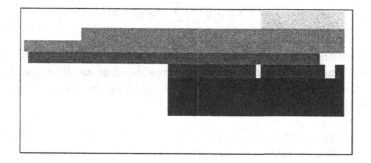

Fig. 4. Assigned areas for information extraction

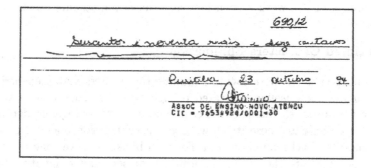

Fig. 5. Interested items extracted from bankcheck

In spite of the redundant information being eliminated in this process, the extracted areas still contain some undesirable items, such as baselines and printed characters strings which overlap the filled-in information. To eliminate these undesirable items we apply two processing techniques as follow.

4.4   Baselines Elimination

The elimination of baselines is based on horizontal projection profiles. Since the baselines extend almost overall bankcheck length, it is expected that its position can be easily identified from the horizontal projection profile. In other words, the maximum values of the horizontal projection profiles occur just in positions where baselines are present. First we compute the horizontal projection profile for $I_{WB}(x, y)$ that is denoted by:

$$PPh_{WB}(y) = \sum_x I_{WB}(x, y) \tag{2}$$

The points with maximum values of projection indicate the position of baselines. We select m maximum points that can be considered as baselines positions. The value of m depends on the bankcheck which are being processed. This operation is denoted by:

    for $MaxPoints = 1$ to m
        find $MaxPPh_{WB}(y)$ and the position (K) where it occurs
        fill $I_{WB}(x, K)$ with white pixels
    end

where $MaxPoint$ is the peaks of projection profile, m is the number of maximum points that will be detected, $MaxPPh_{WB}(y)$ is the value of the maximum point and K is the position where maximum point occurs. Furthermore, the width of baselines, which is generally of 2 pixels, also must be considered when we impose m. In this work we consider that the m can assume values up 12.

4.5   Printed Characters Elimination

In this subsection we show how to eliminate the printed characters strings appearing under the baseline dedicated to signature. In general these strings of characters comprise the name of the customer and his/her register identification number. To eliminate these items we apply a subtraction operation as before to eliminate the background pattern. For each bankcheck, we generate a binary image which contains the printed information present under signature baseline, and uses it to eliminate the printed strings on bankcheck through a subtraction operation. In order to generate this binary image with printed information we seek the following information previously stored in the database: customer's name, register identification number, position, font style and size of each symbol, character and other geometrical object that can be present under signature baseline. The subtraction operation is denoted by:

$$I_{WP}(x,y) = I_{WB}(x,y) - I_{GP}(x,y) \tag{3}$$

where $I_{WB}(x,y)$ is the image resulted from the baselines elimination operation, $I_{GP}(x,y)$ is the generated image with printed characters, and $I_{WP}(x,y)$ is the resulted image. Finally, we obtain an image free of background pattern and printed information, as shown in Fig. 6. This image contains only the information introduced by bank's customer.

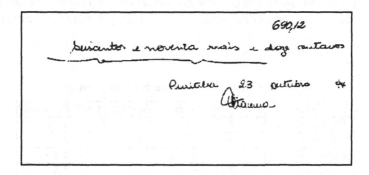

Fig. 6. The bankcheck image resulting from processing

In order to process each item independently, we can segment this image using the template again. Each area assigned with a different gray level in Fig. 5 represents the different items position. Based on this knowledge about items position, we can segment each item from the image. The items segmented from the resulting image are shown in Fig. 7.

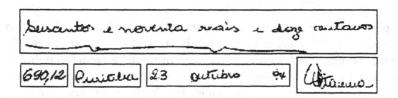

Fig. 7. Extracted information items

5 Experimental Results

The proposed extraction method was tested by real Brazilian bankcheck images with 200 dpi spatial resolution and 256 gray level in BMP format. It is expected that the extraction algorithm provides all filled–in information: digit amount, worded amount, payee's name, city and date, and signature.

In order to evaluate the proposed method, the resulted images were classified by applying a classification rule similar to that proposed by Liu et al. [11]. Tab. 1 summarizes the results of extracting the items from 100 real check images issued by different financial institutions. The first column classify the extracted items according to their information. **Cor_Extr**, **Incor_Extr** and **Unus_Items** denote the percentage of items correctly extracted, incorrectly extracted and correctly extract but presenting some defects which prevents their future use.

Table 1. Extraction results on bankcheck images

Items	Corr_Extr (%)	Incor_Extr (%)	Unus_Items (%)
Digit Amount	98.3	1.7	1.2
Worded Amount	95.3	4.7	3.4
Payee's Name	97.5	2.5	2.9
Place	95.2	4.8	1.5
Date	96.1	3.9	3.3
Signature	88.7	11.3	7.3

We have observed that the items incorrectly extracted is mainly due to the presence of stylist traces that exceed the limits of selected areas. This problem affects especially the signatures. Furthermore, some post-processing on extracted images is needed to improve the quality of images, especially, to recover the lost information due overlap among baselines and handwritten items.

6 Conclusions

This paper presents a technique for extracting the filled–in information from bankchecks based on prior knowledge about their layout structure. We have analyzed the bankcheck characteristics and proposed a model that can be used to locate and extract the filled–in information applicable to any bankcheck. The model is based on prior knowledge about the check layout structure and on the identification of the check by reading the information stored in the MICR line.

The different items of information, such as digit amount, worded amount, payee's name, date, city, and signature are extracted from a digitized image of a bankcheck. The proposed method was applied in extraction of filled–in information from Brazilian bankchecks providing satisfactory results. Experimental results show that the approach presented in this paper is able to read bankchecks and extract all filled–in items. By combining this method with digit recognition, word recognition, and signature verification algorithms, an automatic bankcheck recognition system might be feasible for practical applications.

References

1. Yasuda, Y., Dubois, M., Huang, T. S.: Data compression for check processing machines. Proceedings of the IEEE 68 (1980) 874–885
2. Lethelier, E., Gilloux, M., Leroux, M.: Automatic reading system for handwritten numeral amounts on French cheques. Proc. 3rd Int'l Conf. on Document Analysis and Recognition, Montreal, Canada (1995) 92–97
3. Lam, L., Suen, C. Y., Guillevic, D., Strathy, N. W., Cheriet, M., Liu, K., Said, J. N.: Automatic processing of information on cheques. Proc. Int'l Conf. on Systems Man and Cybernetics, Vancouver, Canada, (1995) 2353–2358
4. Agarwal, A., Granowetter, L., Hussein, K.,Gupta, A.: Detection of courtesy amount block on bank checks. Proc. 3rd Int'l Conf. on Docum. Analysis and Recognition, Montreal, Canada, (1995) 748–751
5. Ueda, K.: Extraction of signature and seal imprint from bankchecks by using color information. Proc. 3rd Int'l Conf. on Document Analysis and Recognition, Montreal, Canada (1995) 665–668
6. Akiyama, A., Hagita, N.: Automated entry for printed documents. Pattern Recognition 23 (1990) 1141–1154
7. Yoshimura, I., Yoshimura, M.: Off-line verification of Japanese signatures after elimination of background. Int'l Journal of Patt. Recogn. and Artif. Intelligence 8 (1994) 693–708
8. Kamel, M., Zhao, A.: Extraction of binary character/graphics images from grayscale document images. CVGIP: Graphic Models and Image Processing 55 (1993) 203–217
9. Doermann, D. S.: Document understanding research at Maryland. Proceedings of ARPA Image Understanding Workshop (1994)
10. Koerich, A. L.: Processing of bankcheck images: information extracting and storage. MSc. Thesis (1997) (in portuguese)
11. Liu, K., Suen, C. Y., Nadal, C.: Automatic extraction of items from cheque images for payment recognition. Int'l Conf. on Pattern Recognition, Vienna, Austria (1995) 798–802

A Simple Methodology to Bank Cheque Segmentation

José Eduardo B. Santos[1], Flávio Bortolozzi[2] and Robert Sabourin[3].
Centro Federal de Educação Tecnológica do Paraná - CEFET-Pr[1,2].
Av. 7 de Setembro, 3165 - 80230-901 - Curitiba, Pr. Brasil
Pontifícia Universidade Católica do Paraná - PUC-Pr.[2]
R. Imaculada Conceição,1155 - 80215-901 - Curitiba-Pr.
École de Technologie Supérieure[3]
1100, rue Notre-Dame Ouest
Montréal (Québec) H3C 1K3, Canadá

jesantos@centerline.com.br[1] fborto@ccet.pucpr.br[2] sabourin@gpa.etsmtl.ca3

Abstract : The segmentation of bank cheque images is a fundamental phase of its automatic processing. In the segmentation phase, one of the most important steps is the background elimination, that has to respect the physical integrity of the rest of the cheque image information. This paper describes a simple and robust solution for the background elimination problem, using a process that involves two stages: the original image enhancement and a posterior global thresholding process.

Keywords : *bank cheque segmentation, background elimination, thresholding, histogram quadratic hyperbolization.*

1 Introduction

By definition, the Document Processing, that involves the Document Analysis and the Document Recognition, has the objective of proceeding the text and the graphical elements recognition, that are in the document image, being able to make an analysis just as a human observer is capable to do. Even though we know that many times a printed document presents a higher quality than that of its electronic similar, there is still a lot of motivations to processing the printed documents, like its re-edition, its re-distribution, its storage and its combination with other information.

Our study is part of a larger research that intends to carry out the identification, recognition and validation of the hand-written elements of bank cheques, using Digital Image Processing and other related tools, as we can see in Fig. 1. The stages that make up the whole process are basically divided by the hand-written elements of the bank cheque, which are: the numerical amount, literal amount, date and signature. The field where the cheque's beneficiary is informed, is not an important information for the electronic document processing, but it's perfectly possible to make it's posterior processing if we find it is necessary.

In this figure we can see that the segmentation phase has a special importance, because all the other steps of verification and recognition depend on the segmentation phase so that they can be executed.

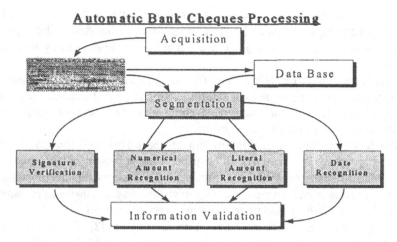

Fig. 1 - Automatic Bank Cheques Processing Phases

Therefore, the segmentation phase has a decisive function in the final process result. The success of the whole process depends on a good quality result of the segmentation phase.

We can identify four jobs in this phase:

- the background elimination;

- the straight lines identification and elimination;

- the pre-printed text identification and elimination;

- the hand-written components identification and separation, sending each one to its adequate verifying or recognition process.

In our study, more attention is given to the first of these problems: the background elimination. This step is not a simple one, especially if we note that the background has to be totally eliminated and the rest of the information must suffer the less degradation possible. In addition, there is another relevant factor that refers to the large diversity that the Brazilian bank cheques present, and that has to be processed with a single automated methodology.

This paper presents a simple and robust solution to solve the background elimination problem, identifying all of the problems and the tools used in our solution.

2   The problem

The elimination of the bank cheques background is an essential part for its processing, when we consider a process based on binary images. This job has many problems that appear before the image processing: for example in its making. However, the document quality is a characteristic that we can't control in many of the projects therefore we search for tools that can electronically give sufficient quality for its processing. This reminds us of two key elements that permit us to see any object in the real world: its luminance and its reflectance. While the first one depends exclusively on the light font and shows the light quantity that strikes the object, the second is determined by object properties and shows the light quantity reflected by the object.

These factors have big relevance in the image processing of bank cheques. The light incidence and the cheque light absorption are factors that characterise the difficulty level registered in the background elimination, because they determine the contrast level between the cheque elements.

The literature shows that this problem can be solved with thresholding techniques, as proposed by Cheriet[1]. In his paper, Cheriet presents a tool based on the Otsu's Method, that obtained good results in the background identification and elimination using a recursive process. However the images shown in the paper are too simple, without the problems observed in the Brazilian cases.

Another interesting proposal comes from Don's work [2], that proposes a new method to document image binarization by a noise image model. In this way, its possible to identify and eliminate many occurrences of text or images that are on the back of the document but that show up on the front. The major problem occurs when the objects of interest in the image do not form relevant peeks in its histogram.

Okada[3] treats the background elimination in a superficial way. In his paper, Okada eliminates not only the background but all of the pre-printed elements, by a morphological subtraction scheme between the original image and a filled document image. Even though this scheme presents good results, this is a solution that doesn't allow the treatment of different document images at the same time, like a very generic system that can treat an image of any bank and not only one specific cheque at a time.

In Brazil, there is a big variety of images and drawings printed on the cheque's background, making it a visit card for each bank. The processing of these cheques is not a trivial job, especially if we remember that we are interested in tools that process any cheque image and not only the cheques of a unique institution. The generalisation of the background elimination process becomes a more complex question. On the other hand, any process that searches for an efficient solution for problems that present a big variety of conditions have to be adopted with vast versatility, always expecting a big number of situations and possibilities.

Another important point in the execution of this job, refers to the physical integrity of the remaining information in the image after the background elimination. Simple thresholding processes can be used in the background elimination. However, we verified that the good results obtained in some cases didn't justify the use of this methodology in all of the images, because there is a big number of cases in which the degradation of the remaining information is too large making the process invalid. Good and bad examples can be viewed in Fig. 2 and Fig. 3, where we show two images that were only processed with a simple global thresholding technique.

3 The Otsu's Method

The method proposed by Otsu[4] has the advantage of not needing any prior knowledge of the image, based only on its gray level histogram. The main idea is to find in the histogram an optimal threshold that divides the image objects by constructing two classes $C_0=\{0,1,...,t\}$ and $C_1=\{t+1,t+2,...,K-1\}$, from any arbitrary gray level, using the discriminant analysis.

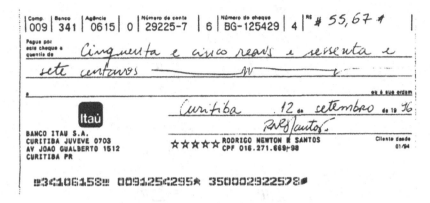

Fig. 2 - A good example of Otsu's Method application.

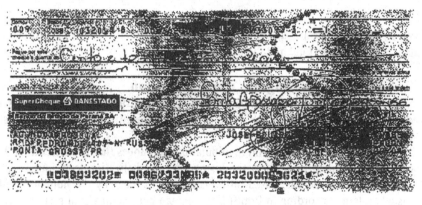

Fig. 3 - A bad example of this methodology.

To find the optimal threshold t, we can use one of the following criteria functions which respect t:

$$\eta = \frac{\sigma_B^2}{\sigma_T^2}, \lambda = \frac{\sigma_B^2}{\sigma_W^2} \text{ or } k = \frac{\sigma_T^2}{\sigma_W^2}$$

We adopted η because it is the simplest one of these functions. σ_T^2, that is the total variance, is independent from the gray level, only being necessary to minimise the function σ_B^2, that is the within-class variance. The optimal threshold t^* will be defined in the following form:

$$t^* = \text{ArgMin } \eta$$

And:

$$\sigma_T^2 = \sum_{i=0}^{l-1} (i - \mu_T)^2 P_i$$

$$\sigma_B^2 = \omega_0 \omega_1 (\mu_1 \mu_0)^2$$

$$\mu_1 = \frac{\mu_T - \mu_t}{1 - \omega_0} \qquad \mu_0 = \frac{\mu_t}{\omega_0}$$

$$\omega_0 = \sum_{i=0}^{t} P_i$$

$$\omega_1 = 1 - \omega_0$$

$$\mu_T = \sum_{i=0}^{l-1} ip_i \qquad \mu_t = \sum_{i=0}^{t} ip_i$$

Where ω_0 and ω_1 correspond to C_0 and C_1 class variance. μ_0 and μ_1 its respective median. μ_T and μ_t the total and the by-class medians.

4   The Quadratic Histogram Hyperbolization

When we use particular laws in the histogram transformation (like Gauss or Normal Distribution, for example) we call histogram specification the method used in its modification.

There is one of these histogram manipulation techniques that is based on the hyperbolization. Proposed by Frei[5], the histogram hyperbolization technique tries to emphasise the image details, based on transformations made by the human peripheral visual system. According to Pratt[6], the human peripheral visual system makes a transformation on the intensity of light that we observe in the image, originating in a non-linear relation between the physical luminous intensity and the subjectively perceived luminous intensity by the human central visual system. This relation doesn't allow us to see the gray levels of this image uniformly distributed. The histogram hyperbolization technique, has as a basic principle, to submit the image to a histogram equalisation followed by the inverse of its transformation made by the peripheral portion of the human visual system. With the result of this operation, the annulment of the effect produced by this non-linear transformation is expected.

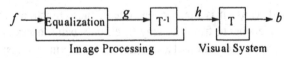

Diagram 1 - Technical Scheme of the Histogram Hyperbolization Technique.

The histogram hyperbolization technique presented by Cobra[7], demonstrates a different model from the one presented by Frei to make the histogram modification. It is based on a new model of the human peripheral visual system, that considers the fact that this system accommodates to the medium intensity of the observed scene, and not to the individual pixel intensity, as shown by the logarithmical model used in the histogram hyperbolization of Frei[5,7,8]. The result of this new model, is a transformation that provides a larger distribution of the gray levels, evicting an excessive concentration in the dark tons, verified in the hyperbolization technique. In this sense, it is possible to obtain an image that makes the human observation a lot easier.

According to Cobra, the logarithmical model for the non-linear characteristic of the human visual system, is based on the approximation

$$g \cong \widetilde{g}$$

that supposes an approximate equality between the value g of any pixel of the image and the medium value of the image $\widetilde{g}$,that is generally not valid. In reality, this approximation doesn't consider that the visual system accommodates to the value of the pixel analysed g. An alternative model, that considers this fact comes from [6]:

$$b(g) = \frac{g(N-1+\widetilde{g})}{g+\widetilde{g}} \qquad\qquad 1$$

Following diagram 1, we intend to find the h image, from the inverse function of (1), applied to h, the results are:

$$g(h) = \frac{\widetilde{g}h}{(\widetilde{g} + N - 1 - h)} \qquad\qquad 2$$

We notice that if $h_{min}=0$ and $h_{max}=N-1$, then $g_{min}=0$ and $g_{max}=N-1$, as we wanted. Therefore, h has it histogram uniformly distributed between 0 and N-1, being obtained from f, which is the value of one pixel from the image, during the transformation

$$h(f) = P_f(f)(N\text{-}1)$$

To obtain the final transformation, all we need is to combine this equation with (2), that results in:

$$g(f) = \frac{\widetilde{g}(N-1)P_f(f)}{\widetilde{g} + (N-1)[1 - P_f(f)]} \qquad\qquad 3$$

We now come across the following problem: the transformation $g(f)$ involves the parameter $\widetilde{g}$ which represents the medium value of the image resultant of the own transformation. This parameter depends on the histogram of the original image and it can't be calculated beforehand, when digital images are considered. However, in the analogic image cases, this dependence is eliminated by the histogram equalisation made to obtain the intermediate image h. By convenience, we suppose that h varies between 0 and N-1, maintaining the superior limit adopted in the discrete case. So if the histogram of h is uniform between 0 and N-1 and the transformation $g(h)$ is known, its possible to determine $\widetilde{g}$, by the relation:

$$\widetilde{g} = \int_0^{N-1} g(h)P_h(h)dh$$

We have then:

$$\widetilde{g} = \int_0^{N-1} \frac{\widetilde{g}h}{\widetilde{g}+N-1-h}\left(\frac{1}{N-1}\right)dh$$

$$\widetilde{g} = \frac{\widetilde{g}}{N-1}\int_0^{N-1} \frac{h}{\widetilde{g}+N-1-h}dh$$

By solving the integral above and simplifying, we obtain:

$$\widetilde{g} = \widetilde{g}\left[\left(1+\frac{\widetilde{g}}{N-1}\right)\ln\left(1+\frac{N-1}{\widetilde{g}}\right)-1\right]$$

Re-organising the terms, we get the final expression:

$$\left(1+\frac{N-1}{\widetilde{g}}\right)\ln\left(1+\frac{N-1}{\widetilde{g}}\right)-2\left(\frac{N-1}{\widetilde{g}}\right)=0$$

When we solve this non-linear equation by using numerical methods, we verify that it is satisfied when

$$\widetilde{g} = 0.255(N-1)$$

Though this result is strictly valid to analogic images, it is fair to use it to give an approximate value for $\widetilde{g}$ in the digital image cases. Typically, when N=256 then $\widetilde{g}$ =65.

Many examples presented in Cobra's work show that a simple histogram equalisation process turns the gray levels in the image, too light, and too dark in the hyperbolization case. More efficiently, the quadratic hyperbolization allows a clear observation of the objects in the image, because they are better contrasted, both in the lighter areas and in the darker areas of the image.

5 Experimental Results

Like we said, the exclusive use of one global thresholding method is not sufficient to solve our problem. Therefore, we have tried to improve the original image quality, by distributing more uniformly and largely distributing, all the pixels of the bank cheque image. What we obtained with this purpose, was an altered image, in which the contrast remained well defined, giving priority to the objects of interest from the image, to the detriment of the background[9].

To test the use of the proposed methodology we used an experimental database, consisting of about 20 images, which showed different backgrounds, contrasts, a variety on the amount of light absorbed and reflected by the paper and different types of handwriting.

Approximately 30% of the images were able to be processed in a satisfactory manner by applying the histogram quadratic hyperbolization only once. In the other cases it was necessary to apply the technique more than once and soon after the Otsu limiarization. However, this percentage increased to as near 95%, when we applied the quadratic hyperbolization process more than once before limiarizing the image.

The visual criteria was used when determining the image quality, taking into consideration the complexity of the quantification that this type of parameter involves. Therefore, we haven't precisely forecast how many interactions will be necessary to obtain the expected results, since it's not possible to quantify how close the original images are to the results.

Even with the possibility of using tools that teach the machine the desired pattern for the image approval we weren't really worried with this aspect in this phase of the process. In the future this will be one of our main concerns wince we believe in the automatic bank cheque treatment project as being totally automated.

The following images shows in Fig. 4, 5 and 6 some situations where our methodology was successfully applied. To prove this, we show an image that had its straight lines perfectly removed before the background elimination.

6 Conclusions

As we presented in section 2, the problem with background elimination from the cheque image, in the Brazilian cases, is a complex job because of the great number of textures background. The proposed solutions seem not to consider the great variety in the processed images.

We introduced a very simple methodology when treating these images that showed highly satisfactory results, since the background was removed and the remaining information didn't suffer much degradation. Eliminating the horizontal lines in Fig. 6 we prove the viability of our methodology in an image segmentation process of cheques or other similar document images.

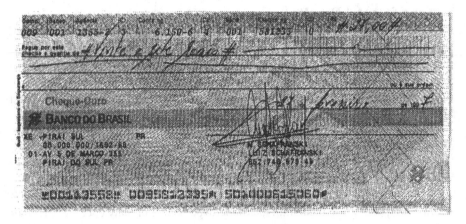

Fig. 4 – Original image cheque

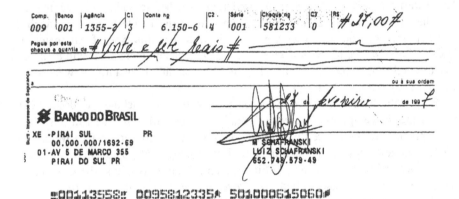

Fig. 5 – Image after application of morphological methods

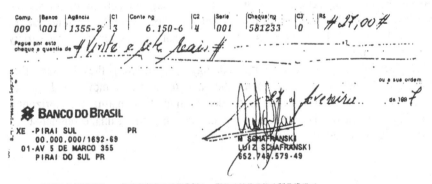

Fig. 6 – Image of straight lines perfectly removed.

References

1. Cheriet, M., Said, J.N. and Suen, C.Y. "A Formal Model for Document Processing of Business Forms." In Proceedings of ICDAR'95. 1995. 210-213.

2. Don, Hon-Son. "A Noise attribute Thresholding Method of Document Image Binarization." In Proceedings of ICDAR'95. 1995. 231-234.

3. Okada, Minoru and Shridhar, Malayappan. "A Morphological Subtraction Scheme for Form Analysis." In Proceedings of ICPR'96. 1996. 190-194.

4. Sahoo, P.K., Soltani, S. and Wong, A.K.C. "A Survey of Thresholding Techniques." Computer Vision, Graphics, and Image Processing. 1988. 41, 233-260.

5. Frei, W. "Image Enhancement by Histogram Hyperbolization."Journal of Computer Graphics and Image Processing. June, 1977. 286-294.

6. Pratt, W.K. "Digital Image Processing." John Wiley & Sons. New York, 1978.

7. Cobra, Daniel T., Costa, José D. e Menezes, Marcelo F. "Realce de Imagens Através de Hiperbolização Quadrática do Histograma." Anais do V SIBGRAPI, 1992. 63-71.

8. Cobra, Daniel T. "A Generalization of the Method of Quadratic Hyperbolization of Image Histograms." In Proceedings of 38th Midwest Symposium on Circuits and Systems. Rio de Janeiro, 1995. 141-144.

9. Santos, José Eduardo B. "Estudo sobre Métodos e Técnicas para a Segmentação de Imagens de Cheques Bancários." Dissertação de Mestrado. CEFET-Pr. 1997.

Author Index

Springer
and the
environment

At Springer we firmly believe that an international science publisher has a special obligation to the environment, and our corporate policies consistently reflect this conviction.
We also expect our business partners – paper mills, printers, packaging manufacturers, etc. – to commit themselves to using materials and production processes that do not harm the environment. The paper in this book is made from low- or no-chlorine pulp and is acid free, in conformance with international standards for paper permanency.

Springer

Lecture Notes in Computer Science

For information about Vols. 1–1265

please contact your bookseller or Springer-Verlag

Vol. 1302: P. Van Hentenryck (Ed.), Static Analysis. Proceedings, 1997. X, 413 pages. 1997.

Vol. 1303: G. Brewka, C. Habel, B. Nebel (Eds.), KI-97: Advances in Artificial Intelligence. Proceedings, 1997. XI, 413 pages. 1997. (Subseries LNAI).

Vol. 1304: W. Luk, P.Y.K. Cheung, M. Glesner (Eds.), Field-Programmable Logic and Applications. Proceedings, 1997. XI, 503 pages. 1997.

Vol. 1305: D. Corne, J.L. Shapiro (Eds.), Evolutionary Computing. Proceedings, 1997. X, 307 pages. 1997.

Vol. 1306: C. Leung (Ed.), Visual Information Systems. X, 274 pages. 1997.

Vol. 1307: R. Kompe, Prosody in Speech Understanding Systems. XIX, 357 pages. 1997. (Subseries LNAI).

Vol. 1308: A. Hameurlain, A M. Tjoa (Eds.), Database and Expert Systems Applications. Proceedings, 1997. XVII, 688 pages. 1997.

Vol. 1309: R. Steinmetz, L.C. Wolf (Eds.), Interactive Distributed Multimedia Systems and Telecommunication Services. Proceedings, 1997. XIII, 466 pages. 1997.

Vol. 1310: A. Del Bimbo (Ed.), Image Analysis and Processing. Proceedings, 1997. Volume I. XXII, 722 pages. 1997.

Vol. 1311: A. Del Bimbo (Ed.), Image Analysis and Processing. Proceedings, 1997. Volume II. XXII, 794 pages. 1997.

Vol. 1312: A. Geppert, M. Berndtsson (Eds.), Rules in Database Systems. Proceedings, 1997. VII, 214 pages. 1997.

Vol. 1313: J. Fitzgerald, C.B. Jones, P. Lucas (Eds.), FME '97: Industrial Applications and Strengthened Foundations of Formal Methods. Proceedings, 1997. XIII, 685 pages. 1997.

Vol. 1314: S. Muggleton (Ed.), Inductive Logic Programming. Proceedings, 1996. VIII, 397 pages. 1997. (Subseries LNAI).

Vol. 1315: G. Sommer, J.J. Koenderink (Eds.), Algebraic Frames for the Perception-Action Cycle. Proceedings, 1997. VIII, 395 pages. 1997.

Vol. 1316: M. Li, A. Maruoka (Eds.), Algorithmic Learning Theory. Proceedings, 1997. XI, 461 pages. 1997. (Subseries LNAI).

Vol. 1317: M. Leman (Ed.), Music, Gestalt, and Computing. IX, 524 pages. 1997. (Subseries LNAI).

Vol. 1318: R. Hirschfeld (Ed.), Financial Cryptography. Proceedings, 1997. XI, 409 pages. 1997.

Vol. 1319: E. Plaza, R. Benjamins (Eds.), Knowledge Acquisition, Modeling and Management. Proceedings, 1997. XI, 389 pages. 1997. (Subseries LNAI).

Vol. 1320: M. Mavronicolas, P. Tsigas (Eds.), Distributed Algorithms. Proceedings, 1997. X, 333 pages. 1997.

Vol. 1321: M. Lenzerini (Ed.), AI*IA 97: Advances in Artificial Intelligence. Proceedings, 1997. XII, 459 pages. 1997. (Subseries LNAI).

Vol. 1322: H. Hußmann, Formal Foundations for Software Engineering Methods. X, 286 pages. 1997.

Vol. 1323: E. Costa, A. Cardoso (Eds.), Progress in Artificial Intelligence. Proceedings, 1997. XIV, 393 pages. 1997. (Subseries LNAI).

Vol. 1324: C. Peters, C. Thanos (Eds.), Research and Advanced Technology for Digital Libraries. Proceedings, 1997. X, 423 pages. 1997.

Vol. 1325: Z.W. Raś, A. Skowron (Eds.), Foundations of Intelligent Systems. Proceedings, 1997. XI, 630 pages. 1997. (Subseries LNAI).

Vol. 1326: C. Nicholas, J. Mayfield (Eds.), Intelligent Hypertext. XIV, 182 pages. 1997.

Vol. 1327: W. Gerstner, A. Germond, M. Hasler, J.-D. Nicoud (Eds.), Artificial Neural Networks – ICANN '97. Proceedings, 1997. XIX, 1274 pages. 1997.

Vol. 1328: C. Retoré (Ed.), Logical Aspects of Computational Linguistics. Proceedings, 1996. VIII, 435 pages. 1997. (Subseries LNAI).

Vol. 1329: S.C. Hirtle, A.U. Frank (Eds.), Spatial Information Theory. Proceedings, 1997. XIV, 511 pages. 1997.

Vol. 1330: G. Smolka (Ed.), Principles and Practice of Constraint Programming – CP 97. Proceedings, 1997. XII, 563 pages. 1997.

Vol. 1331: D. W. Embley, R. C. Goldstein (Eds.), Conceptual Modeling – ER '97. Proceedings, 1997. XV, 479 pages. 1997.

Vol. 1332: M. Bubak, J. Dongarra, J. Waśniewski (Eds.), Recent Advances in Parallel Virtual Machine and Message Passing Interface. Proceedings, 1997. XV, 518 pages. 1997.

Vol. 1333: F. Pichler. R.M. Díaz (Eds.), Computer Aided Systems Theory – EUROCAST'97. Proceedings, 1997. XI, 626 pages. 1997.

Vol. 1334: Y. Han, T. Okamoto, S. Qing (Eds.), Information and Communications Security. Proceedings, 1997. X, 484 pages. 1997.

Vol. 1335: R.H. Möhring (Ed.), Graph-Theoretic Concepts in Computer Science. Proceedings, 1997. X, 376 pages. 1997.

Vol. 1336: C. Polychronopoulos, K. Joe, K. Araki, M. Amamiya (Eds.), High Performance Computing. Proceedings, 1997. XII, 416 pages. 1997.

Vol. 1337: C. Freksa, M. Jantzen, R. Valk (Eds.), Foundations of Computer Science. XII, 515 pages. 1997.

Vol. 1338: F. Plášil, K.G. Jeffery (Eds.), SOFSEM'97: Theory and Practice of Informatics. Proceedings, 1997. XIV, 571 pages. 1997.

Vol. 1339: N.A. Murshed, F. Bortolozzi (Eds.), Advances in Document Image Analysis. Proceedings, 1997. IX, 345 pages. 1997.

Vol. 1340: M. van Kreveld, J. Nievergelt, T. Roos, P. Widmayer (Eds.), Algorithmic Foundations of Geographic Information Systems. XIV, 287 pages. 1997.

Vol. 1341: F. Bry, R. Ramakrishnan, K. Ramamohanarao (Eds.), Deductive and Object-Oriented Databases. Proceedings, 1997. XIV, 430 pages. 1997.

Vol. 1342: A. Sattar (Ed.), Advanced Topics in Artificial Intelligence. Proceedings, 1997. XVIII, 516 pages. 1997.

Vol. 1344: C. Ausnit-Hood, K.A. Johnson, R.G. Pettit, IV, S.B. Opdahl (Eds.), Ada 95 – Quality and Style. XV, 292 pages. 1997.